M000192363

FISH RESPIRATION

This is Volume 17 in the

FISH PHYSIOLOGY series
Edited by Steve F. Perry and Bruce L. Tufts

A complete list of books in this series appears at the end of the volume.

FISH RESPIRATION

Edited by

STEVE F. PERRY

Department of Biology
University of Ottawa
Ottawa, Ontario
Canada

BRUCE L. TUFTS

Department of Biology
Queen's University
Kingston, Ontario
Canada

ACADEMIC PRESS

San Diego   London   Boston   New York   Sydney   Tokyo   Toronto

Front cover photograph: A scanning electron micrograph of a rainbow trout (Oncorhynchus mykiss) gill filament illustrating a repsiratory water channel spanned by two individual lamellae. A mixed population of pavement and chloride cells covers the filamental and lamellar epithelial surfaces. Phototgraph courtesy of Steve F. Perry

This book is printed on acid-free paper. ∞

Academic Press
a division of Harcourt Brace & Company
525 B Street, Suite 1900, San Diego, California 92101-4495, USA
http://www.apnet.com

Academic Press Limited
24-28 Oval Road, London NW1 7DX, UK
http://www.hbuk.co.uk/ap/

Library of Congress Catalog Card Number: 98-84496

International Standard Book Number: 0-12-350441-4

PRINTED IN THE UNITED STATES OF AMERICA
98 99 00 01 02 03 BB 9 8 7 6 5 4 3 2 1

CONTENTS

CONTRIBUTORS ix

PREFACE xi

I. Red Blood Cell Physiology and Biochemistry

1. Hemoglobin Structure and Function
Frank B. Jensen, Angela Fago, and Roy E. Weber

I. Introduction	1
II. Basic Structure of Vertebrate Hemoglobins	2
III. Ligand Binding to the Heme Groups and Its Allosteric Regulation	5
IV. The Bohr Groups	8
V. Hydrogen Ion Equilibria	8
VI. Binding of CO_2	12
VII. Binding of Organic Phosphates	13
VIII. Molecular Basis for the Root Effect	14
IX. Temperature Effect	18
X. Adaptation of Hemoglobin Function	20
XI. Hemoglobin Multiplicity	22
XII. Oxidation of Hemoglobin	27
XIII. Interactions of Hemoglobin with Membrane Proteins	30
References	32

2. Red Blood Cell Metabolism
Patrick J. Walsh, Chris M. Wood, and Thomas W. Moon

I. Introduction	41
II. Methodology	42
III. Metabolism of Mammalian Red Blood Cells	46
IV. Metabolic Poise of Fish Red Blood Cells	49
V. Transport of Metabolic Fuels	54
VI. The Influence of Catecholamines, Oxygen Status, and Intracellular pH Regulation on Red Cell Metabolism	62

VII. Epilogue   68
      References   69

3.   Carbonic Anhydrase and Respiratory Gas Exchange
Raymond P. Henry and Thomas A. Heming

  I. Introduction   75
 II. The Catalytic Mechanism of CA   76
III. Tissue and Isozyme Distribution   77
 IV. CA and Respiratory Gas Exchange: CO_2 Transport and Excretion   83
  V. Ammonia as a Respiratory Gas   88
 VI. Plasma Inhibitors of Carbonic Anhydrase   91
VII. Analytical Techniques   96
VIII. Summary   103
      References   103

II.   Oxygen

4.   The Physiology of the Root Effect
Bernd Pelster and David Randall

  I. Introduction   113
 II. Occurrence of the Root Effect   115
III. Characterization of the Root Effect   117
 IV. Root Effect and Swim Bladder Function   120
  V. Oxygen-Concentrating Mechanisms in the Fish Eye   126
 VI. Adrenergic Effects   131
      References   134

5.   Oxygen Transport in Fish
Mikko Nikinmaa and Annika Salama

  I. Introduction   141
 II. Hemoglobin Function: Basic Principles   143
III. Regulation of Hemoglobin Function by Changes in Erythrocytic Organic
      Phosphate Concentrations   145
 IV. Effects of Cellular Hemoglobin Concentration and Red Cell Volume on
      Oxygen Transport   149
  V. Regulation of Erythrocyte Volume   150

VI. Effects of Protons on Hemoglobin Function   155
VII. Control of Erythrocyte pH   157
VIII. Hemoglobin Oxidation   162
IX. Responses of Hemoglobin Function to Changes in the External and the
    Internal Environment of Fish   163
    References   174

6. Hematocrit and Blood Oxygen-Carrying Capacity
P. Gallaugher and A. P. Farrell

I. Introduction   185
II. The Influence of Sampling Methodology on Hematocrit   187
III. Interspecific Diversity in Hematocrit Values   206
IV. Intraspecific Regulation of Hematocrit   209
V. Critique of the Optimal Hematocrit Theory   215
VI. Conclusions   219
    References   219

III. Carbon Dioxide and Acid–Base Balance

7. Carbon Dioxide Transport and Excretion
Bruce Tufts and Steve F. Perry

I. Introduction   229
II. Carriage of CO_2 in Blood   230
III. Carbon Dioxide Transport and Excretion   236
IV. Future Directions   272
    References   273

8. The Linkage between Oxygen and Carbon Dioxide Transport
C. J. Brauner and D. J. Randall

I. Introduction   283
II. Fundamental Basis for the Linkage   284
III. Hb Characteristics That Influence the Magnitude of the Linkage   288
IV. Physiological Basis for the Linkage   294
V. Conclusion   312
    References   313

9.   Causes and Consequences of Acid–Base Disequilibria
     Kathleen M. Gilmour

     I. Introduction   321
    II. Postbranchial Disequilibria   322
   III. Prebranchial Disequilibria   337
    IV. Consequences of Acid–Base Disequilibria   339
        References   343

INDEX   349

OTHER VOLUMES IN THE FISH PHYSIOLOGY SERIES   355

CONTRIBUTORS

Numbers in parentheses indicate the pages on which the authors' contributions begin.

C. J. BRAUNER *(283), Department of Biology, McMaster University, 1280 Main Street West, Hamilton, Ontario, Canada L8S 4K1*

ANGELA FAGO *(1), Department of Zoophysiology, Aarhus University, DK–8000 Aarhus C, Denmark*

ANTHONY P. FARRELL *(185), Department of Biological Sciences, Simon Fraser University, Burnaby, British Columbia, Canada V5A 1S6*

PATRICIA GALLAUGHER *(185), Continuing Studies in Science, Simon Fraser University, Burnaby, British Columbia, Canada V5A 1S6*

KATHLEEN M. GILMOUR *(321), Division of Environmental and Evolutionary Biology, University of Glasgow, Glasgow, G128 QQ United Kingdom*

THOMAS A. HEMING *(75), Department of Internal Medicine and Physiology and Biophysics, University of Texas Medical Branch, Galveston, Texas 77555*

RAYMOND P. HENRY *(75), Department of Zoology and Wildlife Science, Auburn University, Auburn, Alabama 36849*

FRANK B. JENSEN *(1), Institute of Biology, Odense University, DK–5230 Odense M, Denmark*

THOMAS W. MOON *(41), Department of Biology, University of Ottawa, Ottawa, Ontario, Canada K1N 6N5*

MIKKO NIKINMAA *(141), Department of Biology, University of Turku, FIN–20014 Turku, Finland*

BERND PELSTER *(113), Institut für Zoologie und Limnologie, Universität Innsbruck, A-6020 Innsbruck, Austria*

STEVE F. PERRY *(229), Department of Biology, University of Ottawa, 30 Marie Curie, Ottawa, Ontario, Canada K1N 6N5*

DAVID RANDALL *(113, 283), Department of Zoology, University of British Columbia, Vancouver, British Columbia, Canada V6T 2A9*

ANNIKA SALAMA *(141), Division of Animal Physiology, Department of Biosciences, University of Helsinki, PB 17, FIN-00014 Helsinki, Finland*

BRUCE L. TUFTS *(229), Department of Biology, Queen's University, Kingston, Ontario, Canada K7L 3N6*

PATRICK J. WALSH *(41), Division of Marine Biology and Fisheries, Rosenstiel School of Marine and Atmospheric Science, University of Miami, Miami, Florida 33149*

ROY E. WEBER *(1), Department of Zoophysiology, Aarhus University, DK–8000 Aarhus C, Denmark*

CHRIS M. WOOD *(41), Department of Biology, McMaster University, Hamilton, Ontario, Canada L85 4K1*

PREFACE

The physiology of the respiratory system traditionally has been one of the most dynamic areas in fish physiology—a crucial focal point of fish biology for students and researchers alike. The past decade has witnessed significant advances, including the discovery of fascinating new model systems and the increasing use of molecular techniques. Recognizing these advances, Dave Randall suggested at a 1995 meeting of the Canadian Society of Zoologists that we compile a book that would synthesize the classical literature while highlighting recent developments. This volume, the first in the *Fish Physiology* series to be dedicated entirely to the theme of respiratory physiology, is the result; specific aspects of respiration were covered in previous volumes of *Fish Physiology* (Volumes 4 [1970] and 10 [1984]). We present here nine chapters, arranged to form three sections, which are linked by a unifying theme—the involvement of the respiratory pigment, hemoglobin. Section I, Red Blood Cell Physiology and Biochemistry, sets the stage for later chapters through its detailed coverage of the molecular structure of hemoglobin and the metabolic pathways within the red blood cell that create a crucial link between structure and function. In Section II, Oxygen, the focus is on the mechanisms regulating oxygen uptake at the gill and its delivery to the tissues. Finally, Section III, Carbon Dioxide and Acid–Base Balance, presents a detailed account of blood carbon dioxide transport, the interactions between carbon dioxide and oxygen transport, and the consequences of blood carbon dioxide transport on steady-state and non-steady-state acid–base balance.

In preparing this book we have gained a new appreciation and respect for editors everywhere and have learned that the word "deadline" can be interpreted in many different and often creative ways. We thank all of the contributors for their enthusiastic participation in this project, but most of all we must thank Bill Hoar and Dave Randall for conceiving of this wonderful and ever-evolving series devoted to the endlessly fascinating topic of fish physiology.

STEVE PERRY
BRUCE TUFTS

1

HEMOGLOBIN STRUCTURE AND FUNCTION

FRANK B. JENSEN
ANGELA FAGO
ROY E. WEBER

I. Introduction
II. Basic Structure of Vertebrate Hemoglobins
III. Ligand Binding to the Heme Groups and Its Allosteric Regulation
IV. The Bohr Groups
V. Hydrogen Ion Equilibria
   A. H^+ Equilibria and Blood CO_2 Transport
VI. Binding of CO_2
VII. Binding of Organic Phosphates
VIII. Molecular Basis for the Root Effect
IX. Temperature Effect
X. Adaptation of Hemoglobin Function
   A. Organic Phosphates
   B. Red Cell pH and Volume
   C. Other Effectors
XI. Hemoglobin Multiplicity
   A. Functional Differentiation
   B. Variation in Quaternary Structure
   C. Ontogenetic Changes
   D. Significance and Environmental Influences
XII. Oxidation of Hemoglobin
   A. Autoxidation
   B. Nitrite-Induced Oxidation
   C. Interrelationship between Oxidation and Oxygen Affinity
XIII. Interactions of Hemoglobin with Membrane Proteins
   References

I. INTRODUCTION

Transport of oxygen from the environment to cells and transport of
metabolically produced carbon dioxide and H^+ in the opposite direction

1

Fish Physiology, Volume 17:
FISH RESPIRATION

are essential for vertebrate life. Hemoglobin (Hb) greatly increases the carrying capacity of O_2, CO_2, and H^+ in blood as result of reversible binding of the ligands to the Hb molecule and appropriate allosteric interactions between the binding sites. With the notable exception of Antarctic icefishes, Hb is present in all vertebrates, which strongly decreases the circulatory requirement (i.e., volume of blood pumped by the heart for a unit quantity of O_2 consumed).

Hemoglobin is one of the most intensively studied proteins, which has resulted in a deep understanding of its structure–function relationships. Hb has been termed the "honorary enzyme", since the detailed knowledge of its structure and functions has rendered it a valuable model for studying allosteric interactions in other proteins. Research on vertebrate Hbs, and on fish Hbs in particular, has continued to reveal exciting new aspects of molecular and cellular control mechanisms. Significant advances have been made since fish Hbs were last reviewed in this book series (Riggs, 1970). The present chapter is intended to give an overview of the current knowledge of molecular structure, conformational changes, allosteric interactions, and the multiple functions of Hb in fish.

II. BASIC STRUCTURE OF VERTEBRATE HEMOGLOBINS

Hemoglobin of most vertebrates is a tetrameric globular protein consisting of two α and two β polypeptide chains, each having an oxygen-binding heme (iron protoporphyrin IX). The number of amino acids varies slightly among species. In human Hb, the α chain has 141 amino acids, and the β chain has 146 amino acids. In carp, the numbers are 142 and 147, respectively. The amino acid sequences (primary structures) of a considerable number of Hbs are known, and they reveal homology that reflects phylogenetic relationships. Differences in the number and identity of amino acids result a variation in molecular weight of around 65,000. Each polypeptide chain has a characteristic secondary structure, alternating between α-helical segments (labeled A through H from the N terminus) and in-between nonhelical segments (named with the letters of the adjacent helices, i.e., AB through GH). The ultimate segments at the N- and C-terminal ends are nonhelical and are labeled NA and HC, respectively. Individual amino acid residues are often referred to by their position in these segments (e.g., F8α means the eighth amino acid residue in helix F of the α chain, counting from the amino end). The tertiary structure (folding of individual chains) of vertebrate Hb subunits (Fig. 1) is virtually identical to that of myoglobin. The conformation of α and β subunits differs only by an additional helix

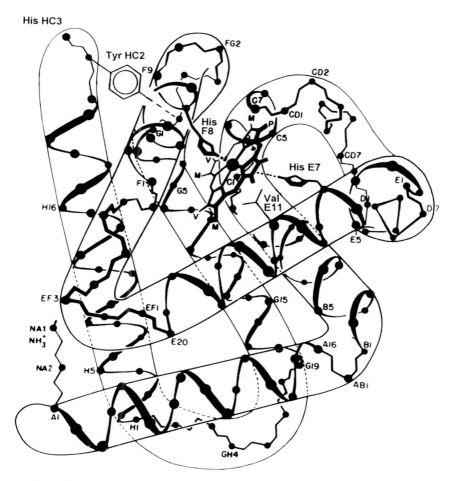

Fig. 1. The β chain of human hemoglobin, illustrating the characteristic three-dimensional fold (tertiary structure) and the alternating helical and nonhelical segments (secondary structure). Some positions of individual amino acid residues are highlighted, e.g., the proximal His F8, the distal His E7 and Val E11 in the heme pocket, and His HC3, which is a major Bohr group in the intact $\alpha_2\beta_2$ tetramer (modified from Perutz, 1990).

(D) in the β subunit. The heme group is buried in a hydrophobic pocket between the E and F helices of each chain and is covalently linked to the proximal histidine (His F8). This histidine and phenylalanine CD1 are present in all Hbs (Perutz, 1990). Phe CD1 is important for providing a hydrophobic environment for the heme. On the distal (ligand) site of the heme, the so-called distal histidine (His E7) and Val E11 are normally present (Fig. 1).

X-ray crystallography of mammalian Hb with different ligands bound to the hemes [HbO$_2$, HbCO, and metHb (where a water molecule is bound)] reveal that the liganded Hbs have a similar overall quaternary structure (three-dimensional arrangement of the four subunits in the tetramer) (Shaanan, 1983), which differs from the quaternary structure of unliganded (deoxygenated) Hb (Fermi *et al.*, 1984). Hemoglobin is in equilibrium between two alternative structures, the T ("tense") structure, characterizing deoxy-Hb, and the R ("relaxed") structure, characterizing oxyHb. In the absence of ligand, the molecule will preferentially be in the T state, which is thermodynamically more stable because of the presence of extra salt bridges and other noncovalent bonds, notably in the interface between the two $\alpha\beta$ dimers. The modification in the tertiary structure upon ligand binding at the heme, as predicted by Koshland *et al.* (1966), progressively loosens the noncovalent interactions that hold the tetramer in the T conformation, allowing a switch to the high-affinity R state (Perutz, 1970). The tetrameric Hb can be considered as being made of two rigid dimeric units ($\alpha_1\beta_1$ and $\alpha_2\beta_2$) held together by looser $\alpha_1\beta_2$ and $\alpha_2\beta_1$ contacts. In the T $\rightarrow$ R transition, the $\alpha_1\beta_1$ dimer rotates relative to the $\alpha_2\beta_2$ dimer by 15°. While the $\alpha_1\beta_1$ and $\alpha_2\beta_2$ interfaces remain rigid, significant changes occur at the looser $\alpha_1\beta_2$ contact, especially between the helix Cα_1 and the corner FGβ_2, known as the "switch" region. The "dove-tailed" shape of this contact allows only two alternative subunit packings, corresponding to those of the T and R states. A detailed analysis of structural changes related to ligand binding is reported by Baldwin and Chothia (1979).

The ability to isolate the intermediates of the ligation process has allowed Ackers and co-workers to establish a sequence of steps in ligand binding. In addition to the two end states (corresponding to the T and R structures), a third and intermediate level was identified (Ackers, 1990). Binding of the first oxygen molecule induces a tertiary modification in the liganded subunit, which is accommodated within the T structure, while the $\alpha_1\beta_2$ interface functions as a constraint. When ligand is bound in at least one subunit of each dimeric half-molecule, the tetramer switches to the R state. The T to R transition releases both the unfavorable "tertiary" and the "quaternary" constraints, represented by the network of salt bridges and noncovalent bonds at the dimer–dimer interface (see Ackers *et al.*, 1992; LiCata *et al.*, 1993; Daughterty *et al.*, 1994).

X-ray crystallographic analyses of tetrameric teleost fish Hbs have been performed recently on the deoxy and CO derivatives of the Hb from the Antarctic fish *Pagothenia bernacchii* (major component; Camardella *et al.*, 1992; Ito *et al.*, 1995) and trout (the "cathodic" Hb I; Tame *et al.*, 1996) and on the carbomonoxy form of the Hb from the spot *Leiostomus xanthurus* (Mylvaganam *et al.*, 1996). The overall three-dimensional structure of the fish Hbs is similar to that of human Hb, although a difference of 3° was observed in the relative orientation of the two $\alpha\beta$ dimers in liganded spot Hb in comparison with human Hb (Mylvaganam *et al.*, 1996). This feature may reflect a different side-chain packing at the $\alpha_1\beta_2$ interface, which may be related to distinctive allosteric effects in fish Hbs (e.g., the Root effect).

Lamprey and hagfish Hbs are monomeric when oxygenated but aggregate to dimers and tetramers upon deoxygenation, upon lowering of pH, and upon elevation of [Hb] (Briehl, 1963; Riggs, 1972; Nikinmaa *et al.*, 1995; Fago and Weber, 1995). The primary structure of the Hbs from cyclostomes is radically different from that of other vertebrate Hbs. The considerable deviation in primary structures from both α- and β-type globins does not permit categorization as either of these (Liljeqvist *et al.*, 1982; Hombrados *et al.*, 1983; Feng and Doolittle, 1987). Nevertheless, the three-dimensional structure of the lamprey *Petromyzon marinus* Hb V includes eight α-helix segments and is similar to the myoglobin fold (Honzatko *et al.*, 1985).

At low concentration (in the μM range and below), tetrameric Hbs dissociate reversibly into $\alpha\beta$ dimers. Fish Hbs are much less liable than mammalian Hbs to dissociate (Edelstein *et al.*, 1976; Kwiatkowski *et al.*, 1994). The dissociation constants for the tetramer–dimer dissociation of liganded carp, menhaden, and blue shark Hbs are approximately 10^{-8} M, which is two orders of magnitude lower than that for human Hb (10^{-6} M) (Edelstein *et al.*, 1976).

Polymerization of teleost Hb occurs rarely, although examples involving intermolecular disulfide bonds between tetramers are known (Fago *et al.*, 1993; Borgese *et al.*, 1994).

III. LIGAND BINDING TO THE HEME GROUPS AND ITS ALLOSTERIC REGULATION

Reversible binding of oxygen to the heme groups demands that heme iron remains in the ferrous state and that binding of oxygen is favored over that of other potential heme ligands. Free heme groups in solution bind carbon monoxide 10^3–10^4 times as strongly as O_2, whereas heme groups in myoglobin and hemoglobin bind CO only some 200 times more tightly than O_2. The distal histidine is involved in this discrimination against CO in Mb and Hb, by sterically hindering bound CO and stabilizing bound O_2 (e.g., Springer *et al.*, 1989; Perutz, 1990). The discrimination between O_2 and CO is physiologically important because CO is endogenously produced.

Thus, breakdown of heme groups [e.g., from senescent red blood cells (RBCs) removed from circulation] is initiated by the conversion of heme to biliverdin, free iron, and CO, in a reaction catalyzed by heme oxygenase (Maines, 1988). If the partition coefficient of Hb between O_2 and CO resembled that in free heme, blood O_2 transport would be severely hindered by this endogenous production of CO.

The oxygen affinity (characterized by P_{50}, the P_{O_2} at 50% O_2 saturation) of Hb varies with globin structure, giving rise to species differences that relate to differences in primary structure. In addition, the O_2 affinity is modulated by allosteric interactions. Binding of allosteric effectors such as protons, chloride, and organic phosphates to specific binding sites on the Hb lowers the affinity of the heme groups for O_2. The heterotropic effectors bind preferentially to the T structure, which they stabilize by introducing additional bonds (Perutz, 1990). The decrease in affinity (increase in P_{50}) with acidification (the alkaline Bohr effect—cf. review by Riggs, 1988) and the decrease in affinity with addition of organic phosphates (e.g., ATP) are illustrated for tench Hb in Fig. 2). The magnitude of the Bohr effect (given by $-d \log P_{50}/d\text{pH}$) varies with pH and the concentration of organic phosphates (Fig. 2b). At physiological pH values, addition of ATP both decreases O_2 affinity (Fig. 2a) and increases the Bohr effect (Fig. 2b), illustrating the complex and reciprocal interaction between O_2, ATP, and H^+ binding sites.

Hb binds oxygen cooperatively (i.e., oxygenation of one subunit increases the oxygen affinity of remaining deoxygenated subunits). This homotropic interaction between the heme groups is reflected in the sigmoid shape of the O_2 equilibrium curve [depicting fractional O_2 saturation (Y) versus P_{O_2}] and can be quantified by the slope n_{50} around half saturation in a plot of $\log(Y/(1-Y))$ versus $\log P_{O_2}$ (Hill plot). Cooperativity is reflected by $n_{50} > 1$. The value is pH independent in mammalian Hbs (Antonini and Brunori, 1971), whereas fish Hbs exhibiting the Root effect are characterized by a decrease in n_{50} at low pH, with values below 1 eventually being reached (Fig. 2c). The presence of organic phosphate can slightly increase n_{50} at physiological pH and shift the decrease in cooperativity with decreased pH to higher pH values (Fig. 2c).

Homotropic and heterotropic effects are often analyzed with the two-state MWC model (Monod *et al.*, 1965), according to which Hb is in equilibrium between two symmetric conformational states (the T and R structures). The cooperative interaction between O_2 binding sites arises from the transition from the T to the R structure during oxygenation, whereas heterotropic effects arise from a displacement of the allosteric T $\leftrightarrow$ R equilibrium (described by the allosteric constant L, which is given by $[T_0]/[R_0]$). Addition of organic phosphates and H^+ (i.e., lowering of pH) to both human and

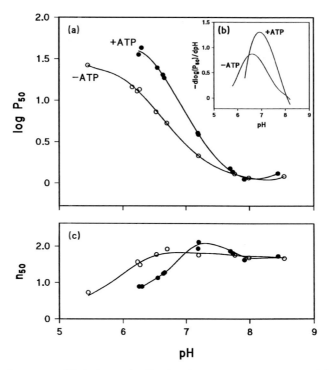

Fig. 2. Oxygen equilibria in tench Hb, showing the pH dependence of (a) log P_{50}, (b) the negative Bohr coefficient, and (c) n_{50} (Hill coefficient at half saturation) in the absence and presence of ATP ([ATP]/[Hb] = 2). The measurements were performed at 15°C in 0.05 M Hepes buffer with 0.1 M KCl (from Jensen and Weber, 1985).

fish Hbs typically increases L and decreases K_T (the association constant for the T state), while K_R is practically unchanged (Imai, 1982; Weber *et al.*, 1987). These effects illustrate that the heterotropic effectors stabilize the T state and postpone the allosteric T → R transition. At high [phosphate]/[Hb] ratios ([NTP]/[Hb] = 10) and [H$^+$] (pH < 7) K_R also falls (Weber *et al.*, 1987). In a strict two-state MWC model, heterotropic effectors should alter L only, and K_T and K_R should be the same for all O_2 binding sites within the T and R states, respectively. Thus, the lowering of the association constants by heterotropic effectors is in principle inconsistent with the model. When the model is extended to include three states (the third state being an altered T state in presence of organic phosphates), O_2 equilibria in tench Hb (under conditions covering natural red cell pH values and [NTP]/[Hb] ratios) can be satisfactorily described in agreement with the basic assumptions of the MWC model (Jensen *et al.*, 1990). Studies on the kinetics of ligand binding also point to the existence of an altered T state in carp Hb (Kwiatkowski *et al.*, 1994). Analyses according to the two-state MWC model, however, remain valuable and provide a straightforward association with the two quaternary Hb structures deduced from X-ray analysis.

IV. THE BOHR GROUPS

The C-terminal His of the β chain (His HC3β) is the most important residue implicated in the pH regulation of Hbs. In human Hb, it is responsible for the Bohr effect observed in the absence of chloride ions and for about 40% of that measured in chloride-containing buffers (Shih *et al.*, 1984, 1993). In the deoxygenated state, the carboxy group of His HC3β makes a salt bridge with the side chain of Lys C5α, while its imidazole ring is salt-bridged to the side chain of Asp FG1β (substituted by a Glu in fish Hbs), which increases its pK_a. In the conformational changes accompanying the T $\rightarrow$ R transition, these salt bridges are broken and protons are released. The importance of His HC3β for the Bohr effect in fish Hbs is clear. This residue is conserved in Hbs showing Bohr or Root effects, and it is replaced by Phe in the cathodic Hbs of trout (Barra *et al.*, 1983) and eel (Fago *et al.*, 1995), whose oxygen affinities are essentially pH-insensitive. His HC3β contributes some 50% to the Bohr effect in fish Hbs, as judged from the halving of the Bohr effect in carp Hb when the C-terminal His is enzymatically removed (Parkhurst *et al.*, 1983).

Another potentially important Bohr group is the N terminus of the α chain (Perutz *et al.*, 1980; Riggs, 1988). In human Hb, Val NA1α accounts for about 20% of the Bohr effect. In teleosts, the N termini of α chains are acetylated (Ac-Ser or Ac-Thr) and cannot contribute to the Bohr effect. A significant fraction of the Bohr effect remains to be accounted for in fish Hb.

V. HYDROGEN ION EQUILIBRIA

The hydrogen ion equilibria of Hb are fundamental to both the structure and the physiological function of the protein. The exchange of H^+ between protein and solvent is important for blood CO_2 transport (by ensuring the necessary binding/release of H^+ for the red cell CO_2 hydration–dehydration reaction), and it makes Hb an effective nonbicarbonate buffer that limits fluctuations in blood pH upon acid or base additions. The H^+ binding properties also determine the grouping of charges on the molecule. The overall charge on Hb controls RBC pH via the Donnan-like distribution of protons across the RBC membrane, and the distribution of charges is important for the formation of inter- and intrasubunit salt bridges and in the binding of ligands such as organic phosphates and chloride.

The basic H^+-binding properties are determined by the type and the amount of dissociable amino acid residues as well as by the molecular microenvironment of these residues, which influences both their pK_a and their accessibility to the solvent. In general, the types of amino acid residues that are titrated can be divided into three classes (Tanford, 1962). The acidic carboxyl groups of glutamic acid and aspartic acid residues are negatively charged at physiological pH and are titrated only when pH falls below 6. Basic groups such as the guanidyl group of arginine and the amino group of lysine side chains are positively charged at physiological pH and release H^+ only when the pH exceeds about pH 9. In the intermediate pH range, it is mainly the imidazole group of histidine residues and the terminal α-amino groups that are titrated.

Information on H^+ equilibria can be obtained by acid–base titration of Hb solutions, using zero net proton charge ($Z_H = 0$) as a reference point (Tanford, 1962). An example is given for carp Hb in Fig. 3. From the direct titration curves of oxyHb and deoxyHb (Fig. 3a), the buffer values at constant protein conformation are obtained by differentiation (Fig. 3b). Buffer values vary with protein conformation and pH (Fig. 3b) and between

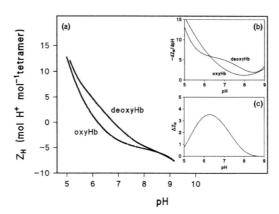

Fig. 3. Hydrogen ion equilibria of stripped carp Hb in 0.1 M KCl, showing (a) net H^+ charge (Z_H, mol H^+/mol tetramer) as a function of pH in oxygenated and deoxygenated Hb solutions. From the direct titration curves information on (b) buffer values ($-dZ_H/dpH$) and (c) the fixed acid Haldane effect (ΔZ_H between deoxyHb and oxyHb) have been obtained. The Hb was kept oxygenated by equilibration with pure O_2 and deoxygenated by pure N_2. Tetrameric [Hb] = 0.21 mM; temperature = 15°C (F. B. Jensen, unpublished data).

species. In general, Hb from teleost fishes have lower buffer values at physiological pH than other groups of fish (e.g., elasmobranchs) and higher vertebrates, which is due to much lower histidine contents and acetylation of the α-amino groups of the α chains in teleost Hbs (Jensen, 1989). Carp Hb, for instance, contains 18 His per tetramer, whereas human Hb contains 38 His residues. These include the proximal and distal histidines of each chain, which do not change ionization significantly at physiological pH (Perutz, 1990). Some fish Hb chains, such as the β chain of cathodic eel Hb (Fago *et al.*, 1995) or lamprey Hb (Hombrados *et al.*, 1983), contain only these two His residues. Histidine residues can also be unavailable for H^+ exchange by being buried inside the molecule. Thus, the number of titratable His residues is lower than the total number of His. In tetrameric carp Hb, about 7 of the 18 His are titrated (Jensen, 1989), whereas 20 of the 38 His are titrated in human Hb (Janssen *et al.*, 1970).

When oxyHb is deoxygenated, the shift in quaternary structure changes the molecular surroundings of certain amino acid residues (the Bohr groups), increasing their pK_a values and thus their H^+ uptake at constant pH. This additional H^+ uptake is known as the fixed acid Haldane effect and is quantified by the difference titration curve (ΔZ_H versus pH; Fig. 3c). The molecular origin of the oxygenation-linked H^+ binding is the same as for the pH dependence of O_2 affinity (the Bohr effect). However, even though the influence of oxygenation on blood H^+/CO_2 equilibration curves could in principle have been foreseen from the discovery of the Bohr effect early in this century, it took a decade before the reciprocal effect—the Haldane effect—was actually discovered (cf. Edsall, 1986). The tight relationship between the Bohr effect and the Haldane effect is illustrated by the classical linkage equation, revealing identity between the Bohr coefficient and the Haldane coefficients (H^+ symbolizing number of bound H^+ per heme and Y being fractional O_2 saturation) (Wyman, 1964):

$$\left(\frac{\partial \log P_{O_2}}{\partial pH}\right)_Y = \left(\frac{\partial H^+}{\partial Y}\right)_{pH}.$$

For a symmetrical O_2 equilibrium curve, the relation reduces to

$$-(\Delta \log P_{50}/\Delta pH) = \tfrac{1}{4}\Delta Z_H.$$

Thus, information on oxygenation-linked H^+ binding can be obtained both from oxygen equilibria (exemplified for tench Hb in Fig. 2b) and from H^+ titration (exemplified for carp Hb in Fig. 3c).

The fixed acid Haldane effect is generally large in teleosts, but differs among species. When solutions of the natural mixture of Hb components are compared in the absence of organic phosphates and at comparable ionic strength and temperature, the maximal ΔZ_H is about 2.7 protons per tetramer in trout and 3.6–3.8 protons per tetramer in carp (Chien and Mayo, 1980; Jensen, 1989; Fig. 3c). Anodic eel Hb has a high maximal ΔZ_H similar to that of carp Hb (Breepoel *et al.*, 1981), but in stripped eel hemolysate, the maximal ΔZ_H is lower (Breepoel *et al.*, 1980) due to the presence of both anodic (Bohr effect) and cathodic (absent or negative Bohr effect) Hbs. In carp, all Hb components possess a Bohr effect (Weber and Lykkeboe, 1978). In the presence of organic phosphates (ATP and GTP), the maximal ΔZ_H increases, as does the pH at which it occurs (Breepoel *et al.*, 1981; Jensen and Weber, 1985), whereby very high ΔZ_H values prevail at natural red cell pH values. The large oxygenation-linked H^+ binding and the low buffer values of the oxy and deoxy conformations of teleost Hbs produce very large changes in red cell pH upon changes in blood O_2 saturation. In teleosts, the difference in red cell pH between deoxygenated and oxygenated RBCs can be 0.35 pH unit (Jensen, 1986), which is more than 10 times that in mammals. One consequence of large changes in red cell pH with changes in O_2 saturation is a lower apparent Hill n value for whole blood O_2 equilibrium curves than for curves measured on Hb solution at a constant buffered pH (cf. Jensen, 1991). In tench, the major change in RBC pH_i occurs between 50 and 100% O_2 saturation, suggesting that the Haldane effect is almost fully exploited within the normal normoxic differences in arterial and venous O_2 saturation (Jensen, 1986). A similar nonlinear relationship between red cell pH and blood O_2 saturation is present in rainbow trout (Brauner and Randall, this volume).

Whereas many teleost Hbs have low buffer values but high Haldane effects, elasmobranch Hbs have high buffer values but small Haldane effects (Jensen, 1989), which correlates with small Bohr effects. The South American lungfish has 46 His residues per tetramer, and the N termini of the α chains are not acetylated (Rodewald *et al.*, 1984), suggesting high buffer values in this representative of dipnoans. Lamprey Hb has few histidines, which is reflected in low buffer values (F. B. Jensen, unpublished), and the Bohr–Haldane effect is large, as in teleosts. In lamprey Hb, the large oxygenation-linked H^+ binding originates in the aggregation of monomers upon lowering of O_2 saturation (e.g., Nikinmaa *et al.*, 1995).

A. H^+ Equilibria and Blood CO_2 Transport

The binding of H^+ to Hb facilitates the uptake in blood of metabolically produced CO_2 by driving the carbonic anhydrase-catalyzed CO_2 hydration inside the RBCs toward bicarbonate formation. Bicarbonate, in turn, is transferred to the plasma by band-3-mediated HCO_3^-/Cl^- exchange across the RBC membrane. Thus, both reaction products are removed, and the equilibrium reaction proceeds farther to the right, increasing CO_2 uptake as the blood passes through the capillaries in the tissues:

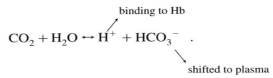

$$CO_2 + H_2O \leftrightarrow H^+ + HCO_3^- \ .$$

The Haldane effect and anion exchange make approximately equal contributions to HCO_3^- formation in humans (Wieth *et al.,* 1982). The large oxygenation-linked H^+ binding in many teleosts suggests that the Haldane effect has greater importance for blood CO_2 transport in teleosts than in mammals (Jensen, 1989). The presence of a large Haldane effect may be related to rate limitations in anion exchange. The Cl^- shift in carp (having a very large Haldane effect) is slower than that in rainbow trout (where the Haldane effect is lower) (Jensen and Brahm, 1995). The Cl^- shift in fish RBCs is relatively rapid (band-3-mediated Cl^- transport in trout at 15°C is as fast as in human RBCs at 37°C—cf. Jensen and Brahm, 1995), but HCO_3^-/Cl^- exchange can nevertheless be rate limiting for CO_2 excretion, notably in situations where blood flow increases and transit times in the capillaries decrease. The formation and concentration of HCO_3^- in the RBCs when blood passes through tissue capillaries are larger with a large Haldane effect and low buffer values than with a small Haldane effect and high buffer values. Upon oxygenation in the gills, the massive Bohr proton release mediates an extensive conversion of red cell HCO_3^- to CO_2 that may compensate for a somewhat delayed entry of HCO_3^- via band 3. Lampreys are the ultimate example of this strategy because they lack band 3 and depend almost exclusively on dehydration of HCO_3^- carried within the RBCs for CO_2 excretion (Tufts and Boutilier, 1989; Nikinmaa and Mattsoff, 1992).

VI. BINDING OF CO_2

CO_2 reacts with uncharged α-amino groups to form carbamic acid. Since α-amino groups may be charged at physiological pH and since carbamic acid dissociates to carbamate, the following equilibria are involved:

$$Hb-NH_3^+ \leftrightarrow Hb-NH_2 + H^+,$$

$$Hb-NH_2 + CO_2 \leftrightarrow Hb-NHCOOH,$$

$$Hb-NHCOOH \leftrightarrow Hb-NHCOO^- + H^+.$$

Carbamate formation is more pronounced in deoxygenated than in oxygenated Hb and is of physiological significance in mammals. In humans, oxygenation-dependent CO_2 binding to Hb accounts for about 13% of the CO_2 exchange (Klocke, 1988). Direct information on oxylabile carbamate formation is scarce in fish. Some insight into the topic may, however, be gained from the linkage equation (Wyman, 1964)

$$\left(\frac{\partial \log P_{O_2}}{\partial \log P_{CO_2}}\right)_Y = \left(\frac{\partial CO_2}{\partial Y}\right)_{P_{CO_2}},$$

which shows that the change in O_2 affinity with a change in $\log P_{CO_2}$ at constant O_2 saturation equals the change in CO_2 bound with changed O_2 saturation at constant P_{CO_2}. The specific CO_2 effect on O_2 affinity (left side of the equation) is generally low in teleost Hbs (e.g., Weber and Lykkeboe, 1978; Farmer, 1979), implying that oxylabile carbamate formation (right side of the equation) is insignificant. The low oxylabile carbamate formation is structurally explained by the acetylation of α-amino groups of the α chains and by the competition of organic phosphates with CO_2 for binding to the α-amino groups in the β chains (cf. Weber and Jensen, 1988). Furthermore, water-breathing fishes have low blood P_{CO_2}, which reduces carbamate formation compared to air breathers with high blood P_{CO_2}.

Preferential carbamate formation in deoxyHb can potentially release up to two protons per CO_2 bound (see chemical equilibria). The insignificant oxylabile carbamino formation in teleost Hbs may have evolved to avoid this H^+ release, which counteracts the large H^+ uptake of deoxy Hb that is characteristic of teleosts.

VII. BINDING OF ORGANIC PHOSPHATES

In anucleated mammalian red cells, the major organic phosphate influencing Hb function is 2,3-DPG, which is produced in anaerobic glycolysis. Nucleated fish red cells, in contrast, contain mitochondria and have an aerobic metabolism producing the nucleoside triphosphates (NTPs) ATP and GTP as potent allosteric effectors of Hb function.

Organic phosphates bind at the entrance to the central cavity between the two β chains in the T structure. Two amino acid substitutions change the site from one fitting 2,3-DPG to one being stereochemically complementary to NTP (Perutz and Brunori, 1982). The residues involved in NTP binding include the N terminus of the β chain, Glu NA2β, Lys EF6β, and Arg H21β (Gronenborn et al., 1984). With GTP an additional hydrogen bond can be formed (with the N-terminal residue of the β chain), which explains why GTP has a greater effect than ATP on HbO_2 affinity (Weber *et al.,* 1975; Lykkeboe *et al.,* 1975). The greater GTP affinity also affects the binding of other ligands, as exemplified by the observation that GTP obliterates the effect of CO_2 on HbO_2 binding in lungfish, *Protopterus amphibius* (which experiences high CO_2 tensions), under conditions where ATP does not (Weber and Johansen, 1979). Curiously, effects of ATP equal to or greater than those of GTP on hemoglobin–O_2 affinity have been observed in some Amazonian fish (Val *et al.,* 1986; Weber, 1996), but the molecular basis remains unknown.

The residues involved in NTP binding are conserved (except for Arg → Lys at position H21β) in cathodic eel Hb (Fago *et al.,* 1995). Addition of NTP to this Hb both decreases the oxygen affinity and obliterates the reverse Bohr effect of the stripped Hb. This suggests that the positively charged residues involved in phosphate binding at the central cavity may act as reverse Bohr groups in fish Hbs when the principal alkaline Bohr groups are replaced (Fago *et al.,* 1995). In human Hb, the reverse Bohr effect observed below pH 6.5 has been ascribed to His H21β (Perutz *et al.,* 1980), which is replaced by Lys in cathodic eel Hb.

VIII. MOLECULAR BASIS FOR THE ROOT EFFECT

The Root effect (reviewed by Brittain, 1987) is characteristic of many (anodic) fish Hbs and is not found in other vertebrate Hbs. At low pH, Root effect Hbs show a large decrease in both oxygen affinity (i.e., a large Bohr effect) and cooperativity. The pH effect is so drastic that complete saturation with oxygen cannot be achieved even at very high oxygen pressures (140 atm; Scholander and Van Dam, 1954). Acidification of blood in the circulatory system will accordingly release large amounts of chemically bound O_2. The physiological role of Root effect Hbs is the secretion of oxygen to the choroid rete of the eye and to the swim bladder (reviewed by Pelster and Weber, 1991; Pelster and Randall, this volume).

The basic molecular mechanism for the Root effect is a debated question, but it is generally assumed that it derives from a strong stabilization of the T state at low pH, which explains the strong reduction in both affinity and cooperativity.

In 1982, Perutz and Brunori proposed the first stereochemical model of the Root effect. According to this model, the replacement of Cys F9β (present in human Hb) with a more hydrophilic Ser residue in many fish Hbs allows formation of two additional hydrogen bonds that increase the stabilization of the T state. Other residues were also considered to be important, leading to the conclusion that "the constellation of polar residues needed to produce the Root effect and the accompanying large Bohr effect in teleost fish seems to consists of Lys EF6, Ser F9, Glu FG1, Arg H21 and His HC3, all on the β chain" (Perutz and Brunori, 1982, p. 426). On the basis of this theoretical model, two mutants of human HbA were produced by site-directed mutagenesis, Hb Nympheas [Cys F9β → Ser] and Daphne [Cys F9β → Ser; His H21β → Arg], to verify induction of the Root effect. However, contrary to the predictions of the model, these Hbs showed a decrease in the Bohr effect and an increased oxygen affinity (Nagai et al., 1985). Crystallographic analysis of these mutants in the T state showed that Ser F9β was bound to Asp FG1β, which thus competes with His HC3β. The partial breakage of the salt bridge between His HC3β and Asp FG1β could account for a less stable T state and the reduction in the Bohr effect observed (Luisi and Nagai, 1986). Introduction of further mutations (e.g., [Cys F9β → Ser; Asp FG1β → Glu]; see Luisi et al., 1987) similarly failed to produce a Root effect in human Hb.

The resolution of the crystallographic structures of the Root effect Hb from the Antarctic fish *Pagothenia bernacchii* in the liganded (Camardella et al., 1992) and unliganded form (Ito et al., 1995) indicates that His HC3β could not function as a Bohr group, as it is free in the solvent in both the T and R structures, which makes the Root effect even more difficult to explain. The only possible oxygen-linked proton-binding site found was that between the two aspartate residues Asp G1α and Asp G3β, which are close to each other in the T state (so that they can share a proton). These two residues are, however, conserved in all known fish Hbs regardless of their functional properties. In Hb I from trout, Asp G3β forms a salt bridge with Arg G6β (and not Asp G1α) in the T state, which is broken upon oxygenation (Tame et al., 1996). The structural basis for the Root effect in *P. bernacchii* Hb has still to be found.

A recently proposed molecular mechanism is based on the crystal structure of the liganded (CO) derivative of spot (*Leiostomus xanthurus*) Hb (Mylvaganam et al., 1996). According to this model, protonation at the positively charged residues in the central cavity between the β chains would *destabilize* the R state and shift the allosteric equilibrium toward the T state at low pH. This behavior would be due mainly to a different orientation of the two $\alpha\beta$ dimers in the liganded spot Hb in comparison with liganded human Hb. Two possible proton-binding sites are suggested: one in the central cavity of the Hb molecule, where one proton could be shared between the N terminus of the β chain and His HC3β, and the other between Asp G1α and Asp G3β, by analogy with *P. bernacchii* Hb.

The number of histidine residues (which can act as proton exchangers in the pH range 6.0–8.0, where the Bohr effect is observed) is lower in

Root effect Hbs than in human Hb. Therefore, factors other than a mere increase in the number of salt bridges must be implicated in the stabilization of the T state of Root effect Hbs. Analysis of phylogenetic relationships in amino acid sequences indicates that Root effect Hbs are "sharply discriminated from mammalian Hbs in several regions of the α and β chains, whereas shark, minor components of teleost fish and amphibian, reptile, and bird Hbs showing no Root effect exhibit a gradual change to mammalian Hb in a straightforward way" (Horimoto *et al.*, 1990, p. 302).

Residues that are important in the regulation of oxygen affinity in Root effect and non-Root effect Hbs are shown in Table 1. The distinction between Bohr effect Hbs (like human Hb), which switch between T and R states at low pH, and Root effect Hbs, which do not, indicates that the $\alpha_1\beta_2$ contact, which is implicated in the cooperative transition, may play a crucial role in the origin of the Root effect. This interface in fish Hbs appears to be significantly different from that of human Hb, as can be seen in Table 1. It appears that a condition for expression of the Root effect is the conservation of amino acid residues with similar properties in positions C3, C6, and CD2 of the α and FG4 of the β subunit at the $\alpha_1\beta_2$ switch interface. It has been hypothesized (Fago, 1995) that, at low pH, this interface is able to accommodate conformational changes occurring in liganded subunits within the T quaternary structure in Root effect Hbs, but not in Hbs with a normal Bohr effect, as in human Hb. This would explain why Root effect Hbs are able to bind ligand while remaining in the T state (with loss of cooperativity) and why transition to the high-affinity state occurs normally upon oxygenation in Bohr effect Hbs. The "overstabilization" of the T state in the Root effect would be due to an ability of the $\alpha_1\beta_2$ region to remain in the low-affinity conformation at low pH. Site-directed mutagenesis studies on cloned anodic (showing Root effect—Weber *et al.*, 1976a) and cathodic (without Root effect—Fago *et al.*, 1995) Hbs from eel are in progress to test this hypothesis.

In Root effect Hbs, the Hill coefficient is commonly reduced to values below unity at low pH (see Fig. 2c for an example), which is related to a marked difference in the O_2 affinity of the α and β subunits. Thus, the functional properties of Root effect Hbs can be explained by a strong stabilization of the T state at low pH, where marked functional chain heterogeneity is expressed. The kinetics of CO binding to Root effect Hbs reveals biphasic curves at low pH, reflecting slow- and fast-reacting molecular species, each contributing 50% to the total reaction. In the low-affinity conformation, the affinity ratio between the chains may be 10 to 100 (Pennelly *et al.*, 1978; Morris and Gibson, 1982) or even up to 500 in some deep-sea fish Hbs (Noble *et al.*, 1986). The affinities of the α and β subunits also differ in T-state human HbA, but only by a factor of 5, as

Table I

Functionally Important Residues in Root Effect and Non-Root Effect Hemoglobins

	α chain			β chain						
	C3	C6	CD2	NA2	EF6	F9	FG1	FG4	H21	HC3
Non-Root Effect Hbs										
Human	Thr	Thr	Pro	His	Lys	Cys	Asp	His	His	His
Lepidosiren paradoxus	Gly	Ser	Pro	His	Lys	Ser	Glu	His	Arg	His
Latimeria chalumnae	Gln	Val	Asp	His	Lys	Phe	His	His	Arg	His
Electrophorus electricus	Glu	Thr	Ala	Glu	Lys	Ser	Glu	His	Lys	His
Gymnodraco acuticeps	Gln	Ile	Ser	Asn	Glu	Ser	Glu	His	Lys	His
Oncorhynchus mykiss HbI	Gln	Thr	Ser	Glu	Leu	Ala	Asn	Phe	Ser	Phe
Anguilla anguilla cathodic Hb	Ala	Val	Ser	Glu	Lys	Asn	Glu	Asn	Lys	Phe
Root Effect Hbs										
Oncorhynchus mykiss HbIV	Gln	Ala	Ser	Glu	Lys	Ser	Glu	His	Arg	His
Cyprinus carpio	Gln	Thr	Ala	Glu	Lys	Ser	Glu	His	Arg	His
Notothenia angustata Hb1	Gln	Thr	Ser	Lys	Ala	Ser	Glu	His	Lys	His
Chelodonichtys kumu	Gln	Thr	Thr	Glu	Lys	Ser	Glu	His	Arg	His
Pagothenia bernacchii Hb1	Gln	Thr	Ser	Glu	Ala	Ser	Glu	His	Lys	His

Note. Sequences were obtained from the Swiss-protein data bank.

demonstrated in oxygen-binding experiments using T-state crystals (Rivetti *et al.*, 1993). The low cooperativity within the T state of these crystals results in a perfectly noncooperative binding curve with $n_{50} = 1$. In the absence of interaction between the subunits, the calculated Hill coefficient for the same ratio of subunit affinities would be lower (Rivetti *et al.*, 1993). In Root effect Hbs, the difference between α and β subunits appears to be so large that it cannot be compensated by the T-state cooperativity. The successive binding of ligand first to the high-affinity subunits and then to the low-affinity subunits produces a pseudo-anticooperative binding in the T state. The presence of cooperativity between subunits in the T state of Root effect Hbs will be masked by their strong functional heterogeneity. In tuna Hb (Morris and Gibson, 1982) and in human HbA (Ribetti *et al.*, 1993), it has been possible to discriminate between the spectral properties of the α and β hemes and to demonstrate a higher ligand affinity in the α subunit. Resonance Raman spectra of the Root effect Hb of a deep-ocean fish suggest that constraints at the proximal side of the β heme (induced by the T conformation of the $\alpha_1\beta_2$ interface) exclude ligand binding to the β subunits (Friedman *et al.*, 1990). The oxygenation process of these Hbs appears therefore to follow the central idea of the α_2-*cooperon* model proposed by Di Cera *et al.* (1987), which assumes that, in the T state, oxygen binds only to the α subunits.

IX. TEMPERATURE EFFECT

Oxygenation of Hb is exothermic, whereby Hb can be considered a heat carrier; the heat absorbed upon oxygen unloading in the tissues is liberated in the gills upon oxygenation. The apparent heat (or enthalpy) of oxygenation ΔH_{app} is usually calculated from the slope of log P_{50} versus $1/T$ plots,

$$\Delta H_{app} = 2.303R \frac{\Delta \log P_{50}}{\Delta(1/T)},$$

where R is the gas constant and T the absolute temperature (Wyman, 1964). ΔH_{app} includes the exothermic intrinsic heat of heme oxygenation but also contributions from heats of ionization and ion (protons and anions) binding/release, the heat of solution of oxygen, and the heat of conformational changes (Atha and Ackers, 1974; Imai, 1982). A temperature increase will decrease oxygen affinity directly because of the exothermic nature of oxygenation. In addition, O_2 affinity will be influenced by the pH decrease associated with the temperature dependence of acid–base status (some -0.01 to -0.02 pH unit $°C^{-1}$), provided the charge of groups contributing to the Bohr effect is changed. The reduced oxygen affinity at high temperatures may increase O_2 unloading to the tissues in parallel with an increased cellular O_2 demand, but may also impair oxygen loading at the gills, thereby decreasing arterial oxygen content. Whether the O_2 affinity change is advantageous or disadvantageous depends on its magnitude and the O_2 tensions at the sites of loading and unloading. Moreover, physiological conclusions based on ΔH determination on stripped Hb can be misleading without knowledge of the influence of cellular cofactors. The numerical value of the intrinsic ΔH in tench Hb is very high, but extensive endothermic release of Bohr protons and NTPs during oxygenation masks the heat released in the exothemic oxygenation process and strongly reduces the temperature sensitivity of the Hb under physiological conditions (Jensen and Weber, 1987).

The evolution of warm-bodied teleosts, such as tuna, is associated with the development of hemoglobins with reduced (or even reversed) temperature effects. The temperature independence of blood–O_2 affinity in tuna Hbs results from a normal temperature dependence at low saturation and a reversed dependence at high saturation levels, whereby O_2 equilibrium curves cross each other near half-saturation (Ikeda-Saito et al., 1983). Endothermic oxygen binding at high saturation appears related to the large number of Bohr protons liberated late in the oxygenation process. In the cathodal Hb from trout, a low temperature effect is achieved by an opposite mechanism: ligand (carbon monoxide) binding is exothermic at high saturation but endothermic at low saturation, which presumably is due to an endothermic conformational change of the T state (Wyman et al., 1977).

A lower enthalpy of oxygenation than in temperate fish Hbs has been considered an "energy saving" adaptive mechanism in the hemoglobins of Antarctic teleosts living at constant ($-1.9°C$) temperature (Di Prisco et al., 1991). The values of the enthalpy of oxygenation are, however, influenced by the temperature of measurements, if the van't Hoff plot deviates significantly from linearity (Weber, 1995). The heat of oxygenation in Antarctic fish Hbs is low at high temperatures, but high (comparable to that of other fish species) at low temperatures (Fago et al., unpublished). This has been related to the heat capacity change of the system, which is neglected when linearity of the van't Hoff plot is assumed (Weber, 1995), but becomes a relevant factor in the presence of strong allosteric effects (e.g., the Root effect) in the hemoglobin molecule.

Thermal acclimation in goldfish does not appear to involve synthesis of new globin chains. Rearrangement of subunits creates new iso-Hb components (Houston, 1980). In trout, no effect of thermal acclimation on the oxygen-binding properties of the hemolysates from warm- and cold-acclimated species could be established, although changes were observed in the iso-Hb pattern (Weber *et al.,* 1976c). A temperature increase favors the formation of asymmetric hybrid Hbs in the hemolysate of the Japanese eel (Shimada *et al.,* 1980).

X. ADAPTATION OF HEMOGLOBIN FUNCTION

The intrinsic (genetically coded) functional properties of hemoglobins determined by globin structure (amino acid sequence) result in species differences that may reflect adaptations to particular environments or metabolic requirements. This is exemplified by the high intrinsic Hb–O_2 affinity in hypoxic-tolerant fish species and low intrinsic O_2 affinity in active fish living in well-aerated water. Within a given species, the ability to transport O_2, CO_2, and H^+ in the circulating blood can be modified by (i) changes in [Hb] (e.g., via release of RBCs from the spleen or stimulation of erythropoiesis), (ii) changes in the erythrocytic concentration of allosteric effectors, and (iii) shifts in synthesis of Hb components with different functional properties (cf. reviews by Weber, 1982, 1990; Jensen, 1991). An evaluation of the physiological significance of such changes must take into consideration the shape and position of blood gas equilibrium curves as well as actual arterial and venous gas tensions and other parameters. In general, the changes are believed to reflect appropriate fine-tunings of O_2 and CO_2 transport properties when external or internal milieu factors vary. A few examples of cofactor modulation of Hb function are given next.

A. Organic Phosphates

Exposure to hypoxia leads to down-regulation of NTP levels in fish (Wood and Johansen, 1972). The response is graded to ambient P_{O_2}, and the time required to attain new steady-state NTP levels varies from hours to days (cf. Weber and Jensen, 1988). The decrease in NTP content raises blood O_2 affinity directly through decreased allosteric interaction, and indirectly via the Bohr effect, due to the increased red cell pH caused by changes in the Donnan-like distribution of H^+ across the RBC membrane (Wood and Johansen, 1973). The O_2 affinity increase elevates the arterial O_2 saturation and blood O_2 capacitance at low P_{O_2}, and contributes to a reduction in the ventilatory requirement. Although ATP is the main phosphate cofactor in some species (such as trout, plaice, and dogfish), others (such as eel, carp, tench, goldfish, and lungfish) also have high concentrations of GTP. Where both cofactors are present, GTP commonly exerts a greater modulator role than ATP, as reflected by greater decreases in concentration compared to ATP under hypoxia, its greater allosteric effect, and a lesser inhibition of its effect by complexing with Mg^{2+} ions (Weber et al., 1976; Weber and Lykkeboe, 1978; review: Weber and Jensen, 1988). The difference in the effects of ATP and GTP on O_2 affinity is consistent with the difference in the free energy contributed to the stability of the T structure by one H bond (Weber and Lykkeboe, 1978; Weber et al., 1987). The decrease in red cell NTP content is universal upon long-term decreases in blood P_{O_2}, and even occurs in Antarctic fish, which are unlikely to encounter hypoxia (Wells et al., 1989). It may be hypothesized that the P_{O_2} modulation of NTP levels is an indigenous feature, which was acquired early in the evolution of fishes. Selection pressure may have worked toward an increase in RBC [NTP] with P_{O_2} increase rather than for a decrease in hypoxia. Early in the evolution of fishes, atmospheric P_{O_2} was low, as may have been red cell [NTP]. As ambient P_{O_2} increased, a lower O_2 affinity and higher [NTP] may have been favored.

Whereas changes in the cellular content of NTPs are important in modulating blood oxygen affinity during hypoxia, the cellular NTP content remains largely unchanged when O_2 shortage results from strenuous exercise (e.g., Weber and Jensen, 1988; Lowe et al., 1995) or severe methemoglobinemia (Jensen et al., 1987). However, the cellular concentration may change as a result of RBC volume changes.

B. Red Cell pH and Volume

The immediate hyperventilation upon exposure to hypoxia limits the fall in arterial P_{O_2}, but also decreases P_{CO_2}, thereby inducing a respiratory alkalosis. Due to the passive distribution of H^+ across the RBC membrane, the alkalosis tends to be transferred to the red cell, increasing O_2 affinity via the Bohr effect. The response may be counteracted by lactacidosis during severe hypoxia or by environmental hypercapnia (increase in P_{CO_2}), which accompanies hypoxia in many habitats (cf. Jensen et al., 1993).

β-Adrenergic stimulation of Na^+/H^+ exchange across fish RBC membranes (Nikinmaa, 1992) is an alternative rapid mechanism for increased Hb–O_2 affinity in hypoxia and other stress situations. The selective elevation of erythrocytic pH is of major importance for the affinity increase (Nikinmaa, 1983), but the increase in cell volume may also contribute by decreasing the cellular concentrations of, and complexing between, Hb and NTP. In carp Hb solutions, P_{50} increases when the Hb concentration is increased toward natural RBC concentrations at constant [NTP]/[Hb] (Lykkeboe and Weber, 1978). A specific influence of cell volume on O_2 affinity is supported by data on intact shrunken (Jensen, 1990) and swollen (Holk and Lykkeboe, 1995) teleost RBCs. In lamprey RBCs, the cellular Hb concentration is also important, the mechanism being related to the aggregation/dissociation behavior of lamprey Hb (Nikinmaa *et al.*, 1995).

C. Other Effectors

Effectors other than NTP and H^+ may be involved in setting whole blood O_2 affinities. In obligatory air-breathing adults of the Amazonian osteoglossid *Arapaima gigas,* the predominant phosphate is inositol penta-phosphate (IPP) (normally present in bird RBCs), whereas NTP dominates in water-breathing juveniles (Isaacks *et al.*, 1977; Val, 1993). IPP also occurs in elasmobranch RBCs, being present at [IPP]/[NTP] ratios of 0.1–0.2 in the dogfish *Squalus acanthias* and the ray *Narcacion nobiliana* (Borgese and Nagel, 1978). The "mammalian" cofactor 2,3-DPG occurs in the facultative gut-breathing *Hoplosterum littorale,* where its concentration increases with ambient temperature (Val, 1993).

The osmotic effector urea modifies Hb–O_2 affinity in elasmobranchs. In *Squalus acanthias,* urea increases O_2 affinity and antagonizes the effect of ATP (Weber *et al.*, 1983), and increased red cell urea concentration raises O_2 affinity during warm acclimation in *Cephaloscyllium isabella* (Tetens and Wells, 1984). Trimethyl amine oxide (which counteracts the destabilizing effect of urea on protein structure; Yancey *et al.*, 1982) does not influence the ATP and urea sensitivities of *S. acanthias* Hb, suggesting that the latter effects are mediated by specific binding rather than quaternary structural changes (Weber, 1983; Weber *et al.*, 1983). In human Hb, high urea concentrations similarly increase O_2 affinity due to formation of carbamylated Hb (Monti *et al.*, 1995).

The discovery that the O_2 affinity of human Hb, in the presence of neutral solutes, decreases linearly with osmotic pressure (Colombo *et al.*, 1992) suggesting modulation of allosteric transitions by water. This type of allosteric modulation has obvious implications for *in vivo* functions, particularly in animals where blood osmotic pressures vary strongly. Such effects have been observed in carp Hb (Kwiatkowski and Noble, 1993). They are, however, absent in cathodic trout Hb I that shows cooperativity (which is considered a requirement for solvation effects—Bellelli *et al.*, 1993).

XI. HEMOGLOBIN MULTIPLICITY

In contrast to humans (and most mammals) that have a single main Hb component, fishes commonly exhibit Hb multiplicity (different "isoHbs" that occur in the same individual at the same or different stages of its development) and Hb polymorphism (different "alloHbs" in genetically different strains of the same species). Hb multiplicity may result from either gene-related heterogeneity (variation in gene activity) or nongenetic heterogeneity (chemical modification *in vivo* or *in vitro*) (Kitchen, 1974; Weber, 1990).

A. Functional Differentiation

Based on multiplicity, teleost fish fall into three major classes (Weber *et al.*, 1976b): class I comprising electrophoretically anodal Hbs that have (normal) Bohr, Root, phosphate, and temperature effects (as in carp; Weber and Lykkeboe, 1978); class II comprising Hbs that consist of both anodal components (with properties as in class I) and cathodal ones, which exhibit high O_2 affinities and small, often reverse Bohr effects (whereby high affinity is maintained at low pH) and low temperature sensitivities (e.g., trout and eel; Binotti *et al.*, 1971; Weber *et al.*, 1976a); and class III Hbs that appear to be sensitive to pH but insensitive to temperature (e.g., tuna; Rossi-Fanelli *et al.*, 1960). The individual Hbs are conventionally labeled with roman numerals starting with the component having the highest isoelectric point (thus in trout, Hbs I and IV are the major cathodal and anodal Hbs, respectively). Multiplicity is also common among elasmobranchs (skates and rays often having 12–13 Hb components; Fyhn and Sullivan, 1975), but marked functional differences as observed in class II teleost Hbs have not been reported.

As earlier reviewed (cf. Weber, 1982, 1990), the cathodal Hbs will assume increasing significance in O_2 transport under hypoxic and acidotic conditions (where O_2 loading to anodal components is compromised due to the Bohr and Root effects) and may function as a circulating O_2 reserve. Predictably, cathodal Hbs will maintain oxygenation in the face of pH decreases associated with acidification of blood in the gas gland of swim bladders and in the eyes, although anodal Hbs leaving the swim bladder will also be able to transport O_2, as their Root effect is switched off as a result of acid movements that alkalize the efferent blood (Pelster and Weber, 1991). The occurrence of cathodal Hbs in catfish species appears to be linked with occupation of fast-flowing streams where frequent acidoses will unload anodal Hbs (Powers, 1972). This aligns with the apparent ubiquity of cathodal components in active salmonids and their absence in inactive flatfish (cf. Weber, 1990).

The cathodal as well as the anodal Hbs may themselves be heterogeneous. The moray eel *Muraena helena* exhibits three Hb phenotypes, having a cathodal component that concurs with either one or both of two anodal components (Pellegrini *et al.*, 1995). Salmonids commonly have four types of chains, α^C, α^A, β^C, and β^A (where C and A denote cathodal and anodal chains), each of which may occur in two electrophoretically distinct forms (Wilkins, 1985). In *Salmo salar,* these 8 chains combine to form 17 distinct (8 cathodal and 9 anodal) Hbs. Sockeye salmon *Onchorhyncus nerka* anodal Hbs are composed of one α and three β chains and the cathodal Hbs of two α and two β chains (Tsuyuki and Ronald, 1971). Electrophoresis indicates the presence of three electrophoretically "slow" (cathodal) and three "fast" (anodal) rainbow trout Hbs that possess allelic variants in the β-type chains (Ronald and Tsuyuki, 1971). In Japanese eel *Anguilla japonica* (Shimada *et al.,* 1980) six different globin chains (α^I, β^I, γ^I, α^{II}, β^{II}, and γ^{II}) form three cathodal (Ia, Ib, and Ic) and three anodal Hb components (IIa, IIb, IIc), where Ib and IIb are asymmetric hybrid tetramers formed by subunit exchange $[\alpha^I_2 \beta^I_2$ (Ia) $+ \alpha^I_2 \gamma^I_2$ (Ic) $\leftrightarrow 2\alpha^I_2 \beta^I \gamma^I$ (Ib) and $\alpha^{II}_2 \beta^{II}_2$ (Ia) $+ \alpha^{II}_2 \gamma^{II}_2$ (IIc) $\leftrightarrow 2\alpha^{II}_2 \beta^{II}\gamma^{II}$ (IIb)]. In the European eel *A. anguilla,* no asymmetric molecules have been detected (Fago *et al.,* 1995). The occurrence of all major trout Hbs in the same red cells (Brunori *et al.,* 1974) suggests that subunit exchange may occur *in vivo.*

The functional differentiation between anodal and cathodal fish Hbs correlates with molecular structure and the available knowledge on the sites implicated in ligand binding in human Hb A (Weber, 1990, 1996). Thus, in cathodal trout Hb I, the absence of the Bohr effect is partly attributable to changes at two sites that account for 84% of the alkaline Bohr effect in human Hb A at physiological pH (Di Cera *et al.,* 1988), viz. acetylation of the N-terminal Val of the α chain and substitution of His HC3β by Phe.

Compared to the anodal components that do respond to organic phosphates, cathodal fish Hbs vary strikingly. Whereas the O_2 affinity of cathodal trout Hb I is insensitive to NTP (Brunori *et al.,* 1975), the ATP and GTP sensitivities of cathodal eel exceed those of anodal eel Hb (Weber *et al.,* 1975, 1976a). In cathodal eel Hb, all the sites implicated in phosphate binding to fish Hbs (Val NA1β, Glu NA2β, Lys EF6β, and Lys H21β) are conserved (Table 1; Fago *et al.,* 1995). In trout Hb I, phosphate insensitivity correlates with Lys EF6$\beta \rightarrow$ Leu and Arg H21$\beta \rightarrow$ Ser replacements, which represents a loss of 3 ATP- and 4 GTP-binding sites (cf. Gronenborn *et al.,* 1984; Weber, 1996). X-ray crystallographic studies of the two quaternary structures of trout Hb I show that minimization of the sensitivities to pH and organic phosphates are associated with a stabilization of the low-affinity T structure in the absence of heterotropic interactions (Tame *et al.,* 1996). In this Hb, the T-state-stabilizing α119Pro-β55Met contact, which is absent in high-affinity Hbs such as altitude-tolerant geese (Weber *et al.,* 1993), appears to be replaced by other contacts (Ile residues at β33 and

α124), causing a low O_2 affinity (Tame *et al.*, 1996) compared to other cathodal Hbs.

Cathodal Hbs from several species exhibit reverse Bohr effects in the absence of phosphates (cf. Weber, 1996). A reversed Bohr effect (which is masked by phosphates) is seen in the cathodal Hbs from trout (Weber *et al.*, 1976c), in the Amazonian teleosts *Mylossoma* sp. and *Hoplosternum littorale* (Martin *et al.*, 1979; Garlick *et al.*, 1979), and in the American and the European eel (Gillen and Riggs, 1973; Weber *et al.*, 1976a). The reverse Bohr effect is associated with the opposite control mechanism from anodal Hbs: in cathodal eel Hb, decreasing pH raises K_T values and destabilizes the T structure (Feuerlein and Weber, 1996).

B. Variation in Quaternary Structure

As mentioned earlier, some fish Hbs dimerize or polymerize under specific conditions. It is not always clear, however, to what extent changes observed *in vitro* occur *in vivo*.

As first observed with lampreys, cyclostome Hb has a lower molecular weight in the oxygenated than in the deoxygenated state (Briehl, 1963). Similarly, dogfish shark (*Squalus acanthias*) Hb appears to be dimeric in the liganded form and tetrameric in the deoxy form, and Hbs from several elasmobranchs polymerize upon oxidation (Fyhn and Sullivan, 1975).

Oxidizing conditions may have a dual effect, i.e., metHb formation (which favors the R state and dissociation) and oxidation of cysteine residues (that favors aggregation through S–S bond formation). Goosefish (*Lophius americanus*) Hb, which readily autoxidizes at neutral pH and has a large number of sulfhydryl groups, polymerizes *in vitro* by disulfide bond formation (Borgese *et al.*, 1988). In the red gurnard *Chelodonichthys kumu*, polydisperse high-order aggregates are formed through S–S bridge formation between externally placed Cys CD8β residues without evidence for altered functional properties (Fago *et al.*, 1993). About 40% of the Hb of *Hoplias malabarica* (particularly the cathodal components) polymerize to octamers and larger aggregates *in vitro* (Reischl, 1976). As observed in the ocean pout polymerization may also be induced by heavy metals such as copper (Borgese *et al.*, 1994).

C. Ontogenetic Changes

Specific fetal isoHbs having higher intrinsic O_2 affinities than adult Hbs occur in the live-bearing elasmobranchs *S. acanthias*, *Raja binoculata*, and *Cephaloscyllium ventriosum* (Manwell, 1958, 1963; King, 1994). In the viviparous teleost *Zoarces viviparus*, which has six major adult and three fetal

Hbs (Hjorth, 1974, 1975), blood NTP levels and ATP sensitivity of Hb oxygenation are similar in both life stages, and maternal–fetal transfer of O_2 is based on a markedly higher intrinsic O_2 affinity and a lower Bohr factor in the fetal Hb (Weber and Hartvig, 1984).

Ontogeny of hemoglobins in *oviparous* fish is well documented, although it is less easily identified than in mammals that have fewer Hb components. Generally, the ratio of cathodal to anodal Hbs (C/A ratio) increases with growth (Hashimoto and Matsuura, 1960). In *Salmo salar* and *S. trutta,* the kinetics of the change (Koch *et al.,* 1966) suggest the absence of cathodal Hb at hatching. The increase in C/A ratios represents an increase in cathodal Hb rather than a decrease in anodal Hb and is influenced by, but not directly attributable to, smoltification, migration, and environmental conditions (cf. Wilkins, 1985). Ontogenetic changes in Hb pattern have been documented in several species. Although alevins of coho salmon (*Onchorhynchus kisutch*) have a single cathodal and 12 anodal Hb components, the fry retains only 3 of these anodal Hbs, and presmolt adults have 5 anodal and 5 cathodal tetramers (Giles and Vanstone, 1976). In Arctic charr *Salvelinus alpinus,* which expresses 17 Hbs during life, embryonic and newly hatched alevins exhibit 10 anodal and 3 cathodal components, only 5 and 1 of which, respectively, appear to be found in older fish (Giles and Rystephanuk, 1989). In rainbow trout, where larval Hb lacks a Bohr effect and has a higher affinity than adult Hb, and where both Hbs appear to be highly polymorphic without having subunits in common (Iuchi, 1973), a major Hb with low isoelectric point ($pI = 5.7$) occurring in 1 g (4 cm) specimens is almost completely replaced by the adult components in 55 g (17 cm) specimens (R. E. Weber, unpublished). The changes are consistent with differences in erythropoeitic events. As indicated by Iuchi (1973), a primitive red cell (L) line synthesized in the yolk sac of prehatch embryos is replaced by the definitive (A) line in the kidney and spleen in the early fry stage.

D. Significance and Environmental Influences

In addition to permitting a division of labor between components that extends the range of conditions under which the composite pigment transports oxygen, Hb multiplicity may have a range of other advantages. Hb multiplicity may safeguard against deleterious mutational changes in the genes that code for specific α and β chains, permit higher total erythrocytic Hb concentration (by the phase rule that protein concentration in a saturated solution is higher in multicompone than single-component solutions), increase the production rate of gene products, and increase the capacity to absorb redox stress (cf. Weber, 1990).

The higher number of Hb components in Amazonian than in Antarctic fish may be related to influences of O_2 availability (which is highly variable in freshwater and constantly high in polar habitats) and O_2 demand (which increases with temperature) on the evolution of heterogeneity patterns, which implies a division of labor between the components (cf. Weber, 1990). The mean number of Hbs reported for temperate species appears to be only marginally higher than that in fishes from thermostable tropical environments (Pérez and Rylander, 1985).

Investigating whether environmental factors influence isoHb abundance within individual fish, Houston and co-workers (Tun and Houston, 1986; Marinsky *et al.*, 1990) found increases in individual components in rainbow trout subjected to variable conditions of temperature, O_2 availability, and photoperiod for 2–3 weeks. The absence of significant isoHb differences in splenectomized and sham-operated trout subjected to gradual warming to stimulate erythropoiesis indicates that such adjustments in isoHb concentrations are not attributable to site-related differences in erythroid cell lineages (Murad and Houston, 1991).

XII. OXIDATION OF HEMOGLOBIN

A. Autoxidation

Autoxidation is the spontaneous oxidation of Hb (heme iron in oxidation state II) to metHb (heme iron in oxidation state III) by molecular oxygen. MetHb cannot bind O_2 and accordingly is functionally inert with respect to oxygen transport. Free heme groups are rapidly and irreversible oxidized by O_2 but the embedding of the heme groups in the globin effectively protects against autoxidation. Oxidation is promoted by anions such as CN^- or N_3^-, which are strong ligands for ferric iron, thus stabilizing the Fe^{3+} state relatively to the Fe^{2+} state (Wallace *et al.*, 1982). Autoxidation of Hb may involve the steps (Wallace *et al.*, 1982)

$$Hb(Fe^{2+})O_2 \leftrightarrow Hb(Fe^{2+}) + O_2$$

$$Hb(Fe^{2+}) + X \leftrightarrow Hb(Fe^{2+})X$$

$$Hb(Fe^{2+})X + O_2 \rightarrow Hb(Fe^{3+})X + O_2^-,$$

where X is a nucleophile (e.g., added CN^- or N_3^-). In the physiological situation, water or OH^- would be the promoting ligand. The reaction scheme suggests that (i) the promoting nucleophile is bound to the iron in metHb, (ii) superoxide is a reaction product, and (iii) autoxidation occurs in the fraction of Hb that is deoxygenated. This latter feature rationalizes why the rate of autoxidation increases in mammalian Hb when P_{O_2} (and thus O_2 saturation) is decreased at high O_2 levels (Antonini and Brunori, 1971; Wallace *et al.*, 1982). A similar effect is seen in carp Hb, which has a higher autoxidation rate when equilibrated with air than with pure O_2 at 40°C (Fig. 4).

The globin provides protection against autoxidation partly by decreasing the accessibility of nucleophiles to the heme pocket. The distal histidine (His E7), which is highly conserved in hemoglobins, contributes significantly to this protection (Springer *et al.*, 1989; Perutz, 1990). This is reflected by the large increase in autoxidation rate when His E7 is replaced by other amino acid residues, using site-directed mutagenesis (Springer *et al.*, 1989). The inhibition of autoxidation by His E7 results from both a restriction of water entry into the heme pocket and a stabilization of bound O_2 (Springer *et al.*, 1989).

The rate of autoxidation increases with an increase in temperature. In fish Hbs, the rates appear to be minimal at the temperatures normally encountered by the species. Thus, when compared at a constant high temperature, cold-adapted species have higher autoxidation rates than warm-adapted ones (Wilson and Knowles, 1987). Figure 4 illustrates the temperature dependence of autoxidation in stripped carp Hb at a typical red cell pH. The autoxidation rate is low between 5 and 25°C, but it increases

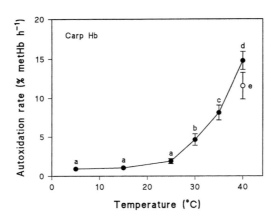

Fig. 4. Rate of autoxidation as function of temperature in stripped carp hemoglobin equilibrated with air (●) or pure O_2 (○). Means ± SEM (N = 4–9). Significant differences between values (analysis of variance followed by the Tukey multiple comparison test) are indicated by different letters at the points. Tetrameric [Hb] = 0.04 mM; pH 7.3; Hepes buffer 0.1 M (F. B. Jensen, C. Tøgersen, and A. Bach, unpublished data).

sharply and significantly for each 5°C increase in temperature between 25 and 40°C (F. B. Jensen *et al.,* unpublished).

Apart from the influences of oxygenation and temperature, autoxidation may depend on pH. A decrease in pH accelerates the oxidation rate of mammalian Hbs (cf. Wallace *et al.,* 1982; Perutz, 1990). Information on this aspect is scant for fish Hbs.

B. Nitrite-Induced Oxidation

Oxidation of Hb is promoted by various chemicals, of which nitrite has physiological relevance in fish. Nitrite is actively taken up across the gills and concentrated in the blood, whereby even low concentrations in the ambient water can result in significant methemoglobinemia. The kinetics of nitrite-induced oxidation of oxyHb differs radically from that of autoxidation. Autoxidation of oxyHb solutions results in a slow and practically linear rise in met Hb content with time. Nitrite-induced oxidation of oxyHb is considerably faster, and it proceeds via an initial lag phase followed by an autocatalytic increase in reaction rate in both mammalian (Kosaka and Tyuma, 1987; Spagnuolo *et al.,* 1987) and fish (Jensen, 1990, 1993) Hbs. The reaction involves several steps in which reactive intermediate species such as H_2O_2, $NO_2\cdot$, and ferrylhemoglobin ($Hb(Fe^{IV})=O$) are produced (Kosaka and Tyuma, 1987; Spagnuolo *et al.,* 1987). The latter two are key species in the free radical chain propagation during the autocatalytic phase. With deoxyHb, nitrite-induced oxidation is slower and proceeds via a reaction mechanism different from that with oxyHb. In fact, a lowering of O_2 saturation via decreased P_{O_2} or addition of ATP or GTP is sufficient to strongly retard metHb formation and to obliterate the autocatalytic phase in rainbow trout Hb (Jensen, 1993). However, this need not imply that oxidation is decreased by a fall in blood O_2 saturation *in vivo,* since nitrite permeates carp RBC membranes much faster at low than at high saturation (Jensen, 1990, 1992). In addition, metHb formation in intact RBCs is countered by metHb reductase systems, principally NADH metHb reductase (Jaffé, 1981; Freeman *et al.,* 1983). The actual degree of methemoglobinemia developed *in vivo* is a complex interplay of numerous factors, including nitrite load, oxygen saturation, concentration of RBC cofactors (e.g., NTP, NADH), and metHb reductase activity.

C. Interrelationship between Oxidation and Oxygen Affinity

Partial oxidation of Hb shifts the T ↔ R equilibrium in favor of the R state, increasing O_2 affinity of remaining ferrous heme groups, as verified by studies on Hb solutions (Falcioni *et al.,* 1977; Kwiatkowski *et al.,* 1994). This effect at the molecular level is, however, not always expressed at the cellular and organismic levels, as exemplified by nitrite-induced methemoglobinemia in carp. During nitrite exposure, a large decrease, rather than an increase, in blood O_2 affinity is observed (Jensen *et al.,* 1987; Jensen, 1990). The decrease in affinity is mainly due to RBC shrinkage, which increases the concentrations and complexing of cellular Hb and NTP (Jensen, 1990). In the Amazonian fish *Semaprochilodus insignis,* blood oxygen affinity has been observed to increase 1 h after an intraperitoneal injection of nitrite (Bartlett *et al.,* 1987), which may reflect the absence of RBC shrinkage in this species under acute and mild methemoglobinemia.

XIII. INTERACTIONS OF HEMOGLOBIN WITH MEMBRANE PROTEINS

In mammalian erythrocytes, Hb binds to membrane components, and the interaction is mainly electrostatic. The most abundant membrane protein is band 3, which is present in about 1 million copies per RBC. Band 3 consists of a membrane domain, which mediates the physiological important anion exchange across the membrane, and a cytoplasmic domain, which is anchored to the cytoskeleton and to which hemoglobin and glycolytic enzymes also bind (Salhany, 1990). Binding of Hb occurs at the N-terminal fragment of the cytoplasmic domain, and deoxyHb binds more tightly than oxyHb (Walder *et al.,* 1984; Tsuneshige *et al.,* 1987). The binding site in Hb is the central cavity between the β chains, the same site as for organic phosphate binding (Walder *et al.,* 1984). The end of the human cytoplasmic band 3 domain contains several negatively charged amino acid residues that seem to correspond stereochemically to the positive charges of the organic phosphate-binding site on the Hb. The primary structure of rainbow trout band 3 reveals a similar cluster of negative charges in the cytoplasmic domain of trout band 3 (Hübner *et al.,* 1992), suggesting that a similar interaction between Hb and the acidic N terminus of the cytoplasmic fragment of band 3 may be present in fish.

The O_2 affinity of human Hb is lowered in the presence of (i) the cytoplasmic domain of band 3, (ii) a synthetic peptide consisting of the first 11 amino acids of the N terminus (Walder *et al.,* 1984), or (iii) RBC membranes (Tsuneshige *et al.,* 1987). This illustrates preferential binding and stabilization of T state Hb but is insignificant for oxygen transport, since only a small fraction of the cellular Hb molecules binds to the membrane. The binding of Hb to the membrane may, however, have other physiological functions. It is possible that competition between Hb and glycolytic enzymes for binding sites on band 3 is involved in the regulation of glycolysis (Walder *et al.,* 1984). Similarly, interaction of Hb with membrane proteins may influence membrane ion transport. Recent data on fish RBCs support the latter idea. Adrenergic Na^+/H^+ exchange is more pronounced in deoxygenated than in oxygenated RBCs (Motais *et al.,* 1987) and is inhibited when deoxyHb is converted to metHb (Nikinmaa and Jensen, 1992). Potassium fluxes similarly respond to changes in Hb oxygenation/conformation. Oxygenation activates a K^+/Cl^- cotransport out of the cells, which is inactivated by deoxygenation (Jensen, 1990, 1992; Nielsen *et al.,* 1992). The K^+ efflux can also be stimulated by nitrite-induced metHb formation in deoxygenated RBCs, which converts a large fraction of T Hb to the R-like conformation, and the transport can be inhibited by a gradual shift of the allosteric R–T equilibrium toward the T state via decreases in pH (Jensen, 1990, 1992). Thus, the relative proportion of T and R structure Hb appears to be involved in the control of ion transport across the RBC membrane. It has been suggested that such effects could originate in the binding of Hb to band 3 by inducing a conformational change in the cytoplasmic domain of band 3 that is transmitted via the cytoskeleton to other membrane ion transport sites (Motais *et al.,* 1987).

If preferential binding of deoxyHb to the cytoplasmic domain of band 3 is involved in the control of Na^+/H^+ exchange and K^+/Cl^- cotransport, then it may also be expected that band-3-mediated anion exchange is influenced. This hypothesis was recently investigated by comparing the kinetics of band-3-mediated Cl^- transport in oxygenated and deoxygenated RBCs from carp, rainbow trout, eel, cod, and human. The rate constant for unidirectional $^{36}Cl^-$ efflux did not differ significantly between oxygenated and deoxygenated RBCs in any of the species (Jensen and Brahm, 1995). The lack of an oxygenation effect may be explained by the flexible link between the cytoplasmic domain (where Hb binds) and the membrane domain (mediating anion transport) of band 3, which results in minimal interactions between the two domains. The results, however, also draw attention to the fact that direct evidence for Hb binding to band 3 and other membrane proteins is lacking in fish. We are presently investigating the interaction between trout Hb and synthetic peptides of the N-terminal fragment of the cytoplasmic domain of trout band 3.

Denatured Hb also binds to the cytoplasmic domain of band 3. In mammalian RBCs, this appears to be important in the removal of old cells from circulation. Thus, copolymerization of denatured Hb (hemichromes) with the cytoplasmic domain of band 3 seems to cause band 3 to form clusters that serve as binding sites for autoantibodies (IgG) that promote removal of the RBCs by the immune system (e.g., phagocytosis by macrophages) (Schlüter and Drenckhahn, 1986; Low, 1991).

ACKNOWLEDGMENTS

The authors acknowledge support from the Danish Natural Science Research Council (Centre for Respiratory Adaptation).

REFERENCES

Ackers, G. K. (1990). The energetics of ligand-linked subunit assembly in haemoglobin require a third allosteric structure. *Biophys. Chem.* **37,** 371–382.

Ackers, G. K., Doyle, M. C., Myers, D., and Daugherty, M. A. (1992). Molecular code for cooperativity in haemoglobin. *Science* **255,** 54–63.

Atha, D. H., and Ackers, G. K. (1974). Calorimetric determination of the heat of oxygenation of human hemoglobin as a function of pH and the extent of reaction. *Biochemistry* **13,** 2376–2382.

Antonini, E., and Brunori, M. (1971). "Haemoglobin and Myoglobin in Their Reactions with Ligands." North-Holland, Amsterdam.

Baldwin, J., and Chothia, C. (1979). Haemoglobin: The structural changes related to ligand binding and its allosteric mechanism. *J. Mol. Biol.* **129,** 175–220.

Barra, D., Petruzzelli, R., Bossa, F., and Brunori, M. (1983). Primary structure of haemoglobin from trout (*Salmo irideus*): Amino acid sequence of the β chain of trout HbI. *Biochim. Biophys. Acta* **742,** 72–77.

Bartlett, G. R., Schwantes, A. R., and Val, A. L. (1987). Studies on the influence of nitrite on methaemoglobin formation in amazonian fishes. *Comp. Biochem. Physiol.* **86C,** 449–456.

Bellelli, A., Brancaccio, A., and Brunori, M. (1993). Hydration and allosteric transitions in hemoglobin. *J. Biol. Chem.* **268,** 4742–4744.

Binotti, I., Giovenco, S., Giardina, B., Antonini, E., Brunori, M., and Wyman, J. (1971). Studies on the functional properties of fish hemoglobins. II. The oxygen equilibrium of the isolated hemoglobin components from trout blood. *Arch. Biochem. Biophys.* **142,** 274–280.

Borgese, T. A., and Nagel, R. L. (1978). Inositol pentaphosphate in fish red blood cells. *J. Exp. Zool.* **205,** 133–140.

Borgese, T. A., Harrington, J. P., Ganjian, I., and Duran, C. (1988). Haemoglobin properties and polymerization in the marine teleost *Lophius americanus* (Goosefish). *Comp. Biochem. Physiol.* **91B,** 663–670.

Borgese, T. A., Bourke, S., Frias, B., Johnson, D., and Harrington, J. (1994). Copper induced polymerisation of haemoglobin from the ocean pout, *Macrozoarces americanus. Biol. Bull.* **187,** 246–247.

Breepoel, P. M., Kreuzer, F., and Hazevoet, M. (1980). Studies of the haemoglobins of the eel (*Anguilla anguilla* L.). I. Proton binding of stripped hemolysate; separation and properties of two major components. *Comp. Biochem. Physiol.* **65A,** 69–75.

Breepoel, P. M., Kreuzer, F., and Hazevoet, M. (1981). Studies of the haemoglobins of the eel (*Anguilla anguilla* L.). III. Proton and organic phosphate binding to the Root effect haemoglobin component. *Comp. Biochem. Physiol.* **69A,** 709–712.

Briehl, R. W. (1963). The relation between the oxygen equilibrium and aggregation of subunits in lamprey hemoglobin. *J. Biol. Chem.* **238,** 2361–2366.

Brittain, T. (1987). The Roof effect. *Comp. Biochem. Physiol.* **86B,** 473–481.

Brunori, M., Giardina, B., Antonini, E., Benedetti, P. A., and Bianchini, G. (1974). Distribution of the haemoglobin components of trout blood among the erythrocytes: Observations by single-cell spectroscopy. *J. Mol. Biol.* **86,** 165–169.

Brunori, M., Falcioni, G., Fortuna, G., and Giardina, B. (1975). Effect of anions on the oxygen binding properties of the hemoglobin components from trout (*Salmo irideus*). *Arch. Biochem. Biophys.* **168,** 512–519.

Camardella, L., Caruso, C., D'Avino, R., Di Prisco, G., Rutigliano, B., Tamburrini, M., Fermi, G., and Perutz, M. F. (1992). Haemoglobin of the Antarctic fish *Pagothenia bernacchii:* Amino acid sequence, oxygen equilibria and crystal structure of its carbonmonoxy derivative. *J. Mol. Biol.* **224,** 449–460.

Chien, J. C. W., and Mayo, K. H. (1980). Carp haemoglobin. II. The alkaline Bohr effect. *J. Biol. Chem.* **255,** 9800–9806.

Colombo, M. F., Rau, D. C., and Parsegian, V. A. (1992). Protein solvation in allosteric regulation: A water effect on hemoglobin. *Science* **256,** 655–659.

Daugherty, M. A., Shea, M. A., and Ackers, G. K. (1994). Bohr effect of the partially ligated (CN-Met) intermediates of haemoglobin as probed by quaternary assembly. *Biochemistry* **33,** 10345–10357.

Di Cera, E., Robert, C. H., and Gill, S. J. (1987). Allosteric interpretation of the oxygen-binding reaction of human haemoglobin tetramers. *Biochemistry* **26,** 4003–4008.

Di Cera, E., Doyle, M. L., and Gill, S. J. (1988). Alkaline Bohr effect of human hemoglobin A°. *J. Mol. Biol.* **200,** 593–599.

Di Prisco, G., Condò, S. G., Tamburrini, M., and Giardina, B. (1991). Oxygen transport in extreme environments. *Trends Biochem. Sci.* **16,** 471–474.

Edelstein, S. J., McEwen, B., and Gibson, Q. H. (1976). Subunit dissociation in fish haemoglobins. *J. Biol. Chem.* **251,** 7632–7637.

Edsall, J. T. (1986). Understanding blood and haemoglobin: an example of international relations in science. *Persp. Biol. Med.* **29,** S107–S123.

Fago, A. (1995). "Structural Basis for Allosteric Effects of Fish Hemoglobins." Ph.D. thesis. Aarhus University.

Fago, A., and Weber, R. E. (1995). The haemoglobin system of the hagfish *Myxine glutinosa:* Aggregation state and functional properties. *Biochim. Biophys. Acta* **1249,** 109–115.

Fago, A., Romano, M., Tamburrini, M., Coletta, M., D'Avino, R., and Di Prisco, G. (1993). A polymerising Root-effect fish haemoglobin with high subunit heterogeneity: Correlation with primary structure. *Eur. J. Biochem.* **218,** 829–835.

Fago, A., Carratore, V., di Prisco, G., Feuerlein, R. J., Sottrup-Jensen, L., and Weber, R. E. (1995). The cathodic haemoglobin of *Anguilla anguilla:* Amino acid sequence and oxygen equilibria of a reverse Bohr effect haemoglobin with high oxygen affinity and high phosphate sensitivity. *J. Biol. Chem.* **270,** 18897–18902.

Falcioni, G., Fortuna, G., Giardina, B., Brunori, M., and Wyman, J. (1977). Functional properties of partially oxidized trout hemoglobins. *Biochim. Biophys. Acta* **490,** 171–177.

Farmer, M. (1979). The transition from water to air breathing: Effects of CO_2 on hemoglobin function. *Comp. Biochem. Physiol.* **62A,** 109–114.

Feng, D.-F., and Doolittle, R. F. (1987). Progressive sequence alignment as a prerequisite to correct phylogenetic trees. *J. Mol. Evol.* **25,** 351–360.

Fermi, G., Perutz, M. F., Shaanan, B., and Fourme, R. (1984). The crystal structure of human deoxyhaemoglobin at 1.74 Å resolution. *J. Mol. Biol.* **175,** 159–174.

Feuerlein, R. J., and Weber, R. E. (1996). Oxygen equilibria of cathodic eel hemoglobin analysed in terms of the MWC model and Adair's successive oxygenation theory. *J. Comp. Physiol. B* **165,** 597–606.

Freeman, L., Beitinger, T. L., and Huey, D. W. (1983). Methaemoglobin reductase activity in phylogenetically diverse piscine species. *Comp. Biochem. Physiol.* **75B**, 27–30.

Friedman, J. M., Campbell, B. F., and Noble, R. W. (1990). A possible new control mechanism suggested by resonance Raman spectra from a deep ocean fish haemoglobin. *Biophys. Chem.* **37**, 43–59.

Fyhn, U. E. H., and Sullivan, B. (1975). Elasmobranch hemoglobins: dimerization and polymerization in various species. *Comp. Biochem. Physiol.* **50B**, 119–129.

Garlick, R. L., Bunn, H. F., Fyhn, H. J., Fyhn, U. E. H., Martin, J. P., Noble, R. W., and Powers, D. (1979). Functional studies on the separated hemoglobin components of an air-breathing catfish, *Hoplosternum littorate* (Hancock). *Comp. Biochem. Physiol.* **62A**, 219–226.

Giles, M. A., and Rystephanuk, D. M. (1989). Ontogenetic variation in the multiple hemoglobins of Arctic charr, *Salvelinus alpinus. Can. J. Fish. Aquat. Sci.* **46**, 804–809.

Giles, M. A., and Vanstone, W. E. (1976). Ontogenetic variation in the multiple hemoglobins of coho salmon (*Oncorhynchus kisutch*) and effect of environmental factors on their expression. *J. Fish. Res. Bd. Canada* **33**, 1144–1149.

Gillen, R. G., and Riggs, A. (1973). Structure and function of the isolated hemoglobins of the American eel, *Anguilla rostrata. J. Biol. Chem.* **248**, 1961–1969.

Gronenborn, A. M., Clore, C. M., Brunori, M., Giardina, B., Falcioni, G., and Perutz, M. F. (1984). Stereochemistry of ATP and GTP bound to fish haemoglobins. *J. Mol. Biol.* **178**, 731–742.

Hashimoto, K., and Matsuura, F. (1960). Comparative studies on two hemoglobins of salmon. V. Change in proportion of two hemoglobins with growth. *Bull. Jpn. Soc. Sci. Fish.* **26**, 931–937.

Hjorth, J. P. (1974). Genetics of *Zoarces* populations. VII. Fetal and adult hemoglobins and a polymorphism common to both. *Hereditas* **78**, 69–72.

Hjorth, J. P. (1975). Molecular and genetic structure of multiple hemoglobins in the eelpout, *Zoarces viviparus* L. *Biochem. Genet.* **13**, 379–391.

Holk, K., and Lykkeboe, G. (1995). Catecholamine-induced changes in oxygen affinity of carp and trout blood. *Respir. Physiol.* **100**, 55–62.

Hombrados, I., Rodewald, K., Neuzil, E., and Braunitzer, G. (1983). Haemoglobins. LX. Primary structure of the major haemoglobin of the sea lamprey *Petromyzon marinus* (*var.* Garonne, Loire). *Biochimie* **65**, 247–257.

Honzatko, R. B., Hendrickson, W. A., and Love, W. E. (1985). Refinement of a molecular model of lamprey haemoglobin from *Petromyzon marinus. J. Mol. Biol.* **184**, 147–164.

Horimoto, K., Suzuki, H., and Otsuka, J. (1990). Discrimination between adaptive and neutral amino acid substitutions in vertebrate haemoglobins. *J. Mol. Evol.* **31**, 302–324.

Houston, A. H. (1980). Components of the hematological response of fishes to environmental temperature change: A review. *In* "Environmental Physiology of Fishes" (Ali, M. A., ed.), pp. 241–298. Plenum, New York.

Hübner, S., Michel, F., Rudloff, V., and Appelhans, H. (1992). Amino acid sequence of band-3 protein from rainbow trout erythrocytes derived from cDNA. *Biochem. J.* **285**, 17–23.

Ikeda-Saito, M., Yonetani, T., and Gibson, Q. H. (1983). Oxygen equilibrium studies on hemoglobin from the bluefin tuna (*Thunnus thynnus*). *J. Mol. Biol.* **168**, 673–686.

Imai, K. (1982). "Allosteric Effects in Haemoglobin." Cambridge Univ. Press, Cambridge.

Isaacks, R. E., Kim, H. D., Bartlett, G. R., and Harkness, D. R. (1977). Inositol pentaphosphate in erythrocytes of a freshwater fish, piraracú (*Arapaima gigas*). *Life Sci.* **20**, 987–990.

Ito, N., Komiyama, N. H., and Fermi, G. (1995). Structure of deoxyhaemoglobin of the Antarctic fish *Pagothenia bernacchii* with an analysis of the structural basis of the Root effect by comparison of the liganded and unliganded haemoglobin structures. *J. Mol. Biol.* **250**, 648–658.

Iuchi, I. (1973). Chemical and physiological properties of the larval and the adult hemoglobins in rainbow trout, *Salmo gairdneri irideus*. *Comp. Biochem. Physiol.* **44B**, 1087–1101.

Jaffé, E. R. (1981). Methaemoglobinaemia. *Clin. Haematol.* **10**, 99–122.

Janssen, L. H. M., De Bruin, S. H., and van Os, G. A. J. (1970). H⁺ titration studies of human haemoglobin. *Biochim. Biophys. Acta* **221**, 214–227.

Jensen, F. B. (1986). Pronounced influence of Hb–O₂ saturation on red cell pH in tench blood in vivo and in vitro. *J. Exp. Zool.* **238**, 119–124.

Jensen, F. B. (1989). Hydrogen ion equilibria in fish haemoglobins. *J. Exp. Biol.* **143**, 225–234.

Jensen, F. B. (1990). Nitrite and red cell function in carp: control factors for nitrite entry, membrane potassium ion permeation, oxygen affinity and methaemoglobin formation. *J. Exp. Biol.* **152**, 149–166.

Jensen, F. B. (1991). Multiple strategies in oxygen and carbon dioxide transport by haemoglobin. In "Physiological Strategies for Gas Exchange and Metabolism" (Woakes, A. J., Grieshaber, M. K., and Bridges, C. R., eds.), pp. 55–78. Cambridge Univ. Press, Cambridge.

Jensen, F. B. (1992). Influence of haemoglobin conformation, nitrite and eicosanoids on K⁺ transport across the carp red blood cell membrane. *J. Exp. Biol.* **171**, 349–371.

Jensen, F. B. (1993). Influence of nucleoside triphosphates, inorganic salts, NADH, catecholamines, and oxygen saturation on nitrite-induced oxidation of rainbow trout haemoglobin. *Fish Physiol. Biochem.* **12**, 111–117.

Jensen, F. B., and Brahm, J. (1995). Kinetics of chloride transport across fish red blood cell membranes. *J. Exp. Biol.* **198**, 2237–2244.

Jensen, F. B., and Weber, R. E. (1985). Proton and oxygen equilibria, their anion sensitivities and interrelationships in tench hemoglobin. *Mol. Physiol.* **7**, 41–50.

Jensen, F. B., and Weber, R. E. (1987). Thermodynamic analysis of precisely measured oxygen equilibria of tench (*Tinca tinca*) hemoglobin and their dependence on ATP and protons. *J. Comp. Physiol. B* **157**, 137–143.

Jensen, F. B., Andersen, N. A., and Heisler, N. (1987). Effects of nitrite exposure on blood respiratory properties, acid-base and electrolyte regulation in the carp (*Cyprinus carpio*). *J. Comp. Physiol. B* **157**, 533–541.

Jensen, F. B., Pedersen, J. B., and Garby, L. (1990). A three-state MWC analysis of oxygenation in tench (*Tinca tinca*) hemoglobin. *J. Comp. Physiol. B* **160**, 407–411.

Jensen, F. B., Nikinmaa, M., and Weber, R. E. (1993). Environmental perturbations of oxygen transport in teleost fishes: causes, consequences and compensations. In "Fish Ecophysiology" (Rankin, J. C., and Jensen, F. B., eds.), pp. 161–179. Chapman & Hall, London.

King, L. A. (1994). Adult and fetal hemoglobins in the oviparous swell shark, *Cephaloscyllium ventriosum*. *Comp. Biochem. Physiol.* **109B**, 237–243.

Kitchen, H. (1974). Animal hemoglobin heterogeneity. *Ann. N. Y. Acad. Science* **241**, 12–24.

Klocke, R. A. (1988). Velocity of CO₂ exchange in blood. *Annu. Rev. Physiol.* **50**, 625–637.

Koch, H. J. A., Bergström, E., and Evans, J. C. (1966). A size correlated shift in the proportion of the haemoglobin components of the Atlantic salmon (*Salmo salar* L.) and of the sea trout (*Salmo trutta* L.). *Mededel. Vlaamse Acad. Kl. Wet.* **11**, 3–20.

Koshland, D. E., Nemethy, G., and Filmer, D. (1966). Comparison of experimental binding data and theoretical models in proteins containing subunits. *Biochemistry* **5**, 365–385.

Kosaka, H., and Tyuma, I. (1987). Mechanism of autocatalytic oxidation of oxyhaemoglobin by nitrite. *Environ. Health Perspect.* **73**, 147–151.

Kwiatkowski, L. D., and Noble, R. W. (1993). The effect of 75% glycerol on the oxygen binding properties of carp hemoglobin. *Biochem. Biophys. Res. Commun.* **195**, 1218–1223.

Kwiatkowski, L. D., DeYoung, A., and Noble, R. W. (1994). Isolation and stability of partially oxidized intermediates of carp haemoglobin: Kinetics of CO binding to the mono- and triferric species. *Biochemistry* **33,** 5884–5893.

LiCata, V. J., Dalessio, P. M., and Ackers, G. K. (1993). Single site modifications of half-ligated haemoglobin reveal autonomous dimer cooperativity within a quaternary T tetramer. *Proteins Struct. Funct. Genet.* **17,** 279–296.

Liljeqvist, G., Paléus, S., and Braunitzer, G. (1982). Haemoglobins XLVIII. The primary structure of a monomeric haemoglobin from the hagfish, *Myxine glutinosa* L: Evolutionary aspects and comparative studies of the function with special reference to the heme linkage. *J. Mol. Evol.* **18,** 102–108.

Low, P. S. (1991). Role of haemoglobin denaturation and band 3 clustering in initiating red cell removal. *In* "Red Blood Cell Aging" (Magnani, M., and DeFlora, A., eds.), pp. 173–183. Plenum, New York.

Lowe, T. E., Wells, R. M. G., and Baldwin, J. (1995). Absence of regulated blood oxygen transport in response to strenuous exercise by the shovelnosed ray, *Rhinobatos typus*. *Mar. Freshwater Res.* **46,** 441–446.

Luisi, B. F., and Nagai, K. (1986). Crystallographic analysis of mutant human haemoglobins made in *Escherichia coli*. *Nature* **320,** 555–556.

Luisi, B. F., Nagai, K., and Perutz, M. F. (1987). X-ray crystallographic and functional studies of human haemoglobin mutants produced in *Escherichia coli*. *Acta Haematol.* **78,** 85–89.

Lykkeboe, G., and Weber, R. E. (1978). Changes in the respiratory properties of the blood in the carp, *Cyprinus carpio,* induced by diurnal variation in ambient oxygen tension. *J. Comp. Physiol.* **128,** 117–125.

Lykkeboe, G., Johansen, K., and Maloiy, G. M. O. (1975). Functional properties of hemoglobins in the teleost *Tilapia grahami*. *J. Comp. Physiol.* **104,** 1–11.

Maines, M. D. (1988). Haem oxygenase: Function, multiplicity, regulatory mechanisms, and clinical applications. *FASEB J.* **2,** 2557–2568.

Manwell, C. (1958). Ontogeny of hemoglobin in the skate *Raja binoculata*. *Science* **128,** 419–420.

Manwell, C. (1963). Fetal and adult hemoglobins of the spiny dogfish *Squalus suckleyi*. *Arch. Biochem. Biophys.* **101,** 504–511.

Marinsky, C. A., Houston, A. H., and Murad, A. (1990). Effect of hypoxia on hemoglobin isomorph abundances in rainbow trout, *Salmo gairdneri*. *Can. J. Zool.* **68,** 884–888.

Martin, J. P., Bonaventura, J., Brunori, M., Fyhn, H. J., Fyhn, U. E. H., Garlick, R. L., Powers, D. A., and Wilson, M. T. (1979). The isolation and characterization of the hemoglobin components of *Mylossoma* sp., an Amazonian teleost. *Comp. Biochem. Physiol.* **62A,** 155–162.

Monod, J., Wyman, J., and Changeaux, J. P. (1965). On the nature of allosteric transitions: a plausible model. *J. Mol. Biol.* **12,** 88–118.

Monti, J. P., Brunet, P. J., Berland, Y. F., Vanuxem, D. C., Vanuxem, P. A., and Crevat, A. D. (1995). Opposite effects of urea on hemoglobin–oxygen affinity in anemia of chronic renal failure. *Kidney Int.* **48,** 827–831.

Morris, R. J., and Gibson, Q. H. (1982). Cooperative ligand binding to haemoglobin: Effects of temperature and pH on a haemoglobin with spectrophotometrically distinct chains (*Tunnus thynnus*). *J. Biol. Chem.* **257,** 4869–4874.

Motais, R., Garcia-Romeu, F., and Borgese, F. (1987). The control of Na^+/H^+ exchange by molecular oxygen in trout erythrocytes: A possible role of hemoglobin as a transducer. *J. Gen. Physiol.* **90,** 197–207.

Murad, A., and Houston, A. (1991). Haemoglobin isomorph abundances in splenectomized rainbow trout, *Oncorhynchus mykiss* (Walbaum). *J. Fish Biol.* **38,** 641–651.

Mylvaganam, S. E., Bonaventura, C., Bonaventura, J., and Getzoff, E. D. (1996). Structural basis for the Root effect in haemoglobin. *Nature Struct. Biol.* **3,** 275–283.

Nagai, K., Perutz, M. F., and Poyart, C. (1985). Oxygen binding properties of human mutant haemoglobins synthesized in *Escherichia coli. Proc. Natl. Acad. Sci. USA* **82,** 7252–7255.

Nielsen, O. B., Lykkeboe, G., and Cossins, A. R. (1992). Oxygenation-activated K fluxes in trout red blood cells. *Am. J. Physiol.* **263,** C1057–C1064.

Nikinmaa, M. (1983). Adrenergic regulation of haemoglobin oxygen affinity in rainbow trout red cells. *J. Comp. Physiol.* **152,** 67–72.

Nikinmaa, M. (1992). Membrane transport and control of hemoglobin–oxygen affinity in nucleated erythrocytes. *Physiol. Rev.* **72,** 301–321.

Nikinmaa, M., and Jensen, F. B. (1992). Inhibition of adrenergic proton extrusion in rainbow trout red cells by nitrite-induced methaemoglobinaemia. *J. Comp. Physiol. B* **162,** 424–429.

Nikinmaa, M., and Mattsoff, L. (1992). Effects of oxygen saturation on the CO_2 transport properties of *Lampetra* red cells. *Respir. Physiol.* **87,** 219–230.

Nikinmaa, M., Airaksinen, S., and Virkki, L. V. (1995). Haemoglobin function in intact lamprey erythrocytes: Interactions with membrane function in the regulation of gas transport and acid–base balance. *J. Exp. Biol.* **198,** 2423–2430.

Noble, R. W., Kwiatkowski, L. D., De Young, A., Davis, B. J., Haedrich, R. L., Tam, L.-T., and Riggs, A. (1986). Functional properties of haemoglobins from deep-sea fish: Correlations with depth distribution and presence of a swimbladder. *Biochim. Biophys. Acta* **870,** 552–563.

Parkhurst, L. J., Goss, D. J., and Perutz, M. F. (1983). Kinetic and equilibrium studies on the role of the β-147 histidine in the Root effect and cooperativity in carp haemoglobin. *Biochemistry* **22,** 5401–5409.

Pellegrini, M., Giardina, B., Olianas, A., Sanna, M. T., Deiana, A. M., Salvadori, S., Di Prisco, G., Tamburrini, M., and Corda, M. (1995). Structure function relationships in the hemoglobin components from moray (*Muraena helena*). *Eur. J. Biochem.* **234,** 431–436.

Pelster, B., and Weber, R. E. (1991). The physiology of the Root effect. *In* "Advances in Comparative and Environmental Physiology," vol. 8, pp. 51–77. Springer-Verlag, Berlin.

Pennelly, R. R., Riggs, A., and Noble, R. W. (1978). The kinetics and equilibria of squirrel-fish haemoglobin: A Root effect haemoglobin complicated by large subunit heterogeneity. *Biochim. Biophys. Acta* **533,** 120–129.

Pérez, J. E., and Rylander, M. K. (1985). Haemoglobin heterogeneity in Venezuelan fishes. *Comp. Biochem. Physiol.* **80B,** 641–646.

Perutz, M. F. (1970). Stereochemistry of cooperative effects in haemoglobin. *Nature* **228,** 726–739.

Perutz, M. F. (1990). Mechanisms regulating the reactions of human haemoglobin with oxygen and carbon monoxide. *Annu. Rev. Physiol.* **52,** 1–25.

Perutz, M. F., and Brunori, M. (1982). Stereochemistry of cooperative effects in fish and amphibian haemoglobins. *Nature* **299,** 421–426.

Perutz, M. F., Kilmartin, J. V., Nishikura, K., Fogg, J. H., Butler, P. J. G., and Rollema, H. S. (1980). Identification of the residues contributing to the Bohr effect of human haemoglobin. *J. Mol. Biol.* **138,** 649–670.

Powers, D. A. (1972). Hemoglobin adaptation for fast and slow water habitats in sympatric catostomid fishes. *Science* **177,** 360–362.

Reischl, E. (1976). The hemoglobins of the fresh-water teleost *Hoplias malabarica* (Bloch, 1794): Heterogeneity and polymerization. *Comp. Biochem. Physiol.* **55B,** 255–257.

Riggs, A. (1970). Properties of fish hemoglobins. *In* "Fish Physiology," (Hoar, W. S., and Randall, D. J., eds.), Vol. IV, pp. 209–252. Academic Press, New York.

Riggs, A. (1972). The haemoglobins. *In* "The Biology of Lampreys," (Hardisty, M. W., and Potter, I. C., eds.), Vol. 2, pp. 261–286. Academic Press, London.

Riggs, A. F. (1988). The Bohr effect. *Annu. Rev. Physiol.* **50,** 181–204.

Rivetti, C., Mozzarelli, A., Rossi, G. L., Henry, E. R., and Eaton, W. A. (1993). Oxygen binding by single crystals of haemoglobin. *Biochemistry* **32,** 2888–2906.

Rodewald, K., Stangl, A., and Braunitzer, G. (1984). Primary structure, biochemical and physiological aspects of hemoglobin from South American lungfish (*Lepidosiren paradoxus,* Dipnoi). *Hoppe-Seyler's Z. Physiol. Chem.* **365,** 639–649.

Ronald, A. P., and Tsuyuki, H. (1971). The subunit structures and the molecular basis of the multiple hemoglobins of two species of trout, *Salmo gairdneri* and *S. clarki clarki. Comp. Biochem. Physiol.* **39B,** 195–202.

Rossi-Fanelli, A., Antonini, E., and Giuffrè, R. (1960). Oxygen equilibrium of haemoglobin from *Thunnus thynnus. Nature* **186,** 895–897.

Salhany, J. M. (1990). "Erythrocyte Band 3 Protein." CRC Press, Boca Raton, FL.

Schlüter, K., and Drenckhahn, D. (1986). Co-clustering of denatured haemoglobin with band 3: Its role in binding of autoantibodies against band 3 to abnormal and aged erythrocytes. *Proc. Natl. Acad. Sci. USA* **83,** 6137–6141.

Scholander, P. V., and Van Dam, L. (1954). Secretion of gases against high pressure in the swimbladder of deep sea fishes. I. Oxygen dissociation in the blood. *Biol. Bull. Mar. Biol. Lab. Woods Hole* **107,** 247–259.

Shaanan, B. (1983). Structure of human oxyhaemoglobin at 2.1 Å resolution. *J. Mol. Biol.* **171,** 31–59.

Shih, D. T.-b, Jones, R., Bonaventura, J., Bonaventura, C., and Schneider, R. G. (1984). Involvement of His HC3(146)β in the Bohr effect of human haemoglobin. *J. Biol. Chem.* **259,** 967–974.

Shih, D. T.-b, Luisi, B. F., Myazaki, G., Perutz, M. F., and Nagai, K. (1993). A mutagenic study of the allosteric linkage of His (HC3)146β in haemoglobin. *J. Mol. Biol.* **230,** 1291–1296.

Shimada, T., Okihama, Y., Okazaki, T., and Shukuya, R. (1980). The multiple hemoglobins of the Japanese eel, *Anguilla japonica:* Molecular basis for hemoglobin multiplicity and the subunit interactions. *J. Biol. Chem.* **255,** 7912–7917.

Spagnuolo, C., Rinelli, P., Coletta, M., Chiancone, E., and Ascoli, F. (1987). Oxidation reaction of human oxyhaemoglobin with nitrite: A reexamination. *Biochim. Biophys. Acta* **911,** 59–65.

Springer, B. A., Egeberg, K. D., Sligar, S. G., Rohlfs, R. J., Mathews, A. J., and Olson, J. S. (1989). Discrimination between oxygen and carbon monoxide and inhibition of autooxidation by myoglobin. *J. Biol. Chem.* **264,** 3057–3060.

Tame, J. R. H., Wilson, J. C., and Weber, R. E. (1996). The crystal structure of trout Hb I in the deoxy and carbonmonoxy forms. *J. Mol. Biol.* **259,** 749–760.

Tanford, C. (1962). The interpretation of hydrogen ion titration curves of proteins. *Adv. Protein. Chem.* **17,** 69–165.

Tetens, V., and Wells, R. M. G. (1984). Oxygen binding properties of blood and hemoglobin solutions in the carpet shark (*Cephaloscyllium isabella*): Roles of ATP and urea. *Comp. Biochem. Physiol.* **79A,** 165–168.

Tsuneshige, A., Imai, K., and Tyuma, I. (1987). The binding of haemoglobin to red cell membrane lowers its oxygen affinity. *J. Biochem.* **101,** 695–704.

Tsuyuki, H., and Ronald, A. P. (1971). Molecular basis for multiplicity of Pacific salmon hemoglobins: Evidence for *in vivo* existence of molecular species with up to four different polypeptides. *Comp. Biochem. Physiol.* **39B,** 503–522.

Tufts, B. L., and Boutilier, R. G. (1989). The absence of rapid chloride/bicarbonate exchange in lamprey erythrocytes: Implications for CO_2 transport and ion distributions between plasma and erythrocytes in the blood of *Petromyzon marinus. J. Exp. Biol.* **144,** 565–576.

Tun, N., and Houston, A. H. (1986). Temperature, oxygen, photoperiod, and the hemoglobin system of the rainbow trout, *Salmo gairdneri. Can. J. Zool.* **64,** 1883–1888.

Val, A. L. (1993). Adaptations of fishes to extreme conditions in fresh waters. *In* "The Vertebrate Gas Transport Cascade: Adaptations to Environment and Mode of Life" (Bicudo, J. E. P. W., ed.), pp. 43–53. CRC Press, Boca Raton, FL.

Val, A. L., Schwantes, A. R., and Almeida-Val, V. M. F. (1986). Biological aspects of Amazonian fishes. VI. Hemoglobins and whole blood properties of *Semaprochilodus* species (Prochilodontidae) at two phases of migration. *Comp. Biochem. Physiol.* **83B,** 659–667.

Walder, J. A., Chatterjee, R., Steck, T. L., Low, P. S., Musso, G. F., Kaiser, E. T., Rogers, P. H., and Arnone, A. (1984). The interaction of haemoglobin with the cytoplasmic domain of band 3 of the human erythrocyte membrane. *J. Biol. Chem.* **259,** 10238–10246.

Wallace, W. J., Houtchens, R. A., Maxwell, J. C., and Caughey, W. S. (1982). Mechanism of autooxidation for haemoglobins and myoglobins: Promotion of superoxide production by protons and anions. *J. Biol. Chem.* **257,** 4966–4977.

Weber, G. (1995). Van't Hoff revisited: Enthalpy of association of protein subunits. *J. Phys. Chem.* **99,** 1052–1059.

Weber, R. E. (1982). Intraspecific adaptation of hemoglobin function in fish to oxygen availability. *In* "Exogenous and Endogenous Influences on Metabolic and Neural Control" (Addink, A. D. F., and Spronk, N., eds.), pp. 87–102. Pergamon Press, Oxford.

Weber, R. E. (1983). TMAO (trimethylamine oxide)—Independence of oxygen affinity and its urea and ATP sensitivities in an elasmobranch hemoglobin. *J. Exp. Zool.,* **228,** 551–554.

Weber, R. E. (1990). Functional significance and structural basis of multiple hemoglobins with special reference to ectothermic vertebrates. *In* "Animal Nutrition and Transport Processes. 2. Transport, Respiration and Excretion: Comparative and Environmental Aspects" (Truchot, J.-P., and Lahlou, B., eds.), pp. 58–75. Karger, Basel.

Weber, R. E. (1996). Hemoglobin adaptations in Amazonian and temperate fish with special reference to hypoxia, allosteric effectors and functional heterogeneity. *In* "Physiology and Biochemistry of the Fishes of the Amazon" (Val, A. L., Almeida-Val, V. M. F., and Randall, D. J., eds.), pp. 75–90. INPA, Brazil.

Weber, R. E., and Hartvig, M. (1984). Specific fetal hemoglobin underlies the fetal-maternal shift in blood oxygen affinity in a viviparous teleost. *Mol. Physiol.* **6,** 27–32.

Weber, R. E., and Jensen, F. B. (1988). Functional adaptations in haemoglobins from ectothermic vertebrates. *Annu. Rev. Physiol.* **50,** 161–179.

Weber, R. E., and Johansen, K. (1979). Oxygenation-linked binding of carbon dioxide and allosteric phosphate cofactors by lungfish hemoglobin. *In* "Animals and Environmental Fitness" (Gilles, R., ed.), pp. 49–50. Pergamon Press, Oxford.

Weber, R. E., and Lykkeboe, G. (1978). Respiratory adaptations in carp blood: Influences of hypoxia, red cell organic phosphates, divalent cations and CO_2 on hemoglobin–oxygen affinity. *J. Comp. Physiol.* **128,** 127–137.

Weber, R. E., Lykkeboe, G., and Johansen, K. (1975). Biochemical aspects of the adaptation of hemoglobin–oxygen affinity of eels to hypoxia. *Life Sci.* **17,** 1345–1350.

Weber, R. E., Lykkeboe, G., and Johansen, K. (1976a). Physiological properties of eel haemoglobin: Hypoxic acclimation, phosphate effects and multiplicity. *J. Exp. Biol.* **64,** 75–88.

Weber, R. E., Sullivan, B., Bonaventura, J., and Bonaventura, C. (1976b). The hemoglobin system of the primitive fish *Amia calva:* Isolation and functional characterization of the individual hemoglobin components. *Biochim. Biophys. Acta* **434,** 18–31.

Weber, R. E., Wood, S. C., and Lomholt, J. P. (1976c). Temperature acclimation and oxygen-binding properties of blood and multiple haemoglobins of rainbow trout. *J. Exp. Biol.* **65,** 333–345.

Weber, R. E., Wells, R. M. G., and Rossetti, J. E. (1983). Allosteric interactions governing oxygen equilibria in the hemoglobin system of the dogfish, *Squalus acanthias. J. Exp. Biol.,* **103,** 109–120.

Weber, R. E., Jensen, F. B., and Cox, R. P. (1987). Analysis of teleost hemoglobin by Adair and Monod-Wyman-Changeux models: Effects of nucleoside triphosphates and pH on oxygenation of tench hemoglobin. *J. Comp. Physiol. B* **157,** 145–152.

Weber, R. E., Jessen, T.-H., Malte, H., and Tame, J. (1993). Mutant hemoglobins (α^{119}-Ala and β^{55}-Ser): Functions related to high-altitude respiration in geese. *J. Appl. Physiol.* **75,** 2646–2655.

Wells, R. M. G., Grigg, G. C., Beard, L. A., and Summers, G. (1989). Hypoxic responses in a fish from a stable environment: Blood oxygen transport in the Antarctic fish *Pagothenia borchgrevinki. J. Exp. Biol.* **141,** 97–111.

Wieth, J. O., Andersen, O. S., Brahm, J., Bjerrum, P. J., and Borders, C. L., Jr. (1982). Chloride–bicarbonate exchange in red blood cells: Physiology of transport and chemically modification of binding sites. *Phil. Trans. R. Soc. Lond. B* **299,** 383–399.

Wilkins, N. P. (1985). Ontogeny and evolution of salmonid hemoglobins. *Int. Rev. Cytol.* **94,** 269–298.

Wilson, R. R., Jr., and Knowles, F. C. (1987). Temperature adaptation of fish haemoglobins reflected in rates of autoxidation. *Arch. Biochem. Biophys.* **255,** 210–213.

Wood, S. C., and Johansen, K. (1972). Adaptation to hypoxia by increased HbO_2 affinity and decreased red cell ATP concentration. *Nature New Biol.* **237,** 278–279.

Wood, S. C., and Johansen, K. (1973). Organic phosphate metabolism in nucleated red cells: Influence of hypoxia on eel HbO_2 affinity. *Neth. J. Sea Res.* **7,** 328–338.

Wyman, J. (1964). Linked functions and reciprocal effects in haemoglobin: A second look. *Adv. Protein Chem.* **19,** 223–286.

Wyman, J., Gill, S. J., Noll, L., Giardina, B., Colosimo, A., and Brunori, M. (1977). The balance sheet of a hemoglobin: Thermodynamics of CO binding by hemoglobin Trout I. *J. Mol. Biol.* **109,** 195–205.

Yancey, P. H., Clark, M. E., Hand, S. C., Bowlus, R. D., and Somero, G. N. (1982). Living with water stress: Evolution of osmolyte systems. *Science* **217,** 1214–1222.

2

RED BLOOD CELL METABOLISM

PATRICK J. WALSH
CHRIS M. WOOD
THOMAS W. MOON

I. Introduction
II. Methodology
    A. Harvesting of Cells
    B. Open versus Closed Metabolic Systems
    C. Transport Studies
    D. Enzyme Activity Measurements
    E. Normalization of Metabolic Parameters
III. Metabolism of Mammalian Red Blood Cells
    A. General Description
    B. Metabolic Regulation
IV. Metabolic Poise of Fish Red Blood Cells
    A. General Description
    B. Enzyme Activities
    C. Metabolic Pathways and Fuel Preferences
V. Transport of Metabolic Fuels
    A. Glucose Transport
    B. Monocarboxylate Transport
    C. Amino Acid Transport
VI. The Influence of Catecholamines, Oxygen Status, and Intracellular pH Regulation on Red Cell Metabolism
    A. The Red Cell pH_i Regulatory Response
    B. Mechanistic Studies *in Vitro*
    C. Mechanistic Studies under *in Vivo* Conditions
VII. Epilogue
    References

I. INTRODUCTION

As evidenced by the other chapters in this volume, the blood of fish has received considerable attention for many years for its gas transport

Fish Physiology, Volume 17:
FISH RESPIRATION

capabilities. However, it is only relatively recently that the blood of fish has attracted detailed research into its metabolic properties. In contrast to mammalian erythrocytes, early studies demonstrated that fish blood consumes O_2 at a significant rate (e.g., Tipton, 1933; Eddy, 1977; Hunter and Hunter, 1957). Other early studies linked cell ATP content (obviously a consequence of metabolism) with hemoglobin (Hb) function (e.g., Greaney and Powers, 1978). Another significant early finding was that red blood cells (RBCs) appeared to possess substantial pentose-phosphate shunt and glutathione reductase enzyme activities, and hence the likely ability to maintain high stores of NADPH and reduced glutathione (GSH) (Bachand and Leray, 1975). More mechanistically oriented studies of fish RBC metabolism began to appear in the mid-1980s in response to the excitement over the linkage of intracellular pH regulation (and its ultimate dependence on active sodium transport) and oxygen transport capacity and the key role of catecholamines in coordinating these processes during respiratory stress (e.g., hypoxia, exercise) (Nikinmaa, 1982, 1983; Nikinmaa *et al.*, 1984; Cossins and Richardson, 1985; Milligan and Wood, 1987). Despite this recent more intensive investigation of fish RBC metabolism, and the transport of the needed metabolic substrates across the fish RBC membrane, there still remain substantial gaps even in our descriptive understanding of these processes. In this review, we point out these gaps to identify key areas for research. Much of the earlier literature on fish RBC metabolism is placed into an excellent comparative perspective by Nikinmaa (1990).

II. METHODOLOGY

Researchers examining the gas transport and acid–base physiology of fish RBCs have made rapid progress understanding mechanisms of these processes, in large measure by adhering to a relatively uniform set of methods, ones that attempt to duplicate *in vivo* conditions seen by RBCs as closely as possible *in vitro*. Adherence to these methods by researchers attacking RBC metabolism and substrate transport by traditional biochemical approaches has been less uniform. For these reasons, we believe an initial discussion of the appropriate methods to use for metabolism and transport studies might lead to a more critical evaluation of existing data and better future uniformity in the field.

A. Harvesting of Cells

Unfortunately, the simpler methods of obtaining blood from fish (e.g., cardiac or caudal puncture) have at least three potential problems. First, it is not often clear whether a venous, arterial, or mixed sample is obtained by such procedures, and the various substrate and effector levels in blood, and therefore the prehistory of the cells, from these sources can vary tremendously (Mommsen, 1984). Second, often a chemical anesthetic (or an ice bath) is required to decrease struggling in order to obtain samples by puncture, and at least one common anesthetic, MS-222, leads to a pronounced acidosis of the blood (Iwama et al., 1989), as would air exposure of the fish or any struggle by the fish during sampling in the absence of anesthetic (Ferguson and Tufts, 1992). Third, massive catecholamine release can be expected with all the aforementioned approaches (Gamperl et al., 1994; McDonald and Milligan, 1997). Since both acid–base disturbance and catecholamine release clearly affect pH_i, and both have been shown to markedly affect the overall metabolic poise of RBCs (e.g., Bourne and Cossins, 1982; Ferguson and Boutilier, 1988; Ferguson et al., 1989; Wood et al., 1990), and some of these effects, especially hormonal ones, can persist even after washing, blood harvesting by puncture is not an appropriate method for most metabolic studies. Fish should be implanted with chronic indwelling catheters (the buccal dorsal aortic catheter is especially useful for salmonids; Soivio et al., 1972), and blood should be sampled slowly, and only until the animal begins to show the first signs of struggle or disorientation. Use of larger fish for blood donors or pooling of blood samples from several similarly treated smaller fish can yield large enough volumes for multifactorial experimental designs, thereby factoring out one potential source of variability.

Once the blood is obtained, a decision must be made either to focus on whole blood, with its multiple blood cell types, or to further process the blood to remove certain cell types, or native substrates and effectors, or the heparin used in the harvesting of blood; to our knowledge, no systematic studies of the effects of heparin (which releases lipoprotein lipase) on fish RBC metabolism have been carried out, a potentially interesting line of research. The decision to use whole blood or to further process the blood depends upon the nature of the question being addressed, so we can suggest only some general cautionary notes. First, under some circumstances and in some species, RBCs seem to be easily damaged by repeated washing in physiological salines, or RBCs may either lose or gain metabolic and transport activity (Bourne and Cossins, 1982; Tufts and Randall, 1989; Marshall et al., 1990; Cossins, 1989). However, in some cases, washing cannot be avoided. For example, if an investigator wishes to examine the K_m for glucose transport or metabolism, the typical glucose concentration in fish blood of several millimolar could be problematic in examining glucose concentrations lower than native values without first washing the blood cells; obviously, higher values can be obtained by simply adding small volumes of concentrated glucose stock solutions. In general, to minimize any potential problems, the metabolic competency of cells, washed or unwashed, should always be examined by viability criteria similar to those used in respiratory/acid–base studies of fish blood and for other cell types (e.g., hepatocytes, Mommsen *et al.,* 1994), namely, stable NTP values that closely match *in vivo* values.

Although RBCs are by mass/volume the dominant cell type in blood, recent evidence demonstrates that white blood cells (WBCs) can have extraordinarily high mass-specific O_2 consumption rates (Wang *et al.,* 1994), likely associated with their "killing" abilities, and thus they can potentially contribute a disproportionate share to whole blood metabolic rate measurements. For example, in rainbow trout blood with a hematocrit (RBC fraction) of 20% and a leukocrit (WBC fraction) of 1%, Wang *et al.* (1994) estimated that WBCs accounted for about 50% of the O_2 consumption and ATP turnover (Fig. 1), despite their sparse occurrence. A doubling of the leukocrit due to chronic low-level infection in experimental fish might well go unnoticed by an investigator, but would have a profound impact on whole blood metabolic rate. Indeed this factor may explain the rather large variations reported for whole blood O_2 consumptions by various workers (see part IVA). For these reasons, we recommend that all further research

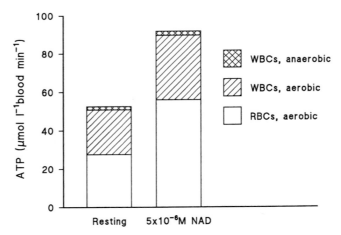

Fig. 1. Estimated partitioning of total ATP turnover between red blood cells (RBCs) and white blood cells (WBCs) in rainbow trout blood at 15°C, assuming a hematocrit of 20% and a leukocrit of 1%. Adrenergic stimulation was achieved by adding 5×10^{-6} M noradrenaline. Aerobic metabolism was estimated from O_2 consumption measurements (assumed P/O ratio = 3) and anaerobic metabolism from lactate production measurements (assumed substrate = glucose). RBCs did not produce significant amounts of lactate. Redrawn from Wang *et al.* (1994).

into RBC metabolism be done on preparations where the WBC layer ("buffy coat") has been removed mechanically after gentle centrifugation. With careful pipetting, this step can be accomplished with minimal removal of plasma and RBCs. Great care should be taken in this separation to avoid removing the youngest, lightest RBCs, which are found immediately below the buffy coat and which may have metabolic properties different from those of older RBCs (Tipton, 1933; Lane et al., 1982; Speckner et al., 1989; Pesquero et al., 1992). Clearly, an obvious further line of research is into the metabolism of the WBCs themselves and into the metabolism of RBCs of different ages.

B. Open versus Closed Metabolic Systems

In vivo, blood is essentially an open system, with metabolic and respiratory exchange with the tissues on one hand, and exchange with the environment at the gills on the other hand. Blood respiratory and acid–base physiologists have recognized this concept for decades, and have adopted open systems for measuring parameters routinely. One exception is the closed system used for oxygen consumption determinations in which small changes in P_{O_2} are measured with minimal change to total O_2 content during short periods of closure. Since most metabolic rate measurements used by biochemists, in particular the monitoring of changes in levels of labeled or unlabeled substrates/products, often require incubations ≥ 1 hour, biochemical researchers have tended to opt for the more convenient closed systems. It is clear, however, that blood incubated for long periods in closed systems will result in a constantly changing acid–base and respiratory environment for the cells, and ultimately set up conditions for metabolic studies that are not reflective of *in vivo* states, nor of the conditions used by respiratory physiologists. Since one important goal of metabolic studies of blood is to reflect *in vivo* reality, and a second important goal is to link respiration/metabolism data with the literature on acid-base balance and hormonal control, we strongly recommend that metabolic studies adhere to the open system design preferred by respiratory physiologists. One example of an open metabolic measurement system is given by Walsh et al. (1990) and Wood et al. (1990).

C. Transport Studies

The current consensus for procedures to be used to measure substrate uptake in fish RBCs was recently reviewed (Moon and Walsh, 1994). Since transport is typically examined on the time scale of seconds to minutes, potential variation of acid–base and respiratory state during the uptake measurement is not as problematic as in metabolic measurements. Nonetheless, prior to the use of cells for transport measurements, harvesting and preincubation should take into account the concerns raised here.

D. Enzyme Activity Measurements

In addition to the concerns raised, there are a few other concerns measuring blood enzyme activities. Obviously, blood cells should be separated from plasma immediately and then frozen, if necessary, prior to assay. One particularly interesting area may be plasma enzyme activity (e.g., Wells *et al.,* 1986). If this is the case, great care should be taken to avoid and quantitate plasma contamination with lysed cells, by measuring plasma Hb content. A second potential problem in measuring blood enzyme activities is the strong spectrophotometric signal of Hb. For these reasons, dilute homogenates and longer than typical reaction times may be needed if the spectrophotometer used is not capable of subtracting out the Hb signal.

E. Normalization of Metabolic Parameters

Metabolic biochemists typically normalize metabolic and enzymatic rates to grams of cells or tissue wet weight or milligrams of protein. Respiratory physiologists typically normalize to hemoglobin content, packed cell volume, or hematocrit. We suggest that investigators working on RBC metabolism present data normalized in two ways, one appropriate to each subdiscipline, so that investigators in both fields can easily assimilate the data. At the very least, interconversion factors for each study should be given; typically, 1 g Hb is equivalent to 4 to 5 ml or g packed red cells (e.g., Bachand and Leray, 1975; Walsh *et al.,* 1990).

III. METABOLISM OF MAMMALIAN RED BLOOD CELLS

The understanding of metabolic pathways in fish red blood cells remains less well understood than that in the blood of mammals. For this reason, it is instructive to first briefly summarize the metabolic poise of mammalian red blood cells, a system that has been studied extensively, no doubt due in part to the importance of this information to blood-banking activities. However, the reader should keep in mind that mammalian red blood cells, which are nonnucleated, present a potentially simpler, stripped-down version of metabolism relative to the nucleated red blood cells of fish and other nonmammalian vertebrates.

A. General Description

The metabolic pathways of RBCs of humans and other mammals are very well understood (for review, see e.g., Joshi and Palsson, 1989; Nikinmaa, 1990; Rapoport, 1986; Schweiger, 1962) and are presented schematically in Fig. 2. There are three main pathways of metabolism, namely, anaerobic glycolysis, the pentose phosphate shunt (PPS), and nucleotide metabolism. The bulk of energy production comes from anaerobic glycolysis (from extracellular glucose), which is estimated to consume 1 to 2 mmol glucose (and produce 2 to 4 mmol ATP) L^{-1} cells h^{-1} (Schweiger, 1962). In fact, aerobic metabolism is virtually absent, with the small amount of oxygen consumption (0.3 mmol O_2 L^{-1} cells h^{-1}) being attributed to the PPS and related to GSH metabolism (Schweiger, 1962). Carbon flow can be diverted from the main glycolytic pathway in two places. First, under resting conditions, 15 to 25% of flow can be diverted to the synthesis of 2,3-DPG, an important allosteric regulator of mammalian Hb function (see Nikinmaa, 1990). Second, an additional 10% of carbon flow can be diverted

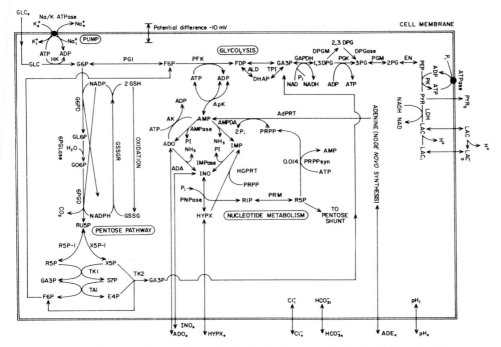

Fig. 2. Schematic diagram of metabolism in the human red blood cell. From Joshi and Palsson (1989), with permission.

to the PPS. This pathway produces two important metabolites. First, it generates reducing power, in the form of NADPH, to maintain high intracellular stores of reduced glutathione (GSH), a compound that is necessary to avoid oxidative damage to macromolecules from the high concentrations of oxygen due to Hb carriage, as well as the highly oxidative environment produced by neighboring macrophages (see, e.g., Low *et al.,* 1995; Rose, 1971). The second important product of the PPS, ribulose 5-phosphate (R5P), links the pathway with nucleotide metabolism, providing R5P or consuming it as needed. The PPS can rejoin glycolysis at two points (fructose 6-phosphate, F6P; and glyceraldehyde 3-phosphate, GA3P), depending upon the activity of transketolases (TK) and transaldolases (TA). From a methodological standpoint, it is important to note that the CO_2 generated from the PPS derives only from the C1 of G6P, whereas those generated in the tricarboxylic acid cycle (TCA) are produced from all glucose carbons. This characteristic enables specific isotopic tests (e.g., ratio of $^{14}CO_2$ production from C1/C6 labeled glucose; e.g., Wood and Katz, 1958) to establish the relative activity of these two pathways, techniques that contributed to the view of the human red blood cell as a nonoxidative tissue, with an active PPS.

Nucleotide metabolism acts largely to regenerate intracellular stores of ATP, which undergo normal turnover to adenosine and inosine. The human red cell, however, does not synthesize adenosine *de novo,* but rather takes up adenosine from the plasma and synthesizes AMP, ADP, and ATP from adenosine in several simple metabolic steps (Fig. 2). The mature human red cell is also not capable of protein or lipid synthesis (Bishop, 1964); therefore, its major ATP-consuming pathways are active transport to maintain ion gradients (e.g., the Na^+/K^+-ATPase is estimated to account for 25% of ATP turnover; Nikinmaa, 1990) and phosphorylation–dephosphorylation reactions (another 25% of ATP turnover; Nikinmaa, 1990) that are important to the flexibility of cell shape (Kodicek *et al.,* 1987) (critical to capillary passage) and to the regulation of enzyme activities (Boivan *et al.,* 1986; Low *et al.,* 1995).

B. Metabolic Regulation

Human red blood cells appear to be regulated in large part by classic feedback mechanisms of ATP/ADP acting allosterically on key glycolytic regulatory enzymes (e.g., PFK, PK) (Rapoport, 1986; Reimann *et al.,* 1977). These metabolic aspects are so well understood, in fact, that several computer simulation models exist for human RBC glycolytic flux (e.g., Der Lee and Palsson, 1992; Joshi and Palsson, 1989). However, recently, another mode of regulation for human RBC metabolism has been discovered, namely, the deactivation of metabolic enzymes by their binding to intracellular structures, specifically, transport proteins. Notably, Low and colleagues have recently gathered evidence allowing them to construct a model in which a key glycolytic enzyme, GA3PDH, binds to the dephosphorylated tyrosine residues of the N terminus of the cytoplasmic domain of band 3, an event that inactivates GA3PDH (Low et al., 1995). Inhibition of cytoplasmic tyrosine phosphatase leads to enhanced phosphorylation of the tyrosine residues of band 3, releasing and activating GA3PDH. The inhibition of cytoplasmic tyrosine phosphatase is believed to occur via activation of a p72 kinase, which itself is activated by a transmembrane electron transport pathway stimulated by membrane impermeable oxidants. The importance of this new mode of glycolytic activation may be especially important in the control of reactive oxygen species; in this scheme, the cell can sense the oxidative environment around it, and GA3PDH appears to be specifically activated to accommodate PPS fluxes (which produce NAPDH). Notably GA3PDH is not specifically activated under classic activation of glycolysis by allosteric action of adenylates on PFK.

IV. METABOLIC POISE OF FISH RED BLOOD CELLS

A. General Description

Although details are sparse, it is likely that the overall metabolic setup of fish RBCs is more complicated than that of mammalian RBCs. Fish blood exhibits substantial basal rates of O_2 consumption (1 to 10 mmol L^{-1} cells h^{-1}), with oxygen consumption rates typically doubling upon maximal adrenergic stimulation (e.g., Bushnell et al., 1985; Eddy, 1977; Ferguson and Boutilier, 1988; Hunter and Hunter, 1957; Schweiger, 1962; Walsh et al., 1990; Wood et al., 1990; Sephton et al., 1991; Tufts and Boutilier, 1991). Notably, in the only study to systematically factor out the white versus red cell contribution to oxygen consumption in fish blood, Wang et al. (1994) have estimated that, although white blood cells typically compose only 1/20th of the blood cell volume, they account for about 50% of the oxygen consumption rate for whole blood (Fig. 1)! They also account for all the lactate production under aerobic conditions. These observations are profoundly important in at least two respects. First, most estimates of fish RBC oxygen consumption rates are probably 100% too high. Even so, a "corrected" rainbow trout RBC oxygen consumption rate at 15°C of 1.4 (unstimulated) to 2.9 (stimulated) mmol O_2 L^{-1} h^{-1} (Wang et al., 1994) is still substantially higher (5- to 10-fold) than rates for mammalian RBCs at

37°C and demonstrates that teleost RBCs truly have an aerobic metabolism. Second, much of the literature that we address later on specific pathways and mechanisms does not specifically address RBCs, but a mixed population of cell types; we will try to point out the distinction by referring to RBCs versus blood, but it is often not clear from the methods reported in the literature which specific cell types are under study. Fortunately, metabolic rates in WBCs do not appear to respond to adrenergic stimulation (Wang *et al.,* 1994; Fig. 1).

B. Enzyme Activities

The potential for activity of several pathways in fish blood has been demonstrated in several studies. Parks *et al.* (1973) reported substantial activities of five enzymes of nucleotide metabolism in several fish species. Bachand and Leray (1975) examined activities in washed RBCs (buffy coat removed) obtained from the yellow perch (*Perca flavescens*) by caudal puncture, and they reported activities for the enzymes of anaerobic glycolysis, the pentose shunt (G6PDH and 6PGDH) and glutathione reductase. Walsh *et al.* (1990) reported similar activities of enzymes of anaerobic glycolysis, the pentose shunt and creatine phosphokinase in packed blood cells of the rainbow trout (*Oncorhynchus mykiss*) obtained by dorsal aortic catheter. A more extensive survey was undertaken for rainbow trout by Ferguson and Storey (1991) using washed blood cells (taken from the ventral aorta of noncatheterized fish) and established the potential for anaerobic glycolysis (but not glycogenolysis), TCA cycle activity, glutathione reductase, PPS activity, creatine and adenylate kinase activity, and amino acid catabolism. Notably, lactate dehydrogenase kinetics indicated a "heart" type enzyme, indicating potential preferential function in lactate utilization, rather than production. (Other enzyme kinetic aspects are only scantly studied and are briefly discussed later, see part VI.) Sephton *et al.* (1991) examined enzyme activities in washed erythrocytes taken from the caudal artery (no catheter) of sea ravens (*Hemitripterus americanus*), confirming anaerobic glycolytic and TCA cycle activity for this species. Soengas and Moon (1995) also report LDH activity in erythrocytes of American eel (*Anguilla rostrata*). Taken together, these studies indicate the potential for anaerobic glycolysis, PPS, and glutathione metabolism as in mammalian erythrocytes, but also the potential for oxidation of glucose, lactate, and amino acids, in agreement with the higher than mammalian oxygen consumption rates noted earlier. Also as expected, enzyme activities are typically 1 to 2 orders of magnitude lower than those in more metabolically active tissues in fish (Moon and Foster, 1995).

C. Metabolic Pathways and Fuel Preferences

1. CATABOLIC PATHWAYS

Clearly the main catabolic pathway in fish blood is respiration (reviewed by Boutilier and Ferguson, 1989; Nikinmaa, 1990). Sea raven blood does not appear to produce lactate under aerobic conditions (Sephton *et al.,* 1991; Sephton and Driedzic, 1994a); however, several studies have reported measurable lactate production by salmonid blood even under aerobic conditions (reviewed by Boutilier and Ferguson, 1989), including direct demonstration of lactate production from labeled glucose (Walsh *et al.,* 1990). Additionally, lactate production increases under anaerobic conditions (Boutilier and Ferguson, 1989; Pesquero *et al.,* 1992). However, it is estimated that anaerobic glycolysis contributes minimally to ATP production in red blood cells under both aerobic and anaerobic conditions (Boutilier and Ferguson, 1989). Notably, a more recent study of rainbow trout blood indicates that it is the WBCs, not the RBCs, that produce most of blood lactate at least under aerobic conditions (Wang *et al.,* 1994; Fig. 1).

The proportion of metabolic fuels used for respiration in fish blood is clearly more diverse than that in mammals, but the exact proportions remain somewhat controversial; in fact the proportions may be an actively regulated adaptation to changing states (see part VI). The confusion over metabolic fuel preferences may stem largely from the range of species, physiological states, and types of methodology that have been studied. There are several general a priori clues that fish blood may be capable of using a number of substrates to fuel respiration. The cells of many species are highly impermeable to glucose (e.g., brown trout, *Salmo trutta;* Bolis *et al.,* 1971), and specific membrane transport systems exist for other substrates (e.g., monocarboxylates and amino acids) (see part V). One could argue that these other transport systems are present for osmotic balance, but this argument is not supported by the presence of enzymes of catabolism of lactate and amino acids in blood cells.

Using whole blood from catheterized resting rainbow trout (containing all native substrates) and trace amounts of various ^{14}C-labeled substrates, Walsh *et al.* (1990) concluded that the order of substrate preference for respiration under resting conditions was glucose > lactate > alanine, with nearly nondetectable rates of oxidation of oleate. Lactate and alanine oxidation, but not glucose oxidation, exhibited saturable kinetics. Since the sum of the individual substrate respiration rates accounted for only 10–25% of parallel measurements of total oxygen consumption rates, and given the presence of very low glycogen content, Walsh *et al.* (1990) and Wood *et al.* (1990) concluded that other substrates (e.g., other amino acids, pyruvate) may contribute to respiration. Additionally, these authors demonstrated that substrate utilization was dynamic because experimentally increased lactate concentrations (20 mM) could depress glucose utilization by 27%. More complex changes occurred after exhaustive exercise (see part VIC). Using a similar approach for washed RBCs of brown trout (*Salmo trutta*) obtained by caudal puncture, Pesquero *et al.* (1994) reported that pyruvate was greatly preferred as an aerobic fuel over glucose, even when present at physiological concentrations of only a small fraction of that of glucose. Tihonen and Nikinmaa (1991b), using washed erythrocytes obtained from carp (*Cyprinus carpio*) in a similar manner, reported an order of substrate preference of lactate > pyruvate > glutamine > glucose > glutamate > aspartate > adenosine > isoleucine when the substrates were presented at normal plasma concentrations. There was no difference in order when other substrates were either present or absent. Lactate and pyruvate oxidation both exhibited saturation kinetics (Tiihonen and Nikinmaa, 1993).

In a study of whole blood and washed erythrocytes (obtained by caudal puncture from ice-bath chilled fish) of the sea raven *Hemitripterus americanus,* Sephton *et al.* (1991) compared rates of glucose utilization and O_2 consumption. These authors concluded that in whole blood, the calculated O_2 consumption rates from the measured disappearance of total glucose very closely matched measured O_2 consumption rates. The match of respiration rate and glucose consumption was less precise in washed erythrocytes; O_2 consumption rates of washed, unstimulated erythrocytes were approximately 50% of those for whole blood, but glucose consumption remained the same, even when O_2 consumption was doubled by maximal adrenergic stimulation. These results pointed toward glucose as a primary fuel in sea raven blood under these conditions. However, the potential for lingering effects of catecholamine release during cell harvesting on metabolism (e.g., via a stimulation of glucose transport and/or an inhibition of alternative fuel transport systems, changes in enzyme phosphorylation state) temper these conclusions. In subsequent studies, Sephton and Driedzic (1994a,b) extended their observations for sea raven, and also examined rainbow trout (whole blood in both cases) by methods similar to those of Sephton *et al.* (1991), and concluded that glucose is a primary fuel for blood cells in both species. Indeed in these studies, the decrease of plasma glucose was two to threefold greater than that required to support measured O_2 consumption suggesting an alternative metabolic fate for glucose, yet there was no net synthesis of glycogen or lactate. This significant consumption of glucose contrasts with the data obtained by Walsh *et al.* (1990) for blood from resting, catheterized rainbow trout, indicating a potential hypermetabolic state of the cells in the studies of Sephton and coworkers.

Sephton and coworkers also measured [^{14}C]glucose conversion to $^{14}CO_2$ using a labeling scheme ("C6") that measured only $^{14}CO_2$ evolved from the Krebs cycle (and not from decarboxylation by pyruvate dehydrodgenase or the PPS shunt). Surprisingly, they concluded that in washed sea raven blood $^{14}CO_2$ production measurements underestimated glucose consumption rates by 1000-fold! Initially, they attributed this massive discrepancy to dilution of exogenous glucose (e.g., by incorporation into endogenous glycogen) and/or to incorporation of ^{14}C label into other other intracellular pools (Sephton *et al.*, 1991). In later experiments, they reported a similar discrepancy in washed trout blood (Sephton and Driedzic, 1994a). In both species, some incorporation of the ^{14}C label into protein, lipid, and glycogen pools was found, but these pools accounted for only a small percentage (~5%) of the discrepancy. Instead, the major fraction of missing ^{14}C label (~50%) was found in the "acid-soluble pool." Sephton and Driedzic (1994a) speculated that the ^{14}C label was "held up" in glycolytic and Krebs cycle intermediates (in the acid soluble fraction) and that equilibration times far greater than the 2-h incubation period used would be required to reach steady-state conditions. Interestingly, they reported that the rate of ^{14}C incorporation into the acid-soluble fraction approximated the measured O_2 consumption rate in whole blood (Sephton and Driedzic, 1994a) and that this rate increased threefold during maximal adrenergic stimulation (Sephton and Driedzic, 1994b). Similar unexplained discrepancies have been reported in the RBCs of brown trout (*Salmo trutta;* Pesquero *et al.*, 1992, 1994). Explanations other than a surprisingly slow carbon flux appear possible, such as a partial PPS shunt (i.e. one that does not return carbon flow to glycolysis). Clearly, more work on this problem is required. In the interim, we concur with Sephton and Driedzic (1994a) that $^{14}CO_2$ production is not a measure of absolute glucose utilization in these sorts of experiments, but probably serves as a mirror of relative rates of glucose metabolism.

Using microcalorimetric methods, Pesquero *et al.* (1994) measured respiration rates in washed trout RBCs (obtained by caudal puncture) in the absence of exogenous substrates and suggested that the low levels of glycogen would be sufficient to fuel metabolism for up to 2 h. Soengas and Moon (1995), using washed erythrocytes from decapitated American eels (*Anguilla rostrata*), performed a particularly complete study of both the uptake and the metabolism of glucose, lactate, and alanine, using both radioisotopic and manometric measurements. There were substantial transport activities for all substrates such that transport was not limiting to metabolism (see part V), and the order of substrate preference was lactate > glucose > alanine. There were 1.6- to 2.6-fold differences measured between the O_2 consumption rate and the $^{14}CO_2$ production rates in terms of absolute values, but these differences were much lower than those in prior studies, perhaps owing to the existence of specific transport systems that allowed more complete permeation of external substrates.

Clearly, the issue of blood substrate preference is not resolved, and it is likely to be a dynamic property of blood depending upon species and physiological state. We believe that greater uniformity in experimental approaches may help to bring more uniformity to the data. Additionally, one approach that has not been taken extensively is the use of various enzyme inhibitors (e.g., the transaminase inhibitor amino oxyacetic acid) to test the relative importance of various substrates. Also important may be studies on isolated blood cell mitochondria.

2. ANABOLIC PATHWAYS

Very little is known of synthetic capabilities of fish blood. Substantial PPS activity in fish blood has been demonstrated in only one study of rainbow trout via the use of $[^{14}C]$glucose labeled in different positions (Walsh *et al.*, 1990). Notably, beyond the measurement of glutathione reductase in perch RBCs by Bachand and Leray (1975), and the measurement of high titers of glutathione in fish RBCs (Buckley, 1982; Härdig and Höglund, 1983), glutathione metabolism is virtually unstudied in fish RBCs. Finally, beyond the measurement of a few enzyme activities, and the obviously high ATP and GTP content of fish RBCs (2,3-DPG has been reported in only very few species) (cf. Table 6.3 in Nikinmaa, 1990), little else is known of nucleoside metabolism in fish RBCs apart from the fact that ATP and GTP levels are extremely sensitive to oxygenation status both *in vitro* and *in vivo* (see part VIB). This response, a decrease in RBC NTP concentrations during hypoxemia, has obvious adaptive advantage in increasing the O_2 affinity of hemoglobin by allosteric influence (e.g., Tetens and Lykkeboe, 1981). However, at present both the metabolic mechanisms involved and the relative importance of O_2 depletion itself as a direct stimulus versus hormonal controls (e.g. catecholamines) remain unclear (Nikinmaa, 1990).

V. TRANSPORT OF METABOLIC FUELS

Red cell metabolite transport has been studied in a variety of fish species, representing the entire breadth of phylogenetic groups. The only conclusion that can be reached to date is the huge diversity of processes that have been reported. This area has recently been reviewed by Nikinmaa and Tiihonen (1994) and assessed methodologically by Moon and Walsh (1994). Red cell membrane fluidity is known to change seasonally (Gabbianelli *et al.*, 1996) and with temperature (Raynard and Cossins, 1991; Dey *et al.*, 1993; Fodor *et al.* 1995), yet few studies have examined such changes in the functional characteristics of transport (Raynard and Cossins, 1991; Cossins and Kilbey, 1989; Gabbianelli *et al.*, 1996).

A. Glucose Transport

There are major differences among species in the uptake of D-glucose or its nonmetabolizable analog 3-O-methyl-D-glucose (3-OMG) into fish red cells. Generally, agnathans (hagfish and lamprey) take up glucose quickly, whereas teleost fishes, with a few exceptions, are sluggish in this regard. The precise reason(s) for this difference remains unknown.

Ingermann et al. (1984) first reported rapid (50% of equilibrium within 10 s) uptake of 3-OMG in hagfish (*Eptatretus stouti*), which has now been confirmed by Young et al. (1994). Although not as active, lamprey (*Lampetra fluviatilis*) RBCs demonstrated equilibrium times of 6–8 min at 20°C (Tiihonen and Nikinmaa, 1991). Other fish species show much longer equilibrium times, exceeding 60 min, except for the electric eel (*Electrophorus electrocus*) and the lungfish (*Lepidosiren paradoxa*), which are in the range 20–30 min (Kim and Isaacks, 1978). What is quite clear from Table 1 is that most teleost fishes demonstrate very low glucose uptake rates, with the exception of the eels *Anguilla japonica* (Tse and Young, 1990) and *A. rostrata* (Soengas and Moon, 1995) and the electric eel (Kim and Isaacks, 1978). In all cases, rates are well below those reported for Agnathan and human red cells (Table 1) and rates have been reported to differ significantly between individuals of the same species (Tse and Young, 1990; Tiihonen et al., 1995). A number of other fish species have been investigated (Table 1), and it is apparent that they are deficient in specific glucose transport, including the salmonids that have been studied to date (Pesquero et al., 1992; Tse and Young, 1990). There does not seem to be a phylogenetic explanation for these differences in rates, although there does seem to be a link between fuel preference and uptake. Ingermann et al. (1985) suggested that the ability for the RBC to take up glucose rapidly ensures adequate substrate supply for metabolic energy, the inference being that those cells with low glucose uptake must use other substrates. Unfortunately, only in the case of the American eel (*A. rostrata*) (Soengas and Moon, 1995) do we have data examining both uptake and metabolism of glucose, lactate, and amino acid that can address this hypothesis. This study does show a clear relationship between transport and metabolism, but transport does not limit glucose metabolism in eel RBCs. Certainly, the differences noted are genetic, just as has been reported for mammals (see Tse and Young, 1990).

Classic mammalian red cell glucose transporters (GLUT-1) demonstrate saturation kinetics, are ion-independent, and are inhibited by both cytochalasin B (binds to the glucose carrier on the membrane inner surface) and phloretin (binds to the glucose carrier on the membrane outer surface) (see Tiihonen et al., 1995). Only in the case of red cells from hagfish

Table I
Selective Values of Glucose Uptake into Fish Red Blood Cells

Class/species	K_m	V_{max} (mmol L^{-1} cell water h^{-1})	Substrate[a]	Ref.
Agnatha				
Hagfish (*E. stouti*)	0.88	414 (10°C)	D-Glucose	1
Lamprey (*L. fluviatilus*)	1.6	18.8 (20°C)	3-OMG	2
Elasmobranchi				
Nurse shark (*G. cirratum*)	—	0.097 (22°C)	D-Glucose	3
Dipnoi				
Lungfish (*L. paradoxa*)	—	0.24 (20°C?)[b]	D-Glucose	4
Teleost				
Pirarucu (*A. gigas*)	—	0.04 (20°C?)[b]	D-Glucose	4
Eel (*A. japonica*)	1.22	12.7 (20°C)	3-OMG	5
		8.9 (20°C)	D-Glucose	
		0.02 (20°C)	L-Glucose	
Eel (*A. rostrata*)	10.4	9.6	3-OMG	6
		1.6 (10°C)	D-Glucose	
		0.6 (10°C)		
Paddyfield eel (*M. albus*)	—	0.20 (20°C)	3-OMG	5
		0.06 (20°C)	D-Glucose	
		0.03 (20°C)	L-Glucose	
Electrical eel (E. electrocus)	—	1.4 (20°C?)[b]	D-Glucose	4
Rainbow trout (*O. mykiss*)	—	0.18 (20°C)	3-OMG	5
Brown trout (*S. trutta*)	—	0.15 (15°C)	3-OMG	7
Carp (*C. carpio*)	—	0.29 (20°C)	3-OMG	2
		0.64 (21°C)	3-OMG	8
Sea Raven (*H. americanus*)	—	1.1 (15°C)[c]	D-Glucose	9
Human	0.4	270 (10°C)	D-Glucose	10

Note. Maximum uptake values have been corrected to mmol L^{-1} cell water h^{-1} using the value of 66% cell water (Finchan *et al.*, 1987). References: 1—Young *et al.*, 1994; 2—Tiihonen and Nikinmaa, 1991; 3—Mauro and Isaacks, 1989; 4—Kim and Isaacks, 1978; 5—Tse and Young, 1990; 6—Soengas and Moon, 1995; 7—Pesquero *et al.*, 1992; 8—Tiihonen *et al.*, 1995; 9—Sephton *et al.*, 1991; 10—Lowe and Walmsley, 1986.

[a] Type of substrate used, either glucose or 3-*O*-methylglucose (3-OMG); concentration is 1 or 5 mM depending upon the study.

[b] Values calculated from figure within reference.

[c] Not radioactive uptake, but "glucose consumption" (see 9).

(Ingermann *et al.*, 1984), lamprey (Tiihonen and Nikinmaa, 1991) and *Anguilla* eels (Soengas and Moon, 1995; Tse and Young, 1990) are these characteristics demonstrated. Affinity constants vary among these species (Table 1), but in each case the inhibitors were shown to decrease V_{max}

(maximum rate of transport). Young *et al.* (1994) have characterized the glucose transporter from the Pacific hagfish and found it to be structurally and immunologically homologous to the GLUT-1 of mammalian RBCs. There does not seem to be any reason to suspect the other stereospecific glucose transporters of fish are any different, except for carp (*Cyprinus carpio*). Tiihonen and colleagues (1991, 1995) have reported that carp RBCs do not demonstrate saturation kinetics for 3-OMG up to 50 mM, but transport was inhibited by cytochalasin B and phloretin. Swelling increased transport that was blocked by cytochalasin B, suggesting that a volume-activated channel may be involved in glucose transport in this system. Kirk *et al.* (1992) reported a swelling-induced increase in nonsaturable glucose, taurine, and uridine uptake in flounder (*Platichthys stellatus*) red cells. These volume-activated fluxes showed sensitivities similar to those of various anion-selective blockers, leading to the conclusion that a volume-activated "chloride channel" may be involved. This raises an important question as to the extent of this second pathway for the transport of organic molecules in red cells from other fish species.

Catecholamines have been shown to increase glucose uptake into Japanese eel (Tse and Young, 1990), brown trout (Pesquero *et al.*, 1992), rainbow trout (Sephton and Driedzic, 1994b), and carp (Tiihonen *et al.*, 1995) RBCs. Red cell energy utilization (Ferguson *et al.*, 1989) and cell volume (Perry *et al.*, 1996) also increase under these conditions. This complex process has been discussed elsewhere in the review (part VI), but the precise mechanism to account for this increase in transport is not understood, particularly since both trout and carp are thought to lack a specific carrier protein. In each case, the authors showed that catecholamine-induced transport is independent of cell swelling and the presence of catecholamines during cell preparation. Obviously, this relationship between catecholamine and uptake requires additional study.

The wide variation in uptake rates of glucose between species is intriguing—in particular, why *Anguilla spp.* among the teleosts and hagfish are the only fish to exhibit carrier-specific glucose transport. There appears to be no adequate explanation for this pattern. In only one case, the Pacific hagfish (Young *et al.*, 1994), do we know that the glucose carrier is homologous to that found in mammalian vertebrates. These questions are both evolutionarily and physiologically important and obvious areas for further study.

B. Monocarboxylate Transport

Monocarboxylate transport occurs in mammalian RBCs by a generalized anion exchanger system associated with the band 3 protein and also by a specific monocarboxylate carrier (Poole and Halestrap, 1993). These transporters are distinguished by the use of specific inhibitors. Few studies on monocarboxylate transport in fish RBCs have been reported, but the available data do support a similar system in mammals and fish. However, the physiological purpose of transport is quite different; fish red cells can oxidize monocarboxylates, so they can possibly utilize lactate and/or pyruvate as substrate from the blood, whereas mammalian RBCs can excrete only metabolites to be used by other tissues (Tiihonen and Nikinmaa, 1993).

Teleost RBCs metabolize lactate and/or pyruvate at rates exceeding those of glucose (carp, Nikinmaa and Tiihonen, 1994; Tiihonen and Nikinmaa, 1991b, 1993; brown trout, Pesquero *et al.*, 1992, 1994; American eel, Soengas and Moon, 1995). Walsh *et al.* (1990) found rainbow trout RBCs oxidized lactate poorly, but exhaustive exercise increased lactate oxidation to exceed that of glucose (Wood *et al.*, 1990). The increase in lactate oxidation rate is far above that attributable to the increase in substrate concentration alone (Fig. 3). This implies a catecholamine-induced process may be involved.

Lactate uptake in both carp (Nikinmaa and Tiihonen, 1994; Tiihonen and Nikinmaa, 1993) and American eel (Soengas and Moon, 1995) RBCs is rapid, with eel cells reaching a stable state within 5 min. Uptake is stereospecific, is Na^+-independent, and consists of both saturating and nonsaturating kinetic components (Fig. 4). The saturating kinetic component is sensitive to inhibition by certain sulfhydryl compounds (e.g., *p*-chloromercuriphenylsulfonic acid, PCMBS) and α-cyano-4-hydroxycinnamate (α-CIM), as well as stilbenes (DIDS,SITS), supporting the presence of a H^+/monocarboxylate carrier in these species as in mammalian RBCs (Poole and Halestrap, 1993). Lactate transport is inhibited by pyruvate (Soengas and Moon, 1995) and carp transport lactate and pyruvate at about equal rates (Tiihonen and Nikinmaa, 1993), suggesting that there is a common carrier for both monocarboxylate compounds. Tiihonen and Nikinmaa (1993) concluded that at physiological concentrations of lactate and pyruvate, transport was through the H^+/monocarboxylate transporter, whereas "diffusional" transport and band-3-mediated transport became more apparent at higher plasma concentrations. These authors also reported significant differences between groups of carp, suggesting that monocarboxylate transport was sensitive to acclimatization. As with glucose, lactate uptake rate does not limit oxidation of this compound by red cells. Skipjack tuna (*Katsuwonus pelamis*) RBCs are reported to possess only diffusion-limited lactate transport (Moon *et al.*, 1987), although prolonged incubation times may have obscured the detection of a saturable transport component.

Tiihonen and colleagues (see Tiihonen, 1995) have also studied pyruvate transport in hagfish and starry flounder (*Platichthys stellatus*) red cells.

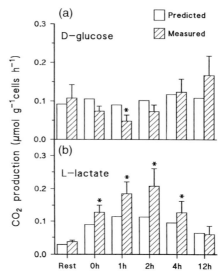

Fig. 3. Rates of $^{14}CO_2$ production from (a) D-[U-^{14}C]glucose and (b) L-[U-^{14}C]lactate in whole blood taken from catheterized rainbow trout at rest and at various times (0–12 h) after exhaustive exercise. At each time, measurements were made using the native, measured *in vivo* substrate levels in the plasma, and the acid–base conditions were maintained at the appropriate measured *in vivo* levels. At each time, the open bar shows the rate predicted simply from the measured substrate concentration, based on kinetic relationships established for resting blood cells under resting acid–base conditions. The cross-hatched bars show the measured rate. Note the depression of measured glucose oxidation rate below predicted levels and the stimulation of measured lactate oxidation rate above predicted levels at 0–2 h postexercise, indicating the influence of factors other than changes in substrate concentration alone. Asterisks indicate significant differences ($P < 0.05$) in measured postexercise rates relative to measured resting rates. Means $\pm$ 1 SEM ($N = 6$–13). Results redrawn from Wood *et al.* (1990) and Walsh *et al.* (1990).

Hagfish RBCs show rapid and concentration-dependent uptake of pyruvate, unlike flounder cells. Transport in hagfish cells was sensitive to PCMBS and to high concentrations of DIDS, but relatively insensitive to α-CIM and specifically dependent upon the presence of extracellular Na$^+$. This is the first example of concentrative pyruvate transport in a system outside the mammalian kidney and intestinal epithelium. Certainly, it does indicate a very different monocarboxylate transporter, the significance of which is yet to be established.

The complexity of monocarboxylate transport precludes an absolute understanding of the extent of monocarboxylates transported by nonspecific and by specific transporters. However, given the wide variation in the ability

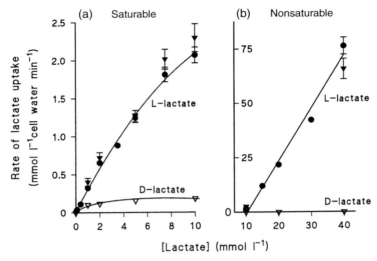

Fig. 4. (a) Saturable uptake kinetics of initial L-lactate transport at low concentration (K_m = 6.7 mmol·L^{-1}; J_{max} = 2.3 mmol lactate·L^{-1} cell water·min^{-1}) and (b) nonsaturable uptake kinetics of initial L-lactate transport at high concentration into washed red blood cells of American eel at 10°C. Means ± 1 SEM. Uptake rates in the presence (filled circles; N = 12) and absence of Na$^+$ (filled triangles; N = 3) in the medium were identical, whereas D-lactate uptake rates (open triangles; N = 4) were negligible, indicating Na$^+$ independence and stereospecificity of transport. Results redrawn from Soengas and Moon (1995).

of fish species to produce and to release lactate into the blood, the RBC may be considered one component of the whole body lactate metabolism. Further studies examining a more diverse group of fish would better define the important components of this transport system.

C. Amino Acid Transport

Amino acid transport has been reviewed (Barker and Ellory, 1990; Christensen, 1990; McGivan and Pastor-Anglada, 1994), and Nikinmaa and Tiihonen (1994) mention studies in fish RBCs. Most studies in fish have dealt with the issue of cell volume regulation and the use of β-amino acids (especially β-alanine and taurine) in this connection (see Nikinmaa and Tiihonen, 1994). More recently, the transport of metabolically important amino acids has been examined to address the evolution of these transport systems in vertebrates.

Fish RBCs metabolize amino acids, although rates are low and generally only L-alanine has been studied (Soengas and Moon, 1995; Walsh *et al.*, 1990). Studies using glutamine/glutamate have demonstrated that both carp

(Nikinmaa and Tiihonen, 1994; Tiihonen and Nikinmaa, 1991) and lungfish (Mauro and Isaacks, 1989) metabolize these amino acids well, with the lungfish demonstrating much higher oxidation rates with glutamate than with glucose. In addition, RBC:plasma distribution ratios generally exceed 1 where reported, including >200 for alanine and >10 for all other amino acids tested in Pacific hagfish (Fincham et al., 1990), 2.8 for L-serine (Gallardo et al., 1992), and 0.6 to 1.5 for L-glycine (Gallardo and Sanchez, 1993) in brown trout. These results suggest specific transporters do exist. A variety of amino acid transporters have been reported that are either Na$^+$ dependent (e.g., systems A, ASC, Gly, and β) or Na$^+$ independent (e.g., systems L, C, y$^+$, and asc) (see Barker and Ellory, 1990; Christensen, 1990, 1996; McGivan and Pastor-Anglada, 1994). The plethora of transporters probably results from the close similarities between many amino acids and their importance to cellular biochemistry. Studies reported in fish seem to support the existence of most of these systems in RBCs, although their specificity and properties differ from those of their homologous mammalian counterparts.

The most active amino acid transporter is that for L-alanine in the Pacific hagfish. Fincham et al. (1990) reported that rates of uptake were 10,000 times higher than similar transport across the mammalian or avian red cell membrane. This high rate was consistent with a very high intracellular content of α-amino acids in these red cells, unlike that reported for teleosts. The transporter was an asc-type system based upon its lack of Na$^+$ dependence and inhibition by other neutral amino acids. Transport was an exchange process with stoichiometry 1:1 with intracellular amino acids. Hagfish RBCs transport L-leucine and L-histidine at rates that exceed those of L-alanine, possibly by a L-type system. Hagfish red cells are unique in that they have little Cl$^-$/HCO$_3^-$ exchange, which means that band 3 transport is of minimal importance.

Gallardo, Sanchez, and colleagues have reported the uptake of L-alanine (Albi et al., 1994), L-glycine (Gallardo and Sanchez, 1993), L-serine (Gallardo et al., 1992), and L-leucine (Gallardo et al., 1996) into brown trout RBCs. These short-chain neutral amino acids are carried by a variety of transporters and probably not by a single transporter. L-Glycine, L-alanine and L-serine are predominantly transported by a Na$^+$-dependent ASC system with variable contributions of a Na$^+$-independent system, including the asc (L-alanine, L-glycine, and L-serine) or the Gly or the L (L-glycine) systems. L-Leucine was reported to be transported predominantly by the Na$^+$-independent L-system. The exact explanation for this variety of transporters is not understood, although this same group has shown that the ASC component of L-serine and L-alanine uptake is reduced in the presence of isoproterenol (Gallardo et al., 1992) and insulin (Canals et al., 1995), respectively. These minor transport components may simply fine-tune the transport processes *in vivo.*

There are significant differences among species in amino acid transport. L-Glycine transport by RBCs of the channel catfish (*Ictalurus punctatus*) is partitioned differently from that of brown trout. The predominant L-glycine transport system is system L (54%), whereas 16.1% is by the *Gly* system, 15.6% by the combination of the ASC and *asc* systems, and 7% by an undefined Na^+-independent system (Angermeier *et al.*, 1996). Additionally, Soengas and Moon (1995) were unable to find saturable uptake of L-alanine into American eel red cells, but L-alanine transport was inhibited by D-alanine, which does not exclude a carrier system. This may mean that yet another transport system exists in eel RBCs.

Studies of amino acids transporters indicate that a variety of carriers exist in fish RBCs. Few specific amino acids have been studied, but those examined are generally carried by more than one transporter type. Whether these carriers are developmentally or environmentally altered has not been established, although some data by Gallardo, Sanchez, and colleagues indicate that distribution ratios for certain amino acids vary seasonally (L-serine, Gallardo *et al.*, 1992; L-glycine, Gallardo and Sanchez, 1993). Certainly, temperature will impact membrane fluidity and transport rates and could alter transport properties, but such changes have yet to be studied. It does not appear, however, that amino acids are particularly important as metabolizable substrates for fish RBCs (see part IV), however, red cells could be important in the interorgan transport of amino acids as previously discussed by Christensen (1990).

VI. THE INFLUENCE OF CATECHOLAMINES, OXYGEN STATUS, AND INTRACELLULAR pH REGULATION ON RED CELL METABOLISM

A. The Red Cell pH_i Regulatory Response

Since the classic discovery by Nikinmaa (1982) that β-adrenergic stimulation causes swelling and a decrease in the pH gradient across salmonid RBCs, a massive amount of evidence has accumulated on both the mechanism and the adaptive significance of the response. This evidence is reviewed in detail by Nikinmaa (1992) and in another chapter in this volume, so it will only be recapped briefly here. It is now clear that under natural conditions of stress, hypoxia, or severe exercise, catecholamines that stimulate β_1-adrenergic receptors on the erythrocyte membrane are mobilized into the blood plasma, resulting in the activation of adenylate cyclase, cAMP production as an intracellular messenger, and cell surface Na^+/H^+ antiporters. The entry of Na^+ along its electrochemical gradient causes a net extrusion of H^+ ions, thereby lowering extracellular pH and raising intracellular pH. The absence of extracellular carbonic anhydrase in the blood plasma ensures that H^+ export will exceed H^+ entry (as CO_2) by the Jacobs–Stewart cycle. However, the presence of intracellular carbonic anhydrase ensures an immediate rise in intracellular HCO_3^- concentration, whose exit in turn forces entry of Cl^- through the band 3 anion exchanger. The increase in intracellular Na^+ and Cl^- levels raises osmolality inside the RBC, causing a net entry of water, erythrocyte swelling, and accompanying dilution of the hemoglobin and other intracellar polyanions such as ATP and GTP ("NTP").

A number of adaptive explanations have been offered for this phenomenon, but the most widely accepted is still that first offered by Nikinmaa (1982). Above all, the response protects RBC O_2 transport at times of hypoxia or acidosis; the increase in pH_i, the dilution of hemoglobin itself, and the decrease in intracellular NTP concentration all serve to elevate hemoglobin O_2 affinity and capacity. In keeping with this role, the responsiveness of the mechanism is enhanced by hypoxia and moderate acidosis; indeed, in some species (e.g., carp, *Cyprinus carpio;* Salama and Nikinma, 1988), such conditions are an absolute requirement for activation of the system. Other potential roles relate to the changing HCO_3^- and P_{CO_2} gradients across the RBC membrane that may alter net CO_2 carriage in the red cell (e.g., Currie and Tufts, 1993) and excretion patterns at the gills (e.g., Perry and Wood, 1989), and the accompanying decrease in pHe, which may serve as a ventilatory stimulus (e.g., Wood and Munger, 1994).

B. Mechanistic Studies *in Vitro*

Although the preceding synthesis has evolved from literally hundreds of studies, there have been only a handful of investigations on the metabolic support of this response. From these few studies, a coherent picture has begun to emerge of the coupling between metabolic energy production and Na^+/H^+ exchange activation, but at the outset, one important *caveat* must be emphasized. With few exceptions, these investigations have used "industrial" levels of catecholamines (e.g., 10^{-5}–10^{-3} M), and often the especially potent synthetic analogue isoprenaline. Such levels are 2 to 4 orders of magnitude greater than the highest concentrations ever likely to occur *in vivo* (Nikinmaa 1992; Gamperl *et al.,* 1994). A second, less serious *caveat* relates to the potential complicating role of the highly metabolically active white blood cells in whole blood studies (Wang *et al.,* 1994; see parts IIA,

IVA). Fortunately, WBC metabolic rate appears to be unresponsive to adrenergic stimulation (Fig. 1).

The first important clue to metabolic coupling of the Na^+/H^+ response was the observation of Nikinmaa (1983) on washed trout RBCs (obtained by cardiac puncture) that intracellular NTP levels decreased more during maximal adrenergic stimulation than could be explained by cell swelling alone. The effect was correlated with a decrease in pyruvate kinase activity, suggesting a decreased glycolytic production of NTP. However, glycolytic rates appear to be low to negligible in teleost erythrocytes under aerobic conditions (Boutilier and Ferguson, 1989; Walsh *et al.*, 1990; Sephton *et al.*, 1991; Sephton and Driedzic, 1994a; Pesquero *et al.*, 1992) and, if anything, are stimulated rather than inhibited by adrenergic stimulation (Pesquero *et al.*, 1992; Sephton and Driedzic, 1994b). Furthermore, erythrocytic NTP levels decline under anaerobic conditions both *in vivo* and *in vitro* (Tetens and Lykkeboe, 1981; Milligan and Wood, 1987; Boutilier *et al.*, 1988).

More detailed studies on whole blood sampled from catheterized Atlantic salmon (*Salmo salar*) were performed by Ferguson and Boutilier (1988). These showed that activation of Na^+/H^+ exchange by maximal adrenergic stimulation under aerobic conditions caused a 15% decrease in erythrocytic NTP concentration (normalized to hemoglobin to correct for RBC swelling), negligible change in lactate production, but a doubling of O_2 consumption rate. Very similar patterns were observed in the blood of rainbow trout (Ferguson *et al.*, 1989). Later work using similar methods showed a comparable increase in the rate of total CO_2 production by the adrenergically stimulated salmon blood cells (Tufts *et al.*, 1991). Aerobic pathways clearly supplied more than 90% of the cells' energy under resting conditions, and an even greater percentage during maximal adrenergic stimulation. The obvious conclusion was that metabolic support for activation of Na^+/H^+ exchange is mainly provided by aerobic respiration. The decrease in NTP either is a contributor to this support or is simply a correlate associated with the strategy of decreasing intracellular polyanions during active pH_i regulation. The former interpretation is supported by comparable experiments on the blood of rainbow trout (Tufts and Boutilier, 1991), which showed that blockade of Na^+/K^+-ATPase by ouabain eliminated the fall in NTP accompanying adrenergic activation of Na^+/H^+ exchange.

Ferguson *et al.* (1989) further examined rainbow trout blood cells under complete anoxia. Intracellular NTP levels declined, as reported previously, but the fall was far less than predicted from the absence of aerobic respiration, as was the modest increase in the rate of lactate production, suggesting that the cells entered a hypometabolic state. More surprisingly, anoxic RBCs remained fully capable of pH_i regulation upon maximal adrenergic stimulation, but now intracellular NTP levels declined by about 50%, whereas lactate production did not change. The decline in NTP (in "O_2 equivalents") upon adrenergic stimulation during anoxia was comparable to the increase in O_2 consumption upon adrenergic stimulation during normoxia. Ferguson *et al.* (1989) concluded that the capability for RBC pH_i regulation was maintained during anoxia, but membrane function became uncoupled from cellular energy-generating processes because of the hypometabolic state.

Ferguson *et al.* (1989) were able to relate high intracellular NTP levels to low intracellular Na^+ (and high intracellular K^+) in trout blood cells. During anoxia, normal ionic gradients were maintained despite the fall in NTP and the apparent hypometabolism. Because even under nonstimulated conditions, cell membrane Na^+/K^+-ATPase may consume 20–25% of normal ATP production (Rapoport, 1986; Tufts and Boutilier, 1991; Wang *et al.*, 1994), they suggested that hypometabolism under anoxia might be achieved by closing membrane leakage channels. Interestingly, after maximal adrenergic stimulation, ionic gradients deteriorated (i.e., intracellular $[Na^+]$ increased greatly while intracellular $[K^+]$ decreased greatly) during anoxia but not during normoxia. The phenomenon was similar to the dissipation of gradients seen when ouabain-treated cells were adrenergically stimulated (Tufts and Boutilier, 1991). It is unclear whether this "breakdown" of regulation during anoxia is simply due to a more potent activation of Na^+/H^+ exchange under these conditions or is in some way linked to the hypothesized uncoupling.

What is the signal that normally accelerates NTP consumption during adrenergic Na^+/H^+ activation? As in many secondary active transport systems, very likely it is the rise in intracellular Na^+ concentration that activates cell membrane-bound Na^+/K^+-ATPase. Bourne and Cossins (1982) recorded a large increase in active K^+ influx at this time in washed erythrocytes. Tufts and Boutilier (1991) found that the ionophore monensin, which in itself effects Na^+/H^+ exchange and a substantial measured Na^+ entry, caused a large decrease in NTP levels in otherwise resting blood cells of rainbow trout. Presumably, the rise in intracellular $[Na^+]$ activated the Na^+/K^+-ATPase pump.

However, the difficulty at present is explaining how O_2 consumption is increased simultaneously. An attractive scenario would be that the fall in [NTP] due to the accelerated consumption of ATP by the pump activates a more or less matching increase in oxidative metabolism so as to reset NTP stores to a new steady-state level. In support of this scenario, Tufts and Boutilier (1991) reported that ouabain blockade of Na^+/K^+-ATPase completely eliminated both the fall in [NTP] and the increase in O_2 consumption accompanying maximal adrenergic stimulation of trout blood cells. Unfortunately, this observation was not confirmed by Wang *et al.*

(1994), who found that elevated O_2 consumption still occurred during maximal adrenergic stimulation of ouabain-treated trout blood, although the response was slightly attenuated. Another difficulty is that Tufts and Boutilier (1991) found no increase in O_2 consumption associated with monensin treatment of resting blood cells, despite the obvious consumption of NTP. Experimental NTP depletion similarly failed to stimulate O_2 consumption. Presumably, mechanisms additional to the consumption of NTP by the membrane-bound Na^+/K^+-ATPase contribute to the elevation in metabolic rate during maximal adrenergic stimulation.

C. Mechanistic Studies under *in Vivo* Conditions

Although the preceding approaches have been very useful in assessing the absolute capabilities of the metabolic machinery in red cells, experiments performed under complete anoxia or with pharmacological levels of catecholamines may not tell us how erythrocytes support adrenergically activated Na^+/H^+ exchange under normal *in vivo* conditions. Metabolic investigations employing realistic levels of catecholamines, hypoxia, and acid–base status are more useful in this regard. Indeed, they are to be encouraged, because the blood is the only tissue in the body where the metabolism of the constituent cells can be examined under the conditions prevailing in the extracellular milieu (i.e., the blood plasma) and where these conditions can be manipulated easily both *in vivo* and *in vitro*.

At present, the only such example is the study of Wood *et al.* (1990) on factors affecting O_2 consumption and oxidative fuel preference in arterial whole blood sampled from catheterized rainbow trout after exhaustive exercise under normoxic conditions. This treatment caused a marked extracellular acid–base disturbance (a large acidosis of mixed respiratory and metabolic origin followed by a small alkalotic overshoot late in recovery), mobilized plasma catecholamines (10^{-7} M range), and activated Na^+/H^+ exchange as shown by RBC pH_i regulation. However, there was no change in blood cell O_2 consumption (measured under the conditions of acid–base, catecholamine, and oxygenation status prevailing *in vivo* at the time of sampling), in contrast to the many previously cited studies that reported increased RBC O_2 consumption under maximal β-adrenergic stimulation *in vitro*. Interestingly, Ferguson *et al.* (1989) reported two postexercise measurements of blood O_2 consumption in trout (sampled by catheter); in one trial it went up and in one it went down, again suggesting a much weaker response than seen with maximal β-adrenergic stimulation *in vitro*.

Factorial analysis largely explained the observed *in vivo* pattern of unchanged blood O_2 consumption after exercise (Wood *et al.*, 1990). When blood from resting trout was subjected to typical postexercise levels of respiratory acidosis and metabolic acidosis, alone or in combination, O_2 consumption was suppressed by about 35%. Conversely, the rate was elevated 30% by metabolic alkalosis. However, when typical postexercise levels of catecholamines (1.5×10^{-7} M adrenaline, 1.5×10^{-8} M noradrenaline) were included in acidotic blood, O_2 consumption remained at resting levels. Catecholamine addition had no effect on the metabolic rate of nonacidotic blood. These results suggest a direct sensitivity of oxidative metabolism to extracellular pH, with realistic levels of catecholamines exerting an appropriate protective effect during acidosis. This explanation does not eliminate the possibility that this "protective effect" represented the aerobic cost of Na^+/H^+ exchange activation (see part VIB), and unfortunately, NTP levels were not measured in this study. Nevertheless, the results suggest that the normal *in vivo* cost of RBC pH_i regulation is much lower than the previously cited studies would indicate.

In parallel experiments, Wood *et al.* (1990) measured $^{14}CO_2$ production from uniformly labeled [^{14}C]glucose and [^{14}C]lactate substrates at the endogenous levels present in the blood at various times after exhaustive exercise. Glucose and lactate have been identified as major aerobic fuels of blood cells under control conditions in trout (Walsh *et al.*, 1990) and several other species (reviewed by Nikinmaa and Tiihonen, 1994). Again, the *in vivo* conditions of acid–base, catecholamine, and oxygenation status recorded at the time of sampling were preserved during the incubations. In view of the mismatch between $^{14}CO_2$ evolution rates and absolute rates of O_2 consumption discussed earlier (see part IVC), data from this approach must be interpreted with caution, as they reflect relative trends rather than absolute values.

The results suggest that fuel usage shifts away from glucose (the preferred fuel at rest), and strongly toward lactate, peaking with a fivefold increase at 2 h postexercise (Fig. 3). Glucose oxidation was depressed by 30–50%, despite more or less unchanged plasma glucose concentrations (Fig. 3a). This effect was partially explained by a direct inhibitory effect of high plasma lactate concentration on glucose utilization (Walsh *et al.*, 1990). In turn, this substantial natural elevation of plasma lactate was partially responsible for the great increase in $^{14}CO_2$ evolution from lactate. However, based on the kinetic concentration-dependence relationships established by Walsh *et al.* (1990) for blood from resting trout, substrate elevation could explain only about 50% of the observed increase in lactate oxidation (Fig. 3b). Factorial analysis revealed the other influences that came into play. Lactate oxidation was stimulated by respiratory acidosis, an effect specifically attributable to the high P_{CO_2} component. Natural catecholamine levels had no effect by themselves, but in the presence of acidosis, they strongly stimulated lactate oxidation. Thus although acid–

base effects on lactate metabolism were very different from acid–base effects on O_2 consumption, the stimulatory role of catecholamines in the presence of acidosis was in accord. Another possible factor contributing to this switch from glucose to lactate was documented by Ferguson and Storey (1991), who found greatly elevated *intracellular* lactate levels but no change in very low glucose levels in the red cells of trout after exhaustive exercise.

Clearly, much more work of this nature is required to understand red cell metabolism under natural stress conditions *in vivo*. For example, how do acid–base, oxygen, and catecholamine status affect the use of other fuels? Both adrenergic stimulation and high pH are reported to alter serine uptake in rainbow trout erythrocytes, but effects on oxidation rate are unknown (Gallardo *et al.*, 1992). Does metabolism change when the red cell β_1-adrenergic pathway becomes either sensitized or densensitized through chronic stress (Perry *et al.*, 1996)? What are the mechanisms of metabolic regulation during natural stresses? To date, most evidence is negative. Pesquero *et al.* (1992), using washed red cells obtained by caudal puncture from brown trout (*Salmo trutta*), found no effect of hypoxic incubation on the activity of key glycolytic enzymes. A common means of regulating glycolytic flux is a change in enzyme binding to cell structure components, but the limited evidence available does not support this as a regulatory mechanism in teleost red cells. Ferguson and Storey (1991) reported low and unchanged binding of key glycolytic enzymes in red cells of rainbow trout obtained by catheter before and after exhaustive exercise. Sephton and Driedzic (1994b) reported higher enzyme binding (probably a technique-related difference) in the blood cells of trout obtained by caudal puncture from ice-bath chilled fish. However, binding was unaffected by maximal adrenergic stimulation despite an apparent increase in glycolytic flux.

VII. EPILOGUE

From this discussion, it is clear that fish blood is a dynamic metabolic system, with many of its basic properties still unexplored. Two additional areas will likely be important areas for research, owing mainly to the blood's position as a first line of defense relative to environmental perturbations. First, the effects of variables such as temperature and salinity will likely yield interesting metabolic adaptations. Second, in addition to blood's routine role as a vital respiratory tissue, it may be an important first line of defense against harmful xenobiotic compounds. As shown in two studies, xenobiotics can form adducts with fish Hb (Shugart *et al.*, 1987; Kennedy *et al.*, 1991), a process that presumably impairs Hb function. In this regard, fish blood is able to transform and detoxify the xenobiotics molinate (Tjeerdema and Crosby, 1988) and benzo[*a*]pyrene (Kennedy *et al.,* 1991) through low levels of its own Phase I and Phase II detoxification enzymes. Understanding the interaction of these pathways with routine metabolism (especially those pathways involving glutathione-based protection; Marshall *et al.,* 1990) will present an additional challenge to researchers studying fish blood function in a warmer and more polluted aquatic world.

VIII. REFERENCES

Albi, J. L., Canals, P., Gallardo, M. A., and Sanchez, J. (1994). Na$^+$-independent L-alanine uptake by trout cells: Evidence for the existence of at least two functionally different *asc* systems. *J. Membr. Biol.* **140,** 189–196.

Angermeier, S. M., Shepard, M. D., and Tunnicliff, G. (1996). Glycine transport by the red cells of channel catfish. *Can. J. Zool.* **74,** 688–692.

Bachand, L., and Leray, C. (1975). Erythrocyte metabolism in the yellow perch (*Perca flavescens* Mitchill). I. glycolytic enzymes. *Comp Biochem. Physiol. B* **50,** 567–570.

Barker, G. A., and Ellory, J. C. (1990). The identification of neutral amino acid transport systems. *Exp. Physiol.* **75,** 3–26.

Bishop, C. (1964). Overall red cell metabolism, pages 148–188, *In* "The Red Blood Cell" (C. Bishop, and D. M. Surgenor, eds.). Academic Press. New York.

Boivan, P. Galand, C., and Bertrand, O. (1986). Properties of a membrane-bound tyrosine kinase phosphorylating the cytosolic fragment of the red cell membrane band 3 protein. *Biochim. Biophys. Acta* **860,** 243–252.

Bolis, L., Luly, P., and Baroncelli, V. (1971). D(+)-glucose permeability in brown trout. *Salmo trutta* L. *J. Fish Biol.* **3,** 273–275.

Bourne, P. K., and Cossins, A. R. (1982). On the instability of K$^+$ influx in erythrocytes of the rainbow trout, *Salmo gairdneri,* and the role of catecholamine hormones in maintaining *in vivo* influx activity. *J. Exp. Biol.* **101,** 93–104.

Boutilier, R. G., Dobson, G., Hoeger, U., and Randall, D. J. (1988). Acute exposure to graded levels of hypoxia in rainbow trout (*Salmo gairdneri*): Metabolic and respiratory adaptations. *Respir. Physiol.* **71,** 69 –82.

Boutilier, R. G. and Ferguson, R. A. (1989). Nucleated red cell function: Metabolism and pH regulation. *Can. J. Zool.* **67,** 2989–2993.

Buckley, J. A. (1982). Hemoglobin-glutathione relationships in trout erythrocytes treated with monochloramine. *Bull. Environ. Contam. Toxicol.* **29,** 637–644.

Bushnell, P. G., Nikinmaa, M., and Oikari, A. (1985). Metabolic effects of dehydroabietic acid on rainbow trout erythrocytes. *Comp. Biochem. Physiol. C* **81,** 391–394.

Canals, P., Gallardo, M. A., and Sanchez, J. (1995). Effects of insulin on the uptake of amino acids by hepatocytes and red blood cells from trout (*Salmo trutta*) are opposite. *Comp. Biochem. Physiol. C* **112,** 221–228.

Christensen, H. N. (1990). Role of amino acid transport and countertransport in nutrition and metabolism. *Physiol. Rev.* **70,** 43–77.

Cossins, A. R. (1989). Intracellular pH regulation by fish red cells. *Nature* **340,** 20–21.

Cossins, A. R., and Richardson, P. A. (1985). Adrenalin-induced Na$^+$/H$^+$ exchange in trout erythrocytes and its effects upon oxygen-carrying capacity. *J. Exp. Biol.* **118,** 229–246.

Cossins, A. R., and Kilbey, R. V. (1989). The seasonal modulation of Na^+/H^+ exchanger activity in trout erythrocytes. *J. Exp. Biol.* **144**, 463–478.

Currie, S., and Tufts, B. L. (1993). An analysis of carbon dioxide transport in arterial and venous blood of the rainbow trout, *Oncorhynchus mykiss,* following exhaustive exercise. *Fish Physiol. Biochem.* **12**, 183–192.

Der Lee, I., and Palsson, B. O. (1992). A Macintosh software package for simulation of human red blood cell metabolism. *Computer Methods and Programs in Biomedicine* **38**, 195–226.

Dey, I., Szegletes, T., Buda, C., Nemcsok, J., and Farkas, T. (1993). Fish erythrocytes as a tool to study temperature-induced responses in plasma membranes. *Lipids* **28**, 743–746.

Eddy, F. B. (1977). Oxygen uptake by rainbow trout blood, *Salmo gairdneri. J. Fish. Biol.* **10**, 87–90.

Ferguson, R. A., and Boutilier, R. G. (1988). Metabolic energy production during adrenergic pH regulation in red cells of the Atlantic salmon, *Salmo salar. Respir. Physiol.* **74**, 65–76.

Ferguson, R. A., and Tufts, B. L. (1992). Physiological effects of brief air exposure in exhaustively exercised rainbow trout (*Oncorhynchus mykiss*): Implications for "catch and release" fisheries. *Can. J. Fish. Aquat. Sci.* **49**, 1157–1162.

Ferguson, R. A., Tufts, B. L., and Boutilier, R. G. (1989). Energy metabolism in trout red cells: Consequences of adrenergic stimulation in vivo and in vitro. *J. Exp. Biol.* **143**, 133–147.

Ferguson, R. A., and Storey, K. B. (1991). Glycolytic and associated enzymes of rainbow trout (*Oncorhynchus mykiss*) red cells: In vitro and in vivo studies. *J. Exp. Biol.* **155**, 469–485.

Fincham, D. A., Wolowyk, M. W., and Young, J. D. (1987). Volume-sensitive taurine transport in fish erythrocytes. *J. Membr. Biol.* **96**, 45–56.

Fincham, D. A., Wolowyk, M. W., and Young, J. D. (1990). Characterization of amino acid transport in red blood cells of a primitive vertebrate, the Pacific hagfish (*Eptatretus stouti*). *J. Exp. Biol.* **154**, 355–370.

Fodor, E., Jones, R. H., Buda, C., Kitajka, K., Dey, I., and Farkas, T. (1995). Molecular architecture and biophysical properties of phospholipids during thermal adaptation in fish: An experimental and model study. *Lipids* **30**, 1119–1126.

Gabbianelli, R., Falcioni, G., Mazzanti, L., Bertoli, E., and Zolese, G. (1996). Seasonal variations of physical and biochemical membrane properties in trout erythrocytes (*Salmo irideus*). *Comp. Biochem. Physiol. B* **114**, 275–279.

Gallardo, M. A., Albi, J. L., and Sánchez, J. (1996). Uptake of L-leucine by trout red blood cells and peripheral lymphocytes. *J. Membr. Biol.* **152**, 57–63.

Gallardo, M. A., Planas, J., and Sanchez, J. (1992). L-Serine uptake by trout (*Salmo trutta*) red blood cells: The effect of isoproterenol. *J. Exp. Biol.* **163**, 85–95.

Gallardo, M. A., and Sanchez, J. (1993). Glycine uptake by trout (*Salmo trutta*) red blood cells. *J. Membr. Biol.* **134**, 251–259.

Gamperl, A. K., Vijayan, M. M., and Boutilier, R. G. (1994). Experimental control of stress hormone level in fishes: Techniques and applications. *Rev. Fish Biol. Fisheries.* **4**, 215–255.

Greaney, G. S., and Powers, D. A. (1978). Allosteric modifiers of fish hemoglobins: In *vitro* and *in vivo* studies of the effect of ambient oxygen and pH on erythrocyte ATP concentrations. *J. Exp. Zool.* **203**, 339–350.

Härdig, J., and Höglund, L. B. (1983). Seasonal and ontogenetic effects on methaemoglobin and reduced glutathione contents in the blood of reared Baltic salmon. *Comp. Biochem. Physiol. A* **75**, 27–34.

Hunter, A. S., and Hunter, F. R. (1957). A comparative study of erythrocyte metabolism. *J. Cell. Comp. Physiol.* **49**, 479–502.

Ingermann, R. L., Bissonnette, J. M., and Hall, R. E. (1985). Sugar uptake by red blood cells. *In* "Circulation, Respiration, and Metabolism" (R. Gilles, ed.), pp. 290–300. Springer-Verlag, Berlin.

Ingermann, R. L., Hall, R. E., Bissonnette, J. M., and Terwilliger, R. C. (1984). Monosaccharide transport into erythrocytes of the Pacific hagfish *Eptatretus stouti. Mol. Physiol.* **6,** 311–320.

Iwama, G. K., McGeer, J. C., and Pawluck, M. P. (1989). The effects of five anesthetics on acid–base balance, hematocrit, blood gases, cortisol, and adrenaline in rainbow trout. *Can. J. Zool.* **67,** 2065–2073.

Joshi, A., and Palsson, B. O. (1989). Metabolic dynamics in the human red cell. I. A comprehensive kinetic model. *J. Theor. Biol.* **141,** 515–528.

Kennedy, C. J., Gill, K. A., and Walsh, P. J. (1991). In-vitro metabolism of benzo(a)pyrene in the blood of the gulf toadfish, *Opsanus beta. Mar. Env. Res.* **31,** 37–53.

Kim, H. D., and Isaacks, R. E. (1978). The membrane permeability of nonelectrolytes and carbohydrate metabolism of Amazon fish red cells. *Can. J. Zool.* **56,** 863–869.

Kirk, K., Ellory, J. C., and Young, J. D. (1992). Transport of organic substrates via a volume-activated channel. *J. Biol. Chem.* **267,** 23475–23478.

Kodicek, M., Mircevova, L., and Marik, T. (1987). Energy requirements of erythrocytes under mechanical stress. *Biomed. Biochim. Acta* **46,** 103–107.

Lane, H. C., Weaver, J. W., Benson, J. A., and Nichols, H. A. (1982). Some age-related changes of adult rainbow trout, *Salmo gairdneri,* peripheral erythrocytes separated by velocity sedimentation at unit gravity. *J. Fish. Biol.* **21,** 1–13.

Low, P. S., Kiyatkin, A., Li, Q., and Harrison, M. L. (1995). Control of erythrocyte metabolism by redox-regulated tyrosine phosphatases and kinases. *Protoplasma* **184,** 196–202.

Lowe, A. G., and Walmsley, A. R. (1986). The kinetics of glucose transport in human red blood cells. *Biochim. Biophys. Acta* **857,** 146–154.

Marshall, W. S., Bryson, S. E., and Sapp, M. M. (1990). Volume regulation in glutathione-treated brook trout (*Salvelinus fontinalis*) erythrocytes. *Fish Physiol. Biochem* **8,** 19–28.

Mauro, N. A., and Isaacks, R. E. (1989). Relative oxidation of glutamate and glucose by vertebrate erythrocytes. *Comp. Biochem. Physiol. A* **94,** 95–97.

McDonald, D. G., and Milligan, L. (1997). Ionic, osmotic, and acid–base regulation in stress. *In* "Fish Stress and Health in Aquaculture" (G. K. Iwama, J. Sumpter, A. Pickering, and C. B. Schreck, eds.), pp. 119–144. Society of Experimental Biology Seminar Series, Cambridge Univ. Press, Cambridge.

McGivan, J. D., and Pastor-Anglada, M. (1994). Regulatory and molecular aspects of mammalian amino acid transport. *Biochem. J.* **299,** 321–334.

Milligan, C. L., and Wood, C. M. (1987). Regulation of blood oxygen transport and red cell pHi after exhaustive activity in rainbow trout (*Salmo gairdneri*) and starry flounder (*Platichthys stellatus*). *J. Exp. Biol.* **133,** 263–282.

Mommsen, T. P. (1984). Metabolism of the fish gill. *In* "Fish Physiology," Vol. 10B (W. S. Hoar and D. J. Randall, eds.), pp. 203–238. Academic Press, Orlando.

Mommsen, T. P., Walsh, P. J., and Moon, T. W. (1994). Hepatocytes: Isolation, maintenance and utilization. *In* "Biochemistry and Molecular Biology of Fishes" (P. W. Hochachka and T. P. Mommsen, eds.), pp. 355–373. Elsevier Science, Amsterdam.

Moon, T. W., and Foster, G. D. (1995). Tissue carbohydrate metabolism, gluconeogenesis and hormonal and environmental influences. *In* "Biochemistry and Molecular Biology of Fishes" (P. W. Hochachka and T. P. Mommsen, eds.), pp. 65–100. Elsevier Science, Amsterdam.

Moon, T. W., Brill, R. W., Hochachka, P. W., and Weber, J.-M. (1987). L-(+)-Lactate translocation in the red blood cells of the skipjack tuna (*Katsuwonus pelamis*). *Can. J. Zool.* **65,** 2570–2573.

Moon, T. W., and Walsh, P. J. (1994). Metabolite transport in fish red blood cells and hepatocytes. *In* "Biochemistry and Molecular Biology of Fishes" (P. W. Hochachka and T. P. Mommsen, eds.), pp. 615–624. Elsevier Science, Amsterdam.

Nikinmaa, M. (1982). Effects of adrenaline on red cell volume and concentration gradient of protons across the red cell membrane in the rainbow trout, *Salmo gairdneri. Molecular Physiol.* **2,** 287–297.

Nikinmaa, M. (1983). Adrenergic regulation of haemoglobin oxygen affinity in rainbow trout red cells. *J. Comp. Physiol. B* **152,** 67–72.

Nikinmaa, M. (1992). Membrane transport and control of hemoglobin–oxygen affinity in nucleated erythrocytes. *Physiol. Reviews* **72,** 301–321.

Nikinmaa, M. (1990). "Vertebrate Red Blood Cells. Adaptations of Function to Respiratory Requirements." Springer-Verlag, Berlin. 262 pp.

Nikinmaa, M., Cech, J. J., and McEnroe, M. (1984). Blood oxygen transport in stressed striped bass (*Morone saxatalis*): Role of β-adrenergic responses. *J. Comp. Physiol. B* **154,** 365–369.

Nikinmaa, M., and Tiihonen, K. (1994). Substrate transport and utilization in fish erythrocytes. *Acta Physiol. Scand.* **152,** 183–189.

Parks, R. E. Jr., Brown, P. R., Cheng, Y.-C., Agarwal, K. C., Kong, C. M., Agarwal, R. P., and Parks, C. C. (1973). Purine metabolism in primitive erythrocytes. *Comp. Biochem. Physiol. B* **45,** 355–364.

Perry, S. F., and Wood, C. M. (1989). Control and coordination of gas transfer in fishes. *Can. J. Zool.* **67,** 2961–2970.

Perry, S. F., Reid, S. G., and Salama, A. (1996). The effects of repeated physical stress on the β-adrenergic response of the rainbow trout red blood cell. *J. Exp. Biol.* **199,** 549–562.

Pesquero, J., Albi, J. L., Gallardo, M. A., Planas, J., and Sanchez, J. (1992). Glucose metabolism by trout (*Salmo trutta*) red blood cells. *J. Comp. Physiol. B* **162,** 448–454.

Pesquero, J., Roig, T., Bermudez, J., and Sanchez, J. (1994). Energy metabolism by trout red blood cells: Substrate utilization. *J. Exp. Biol.* **193,** 183–190.

Poole, R. C., and Halestrap, A. P. (1993). Transport of lactate and other monocarboxylates across mammalian plasma membranes. *Am. J. Physiol.* **264,** C761–C782.

Rapoport, S. M. (1986). "The reticulocyte." CRC Press, Boca Raton, Fl. 238 pp.

Raynard, R. S., and Cossins, A. R. (1991). Homeoviscous adaptation and thermal compensation of sodium pump of trout erythrocytes. *Am. J. Physiol.* **260,** R916–R924.

Reimann, B., Kuttner, G., Maretzki, D., and Rapoport, S. M. (1977). Experimental proof of the control function of glycolysis enzymes in erythrocytes. *Ergeb. Exp. Med.* **24,** 177–180.

Rose, I. A. (1971). Regulation of human red cell glycolysis: A review. *Exp. Eye Res.* **11,** 264–272.

Schweiger, H. A. (1962). Pathways of metabolism in nucleate and anucleate erythrocytes. *Int. Rev. Cytol.* **13,** 135–201.

Sephton, D. H., Macphee, W. L., and Driedzic, W. R. (1991). Metabolic enzyme activities, oxygen consumption and glucose utilization in sea raven (*Hemitripterus americanus*) erythrocytes. *J. Exp. Biol.* **159,** 407–418.

Sephton, D. H., and Driedzic, W. R. (1994a). Glucose metabolism by sea raven (*Hemitripterus americanus*) and rainbow trout (*Oncorhynchis mykiss*) erythrocytes. *J. Exp. Biol.* **194,** 167–180.

Sephton, D. H., and Driedzic, W. R. (1994b). Adrenergic stimulation of glycolysis without change in glycolytic enzyme binding in rainbow trout (*Oncorhynchus mykiss*) erythrocytes. *Can. J. Zool.* **72,** 950–953.

Shugart, L., McCarthy, J., Jimenez, B., and Daniels, J. (1987). Analysis of adduct formation in the bluegill sunfish (*Lepomis macrochirus*) between benzo[a]pyrene and DNA of the liver and hemoglobin of the erythrocyte. *Aq. Toxicol.* **9,** 319–325.

Soengas, J. L., and Moon, T. W. (1995). Uptake and metabolism of glucose, alanine and lactate by red blood cells of the American eel *Anguilla rostrata. J. Exp. Biol.* **198,** 877–888.

Soivio, A., Westman, K., and Nyholm, K. (1972). Improved method of dorsal aorta catheterization: Haematological effects followed for three weeks in rainbow trout. *Finnish Fish. Res.* **1,** 11–21.

Speckner, W., Schindler, J. F., and Albers, C. (1989). Age-dependent changes in volume and haemoglobin content of erythrocytes in the carp (*Cyprinus carpio*). *J. Exp. Biol.* **141,** 133–149.

Tetens, V., and Lykkeboe, G. (1981). Blood respiratory properties of rainbow trout, *Salmo gairdneri*: Responses to hypoxia acclimation and anoxic incubation of blood *in vitro*. *J. Comp. Physiol. B* **145,** 117–125.

Tiihonen, K. (1995). Substrate transport and utilization in fish erythrocytes. Ph.D. thesis. Univ. of Helsinki, pp. 1–31.

Tiihonen, K., and Nikinmaa, M. (1991a). D-Glucose permeability in river lamprey (*Lampetra fluviatilis*) and carp (*Cyprinus carpio*) erythrocytes. *Comp. Biochem. Physiol. A* **100,** 581–584.

Tiihonen, K., and Nikinmaa, M. (1991b). Substrate utilization by carp (*Cyprinus carpio*) erythrocytes. *J. Exp. Biol.* **161,** 509–514.

Tiihonen, K., and Nikinmaa, M. (1993). Membrane permeability and utilization of L-lactate and pyruvate in carp red blood cells. *J. Exp. Biol.* **178,** 161–172.

Tiihonen, K., Nikinmaa, M., and Lappivaara, J. (1995). Glucose transport in carp erythrocytes: Individual variation and effects of osmotic swelling, extracellular pH and catecholamines. *J. Exp. Biol.* **198,** 577–583.

Tipton, S. R. (1933). Factors affecting the respiration of vertebrate red blood cells. *J. Cell. Comp. Physiol.* **3,** 410–414.

Tjeerdema, R. S., and Crosby, D. G. (1988). Disposition, biotransformation and detoxication of molinate (Ordram) in whole blood of the common carp (*Cyprinus carpio*). *Pest. Biochem. Physiol.* **31,** 24–35.

Tse, C. M., and Young, J. D. (1990). Glucose transport in fish erythrocytes. Variable cytochalasin-B-sensitive hexose transport activity in the common eel (*Anguilla japonica*) and transport deficiency in the paddyfield eel (*Monopterus albus*) and rainbow trout (*Salmo gairdneri*). *J. Exp. Biol.* **148,** 367–383.

Tufts, B. L., and Randall, D. J. (1989). The functional significance of adrenergic pH regulation in fish erythrocytes. *Can. J. Zool.* **67,** 235–238.

Tufts, B. L., and Boutilier, R. G. (1991). Interactions between ion exchange and metabolism in erythrocytes of the rainbow trout *Oncorhynchus mykiss*. *J. Exp. Biol.* **156,** 139–151.

Tufts, B. L., Tang, Y., Tufts, K., and Boutilier, R. G. (1991). Exhaustive exercise in "wild" Atlantic salmon (*Salmo salar*): Acid–base regulation and blood gas transport. *Can. J. Fish. Aquat. Sci.* **48,** 868–874.

Walsh, P. J., Wood, C. M., Thomas, S., and Perry, S. F. (1990). Characterization of red blood cell metabolism in rainbow trout. *J. Exp. Biol.* **154,** 475–489.

Wang, T., Nielsen, O. B., and Lykkeboe, G. (1994). The relative contribution of red and white blood cells to whole-blood energy turnover in trout. *J. Exp. Biol.* **190,** 43–54.

Wells, R. M. G., McIntyre, R. H., Morgan, A. K., and Davie, P. S. (1986). Physiological stress responses in big gamefish after capture: Observations on plasma chemistry and blood factors. *Comp. Biochem. Physiol. A* **84,** 565–571.

Wood, C. M., and Munger, R. S. (1994). Carbonic anhydrase injection provides evidence for the role of blood acid–base status in stimulating ventilation after exhaustive exercise in rainbow trout. *J. Exp. Biol.* **194,** 225–253.

Wood, C. M., Walsh, P. J., Thomas, S., and Perry, S. F. (1990). Control of red blood cell metabolism in rainbow trout after exhaustive exercise. *J. Exp. Biol.* **154,** 491–507.

Wood, H. G., and Katz, J. (1958). The distribution of C14 in the hexose phosphates and the effect of recycling in the pentose cycle. *J. Biol. Chem.* **233,** 1279–1282.

Young, J. D., Yao, S. Y.-M., Tse, C. M., Davies, A., and Baldwin, S. A. (1994). Functional and molecular characteristics of a primitive vertebrate glucose transporter. Studies of glucose transport by erythrocytes from the Pacific hagfish (*Eptatretus stouti*). *J. Exp. Biol.* **186,** 23–41.

3

CARBONIC ANHYDRASE AND RESPIRATORY GAS EXCHANGE

RAYMOND P. HENRY
THOMAS A. HEMING

   I. Introduction
  II. The Catalytic Mechanism of CA
 III. Tissue and Isozyme Distribution
 IV. CA and Respiratory Gas Exchange: CO_2 Transport and Excretion
  V. Ammonia as a Respiratory Gas
 VI. Plasma Inhibitors of Carbonic Anhydrase
VII. Analytical Techniques
     A. CA Assays
     B. CA Inhibitors
     C. pH Disequilibria
VIII. Summary
     References

I. INTRODUCTION

The enzyme carbonic anhydrase (CA; EC 4.2.1.1) catalyzes the reversible hydration/dehydration of carbon dioxide. The reaction scheme is usually written as

$$CO_2 + H_2O \overset{CA}{\leftrightarrow} H^+ + HCO_3^-.$$

First discovered in mammalian erythrocytes (Meldrum and Roughton, 1933), CA was soon found to be present in a wide variety of tissue and cell types throughout the animal kingdom (for reviews, see Maren, 1967a; Tashian and Hewett-Emmett, 1984). The enzyme was discovered in fish erythrocytes more than 60 years ago (Brinkman *et al.*, 1932). Because of its initial discovery in erythrocytes, CA has traditionally been thought of as an enzyme whose function is primarily in the area of facilitated gas exchange (i.e., the transport and excretion of CO_2 via the circulatory sytem).

75

Fish Physiology, Volume 17:
FISH RESPIRATION

CA has been shown, however, to be a multifunctional enzyme, playing a role in a diverse assemblage of physiological and biochemical processes, including gas exchange, acid–base balance, acid and/or base secretion, ion transport, calcification, muscle contraction, gluconeogenesis, ureagenesis, fatty acid synthesis, and photosynthesis in plants (reviewed by Maren, 1967b; Bauer *et al.*, 1980; Henry, 1984, 1996; Walsh and Henry, 1991; Gros and Dodgson, 1988; Dodgson *et al.*, 1991). CA serves as a molecular link among interrelated physiological and biochemical processes. It also continues to be the focus of widespread and productive research within the area of respiratory gas exchange, primarily with respect to CO_2 transport and excretion.

II. THE CATALYTIC MECHANISM OF CA

Because CA is a central enzyme to so many physiological processes, it has been written about extensively in the physiological literature. Unfortunately, a number of misperceptions concerning both the mechanism and the significance of the catalyzed reaction have persisted, especially in the comparative literature. First, it is important to understand that the uncatalyzed and the catalyzed CO_2 reactions proceed via different pathways. The uncatalyzed hydration reaction is

$$CO_2 + H_2O \leftrightarrow H_2CO_3 \leftrightarrow H^+ + HCO_3^-.$$

Carbon dioxide is hydrated to carbonic acid, which then dissociates to hydrogen and bicarbonate ions. At physiological pH (7.4–8.1), the dissociation constant of the CO_2 system ($pK' = 6.4$; Cameron, 1989) favors the formation of H^+ and HCO_3^-. The uncatalyzed hydration reaction is by itself quite rapid, having a rate constant of 3.5×10^{-2} s^{-1} (Edsall, 1968). For HCO_3^- supply to be limiting to a particular process, that process must proceed at a faster rate. Therefore, it may not always be necessary for CA to play a role in physiological or biochemical processes dependent on H^+ or HCO_3^- formation unless other factors are involved as well. On the other hand, it is the dehydration reaction, with a rate constant of approximately $20\,s^{-1}$ (Edsall, 1968), 3 orders of magnitude slower, that is usually considered to be the rate-limiting step in physiological or biochemical processes in which CA is believed necessary.

A second, extremely rapid reaction,

$$CO_2 + OH^- \leftrightarrow HCO_3^-$$

occurs at high pH and predominates at pH values of 10 or higher. CA takes advantage of this reaction by proceeding in two steps (Coleman, 1980; Tu *et al.*, 1989; Silverman, 1991):

$$CA\text{--}Zn + H_2O \leftrightarrow CA\text{--}ZnOH^- + BH^+ \tag{1}$$

$$CA\text{--}ZnOH^- + CO_2 \leftrightarrow CA\text{--}Zn + HCO_3^- \tag{2}$$

The first step involves the splitting of water at the active site and the creation of a zinc–hydroxyl complex; the resultant proton is shuttled out of the active site by a mobile nonbicarbonate buffer complex, BH^+, after which it dissociates to form a free proton. The second step is a direct nucleophilic attack of the hydroxyl on CO_2, forming HCO_3^-. This reaction is very fast (rate constant $= 8.5 \times 10^3$ s^{-1} M^{-1}), enhancing the uncatalyzed rate by about 5 orders of magnitude. CA therefore catalyzes the reaction by favoring a pathway that occurs naturally at a physiologically lethal pH, but it does so by creating a highly alkaline environment that is restricted to the active site. There are a couple of important implications of this mechanism. First, the catalyzed reaction does not involve the direct production of either carbonic acid or protons (the latter arise from the dissociation of the buffer–proton complex). Second, it is the shuttling of protons in and out of the active site that is the rate-limiting step in the reaction mechanism. This is consistent with data showing CA activity depends on the concentration of mobile buffers (including phosphate-containing compounds such as ATP) (Lindskog, 1980; Tu *et al.*, 1989; Paranawithana *et al.*, 1989); a buffer concentration of 20 mM appears to be critical to keep proton transfer from becoming rate limiting in CO_2 hydration. Conversely, CA activity is an important component of the overall buffering capacity of biological micro-environments and fluid compartments (e.g., Chen and Chesler, 1992).

III. TISSUE AND ISOZYME DISTRIBUTION

From the perspective of respiratory gas exchange in fish, CA is found in tissues at every major step in the process, from actively metabolizing tissue (e.g., muscle) where CO_2 is produced, to elements in the circulatory system where CO_2 transport takes place, to the gills where CO_2 is excreted (Fig. 1).

CA has been found in muscle tissue of mammals and lower vertebrates (reviewed by Gros and Dodgson, 1988; Henry, 1996). In general, the type of CA isozyme in muscle tissue is dependent on the metabolic type of fiber, with the rapid turnover, sulfonamide-sensitive type II isozyme being predominant in fast twitch (white) and a sulfonamide-resistant type III isozyme being present in slow twitch (red) fibers. Furthermore, it appears that there is the type IV (membrane-associated) isozyme localized to the sarcolemma of all muscle fiber types. Very little work has been done on fish muscle, but the available evidence indicates that the distribution of

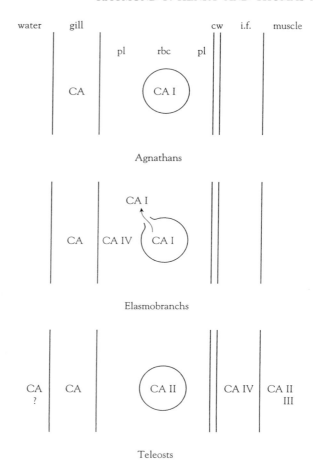

Fig. 1. Schematic representation of the distribution of carbonic anhydrase isozymes in the tissue and fluid compartments associated with CO_2 transport and excretion in fish. CA I–IV represent known kinetic classes of isozymes; CA signifies the presence of enzyme activity that has not yet been characterized as a specific isozyme; a question mark represents reported CA activity that has not been definitively confirmed. rbc, red blood cell; pl, plasma; cw, capillary wall; i.f., interstitial fluid. See text for complete description.

CA in fish muscle is similar to that in other vertebrates. Specifically, there seems to be two cytoplasmic isozymes (II and III) that are distributed according to metabolic fiber type (Sanyal *et al.*, 1982a,b) and a type IV isozyme associated with the sarcolemma (Henry *et al.*, 1997a). Recent work with selectively permeable CA inhibitors indicates that the sarcolemma CA of fish white muscle has an extracellular orientation and functions in transmembrane CO_2 transport in a similar manner to that in mammalian muscle.

As is the case in all vertebrates, CA is present in high concentrations in the erythrocytes of every fish species examined. Electrophoretic and kinetic data have shown that agnathans, elasmobranchs, and teleosts have a single CA isozyme in their red cells (Maynard and Coleman, 1971; Maren *et al.*, 1980; Carlsson *et al.*, 1980; Sanyal *et al.*, 1982c; Kim *et al.*, 1983; Hall and Schraer, 1983; Henry *et al.*, 1993). In the most primitive group, the agnathans, this appears to be a type I isozyme characterized by a relatively low turnover rate and relatively low sensitivity to the sulfonamide CA inhibitors. Teleosts appear to have the high turnover, sulfonamide-sensitive type II isozyme in their erythrocytes. CA in the red cells of elasmobranchs, a side branch of the main phylogenetic line of fish evolution, may be an evolutionary intermediate between the type I and II isozymes, as it possesses traits that are characteristic of both. Kinetically, elasmobranch CA is more similar to the type I isozyme, having a low turnover rate and reduced sulfonamide sensitivity (Maren *et al.*, 1980); however, it resembles the type II isozyme in its primary structure (Maynard and Coleman, 1971; Bergenhem and Carlsson, 1990). See Table 1 for a summary of the kinetic properties of fish erythrocyte CA.

Early reports suggested that fish red cells had two pools of CA, cytoplasmic and membrane associated (Haswell, 1977; Smeda and Houston, 1979; Houston and Mearow, 1979). These results, however, could not be duplicated; Henry *et al.* (1993) found virtually no CA activity in microsomal fractions of erythrocytes from either lamprey or trout. It is possible that putative membrane CA activity was actually cytoplasmic CA that had not been completely washed out of the initial pellet. That pellet, containing whole cells, large cell fragments, and cellular debris, is high in CA activity; however, that activity can be removed via repeated washing and sonication (Henry *et al.*, 1988, 1993). Although it is doubtful that fish erythrocytes contain the true membrane-associated CA IV isozyme, it is possible that cytoplasmic CA in the intracellular fluid (ICF) boundary layer is structurally associated with components of the membrane. In mammalian erythrocytes, soluble CA forms a complex with the cytosolic pole of the band 3 anion exchange protein (Kifor *et al.*, 1993), and this linkage allosterically alters the sulfonamide binding site of the enzyme. Allosteric effects of CA linkage to band 3 (or other membrane constituents) may explain Parkes and Coleman's (1989) observation that the association of CA (human CA I and II, and bovine CA II) with human erythrocyte membranes significantly enhances the enzyme's catalytic activity, presumably by facilitating inter- and/or intramolecular H^+ transfer. It is therefore possible that the CA

Table I

Distribution and Kinetic Properties of Erythrocyte Carbonic Anhydrase in Fish

Species	K_m (mmol L^{-1})	k_{cat} (s^{-1} 10^{-4})	K_i Az (nmol L^{-1})	K_i Cl^{-1} (mmol L^{-1})	pI	Ref.
Agnathans						
Myxine glutinosa	—	1.3	—	—	—	Maren *et al.*, 1980
	—	—	9	300	—	Carlsson *et al.*, 1980
Petromyzon marinus	5	1.5	22	150	—	Henry *et al.*, 1993
Elasmobranchs						
Squalus acanthias	5	2.5	200	—	—	Maren *et al.*, 1980
Galeocerdo cuvieri	—	—	—	—	5.0	Bergenhem and Carlsson, 1990
Carcharhinus leucas	—	—	—	—	4.5	Maynard and Coleman, 1971
Teleosts						
Oncorhynchus mykiss	10	70	—	—	—	Maren *et al.*, 1980
	10	—	2	175	—	Henry *et al.*, 1993
	—	—	—	—	9.3	Hall and Schraer, 1983
Archosargus probatocephalus	25	7.7	30	200	—	Sanyal *et al.*, 1982c

Note. K_m, Michaelis constant; k_{cat}, turnover number; K_i, inhibition constant; pI, isoelectric point.

activity reported for the membrane fraction of fish erythrocytes was, in fact, cytoplasmic CA that was linked to an integral component of the membrane.

Very little detail is known about the evolution of CA outside mammalian systems, but from the available information, it appears that the ancestral form of vertebrate erythrocyte CA is the type I isozyme, with the type II isozyme replacing it in the more highly derived groups. The evolution of CA in general, and erythrocyte CA in particular, remains a virtually un-tapped field of investigation, as the selective pressures that shaped the distribution and characteristics of the various CA isozymes have never been systematically studied.

With one exception (elasmobranchs), CA is absent from fish plasma. Recently, Wood *et al.* (1994) reported finding CA activity in the plasma of the elasmobranch *Scyliorhinus canicula*. This was confirmed for another species, *Squalus acanthias* (Gilmour *et al.*, 1997). Plasma CA activity is very low compared to that in the erythrocytes, and at this point the evidence indicates that it is not a separate isozyme that has been selected for; rather, it appears to be released from erythrocyte lysis as the result of natural, endogenous red cell turnover and hemolysis (Henry *et al.*, 1997b).

CA has also been found in high concentrations in the gill. The gill of aquatic organisms is a multifunctional organ, being the site of gas exchange, ion transport, acid–base balance, and nitrogenous waste (ammonia) excre-tion (see Hoar and Randall, 1984a,b for reviews). As a multifunctional enzyme, CA is believed to play an important role in all those processes (e.g., Perry and Laurent, 1990; Walsh and Henry, 1991; Henry, 1996), a view that was originally difficult to conceptualize. Gas exchange at the gill (i.e., CO_2 excretion) requires the net dehydration of HCO_3^- (generally considered impermeable to membranes in the absence of specific transport proteins) (Effros *et al.*, 1981; Perry *et al.*, 1982; Tang *et al.*, 1992) to freely diffusable CO_2 gas. Conversely, ion transport and acid–base balance require the net hydration of CO_2 to H^+ and HCO_3^-, which serve as counterions for the uptake of Na^+ and Cl^-, respectively (Kirshner, 1979). This apparent paradox can be explained by the physiological roles of CA in the gill being compartmentalized to specific subcellular fractions within the tissue where the net hydration and dehydration reactions can proceed independently. Specifically, soluble (cytoplasmic) CA is believed to function in support of ion transport, and particulate (membrane-associated) CA is believed to function in facilitated CO_2 diffusion (Henry, 1988).

The concept of compartmentalization of branchial CA function is sup-ported by the presence of discrete subcellular pools of CA. The enzyme has been measured in cytoplasmic, mitochondrial, and microsomal fractions of gill homogenates (Henry *et al.*, 1988, 1993, 1997b). Furthermore, histo-chemical localization, using both vital staining and monoclonal antibodies, confirms that branchial CA is heterogenously distributed within the cells of the respiratory epithelium (Rahim et al., 1988).

Although the distribution and function of cytoplasmic CA in the gills is well established, the same cannot be said for membrane-associated CA in fish gills. In mammals and other lower vertebrates, there is a pool of membrane-associated CA on the endothelial membrane of the respiratory epithelium, with the active site accessible to plasma HCO_3^- (e.g., Crandall and O'Brasky, 1978; Klocke, 1978; Stabenau et al., 1996). This appears to be absent in both agnathans and teleosts (Henry et al., 1988, 1993). Physiological studies seem to have confirmed the absence of a plasma accessible, branchial membrane-associated CA in teleosts: the presence of a postbranchial arterial of pH disequilibrium suggests that the plasma CO_2 reactions are not being catalyzed by CA (reviewed by Gilmour, 1998). Recently, however, evidence for endothelial membrane-associated CA, with its active site accessible to plasma HCO_3^-, was found for elasmobranch gills (Swenson et al., 1995, 1996; Henry et al., 1997b).

Some fish are bimodal breathers and exchange respiratory gases with air as well as water, using a primitive lung for aerial gas exchange (reviewed by Rahn and Howell, 1976; Randall et al., 1982; Graham, 1994). The fish lung has also been shown to have significant CA activity (Burggren and Haswell, 1979; Heming and Watson, 1986), but its subcellular distribution and function have not been systematically examined.

Recently, CA was suggested as being present on the external surface of the gill, in the mucus of the boundary layer (Wright et al., 1986). This was based on indirect evidence: downstream changes in the pH of the water flowing over the surface of the gills; the potential problems of this approach are discussed later. CA activity was detected in skin mucus scraped from the bodies of trout with a metal spatula. However, fish capture and handling stress have been shown to cause the appearance of occult hemoglobin in skin mucus (Smith and Ramos, 1976) and have been implicated in causing contamination of skin mucus samples with blood serum components (Hjelmeland et al., 1983). Thus, it is difficult to interpret assays of skin mucus CA activity without concurrent information about the potential contamination of the mucus samples by blood elements (including erythrocyte CA). Examination of the distribution of CA in trout gills via immunocytochemical localization (Rahim et al., 1998) showed that mucus granules in the gill mucus cells were labeled by an antibody against purified trout gill CA, but it is not known whether CA is an actual component of the mucus or part of the metabolic machinery of mucus synthesis. Lacy (1983) found that CA localized histochemically in the opercular mucus cells was largely confined to the cytoplasmic matrix with some CA activity in the mucus granules.

The presence of CA in the actual mucus layer on the surface of the gill was not examined.

IV. CA AND RESPIRATORY GAS EXCHANGE: CO_2 TRANSPORT AND EXCRETION

Metabolically active tissue, such as muscle, is probably the single largest source of CO_2 production in fish. From its origin in the mitochondria, CO_2 must cross a number of membrane barriers and be transported through a number of different fluid compartments before it is eliminated. Biological membranes (e.g., endothelial membranes of lungs and gills, and the sarcolemma of muscle) are considered to be functionally impermeable to HCO_3^-, the predominant chemical species at physiological pH (Effros et al., 1981; Perry et al., 1982; Tang et al., 1992). The rate-limiting step in CO_2 transport and excretion, therefore, is its movement across each membrane barrier, and this in turn depends on the conversion of HCO_3^- to CO_2 gas. Gutknecht et al. (1977) have shown experimentally, that in the absence of CA, the rate-limiting step in CO_2 transport is the uncatalyzed conversion of HCO_3^- to CO_2 gas in the boundary layer of the membrane. With CA present, however, a virtual instantaneous equilibrium between the chemical species is maintained, allowing transport to occur at the rate of CO_2 diffusion through the membrane.

As such, the first step in the overall process of CO_2 excretion is its transport across the sarcolemma from the intracellular fluid (ICF) of muscle to the extracellular fluid (ECF) that ultimately leads to the circulatory system. It was originally believed that the presence of CA in muscle would actually retard CO_2 excretion, i.e., that rapid catalysis of metabolically produced CO_2 would lead to bicarbonate trapping within the cell (Roughton, 1935). However, the rate of the uncatalyzed hydration reaction is very fast, having a rate constant 3 orders of magnitude faster than that of the dehydration reaction (Edsall, 1968). So even in the absence of CA, a significant fraction of CO_2 would be trapped intracellularly at HCO_3^-. The combination of cytoplasmic and membrane-associated CA actually facilitates CO_2 movement across the sarcolemma in both mammals (reviewed by Gros and Dodgson, 1988; Henry, 1996) and lower vertebrates (fish) (Henry et al., 1997a). Cytoplasmic CA maintains chemical equilibrium between HCO_3^- and CO_2 in the ICF, including the intracellular boundary layer of the sarcolemma. This prevents the dehydration reaction from being limiting, and it allows transport to occur at the rate of CO_2 diffusion even at low ICF PCO_2 (i.e., as CO_2 diffuses across the sarcolemma, it is immediately replaced from the large HCO_3^- pool through the rapid action of CA).

Membrane-associated CA, on the extracellular surface of the sarcolemma, would be driven by mass action to catalyze the hydration of outwardly diffusing CO_2. This would serve to maintain low P_{CO_2} in the ECF boundary layer, thus maximizing the P_{CO_2} gradient across the sarcolemma (Henry, 1996; Henry et al., 1997a).

From the interstitial space, CO_2 crosses the capillary wall and enters the circulatory system via diffusion. Although definitive evidence may still be lacking, the available data indicate there do not appear to be significant amounts of CA associated with the capillary wall in fish. Infusion of CA into plasma facilitates CO_2 removal from the blood and also abolishes the small venous pH disequilibrium (Currie et al., 1995; Perry et al., 1997). CO_2 also crosses the erythrocyte membrane via diffusion, because despite reports to the contrary (Haswell, 1977; Smeda and Houston, 1979; Houston and Mearow, 1979), there does not appear to be the membrane-associated CA IV in red cells of either mammals or fish (Randall and Maren, 1972; Henry et al., 1993). Inside the erythrocyte, CO_2 is rapidly hydrated to H^+ and HCO_3^- by the high levels of intracellular CA activity. In the systemic circulation of elamobranches and teleosts, mass action favors hydration, and product inhibition is minimized through the buffering of protons by the high concentrations of hemoglobin within the red cell and by the removal of HCO_3^- via the chloride shift. As a result, the bulk of CO_2 is transported in the plasma as HCO_3^-. Agnathans, however, lack the band 3 protein responsible for rapid erythrocyte anion exchange (Ellory et al., 1987; Nikinmaa and Raillo, 1987; Tufts and Boutilier, (1990a); consequently, the majority of the CO_2 is transported as HCO_3^- win the red cell (Tufts and Boutilier, 1989, 1990b; Ferguson et al., 1992; Tufts et al., 1992; Nikinmaa, 1992). For a more detailed description of the anion exchange capabilities of fish red cells, especially among agnathans, see Chapter 6 of this volume (Tufts and Perry, 1998).

At the site of respiratory gas exchange with the ambient medium (in this case, the gill), the aforementioned processes take place in reverse, and CO_2 is excreted in the gas form. In elasmobranchs and teleosts, plasma HCO_3^- is mobilized via a reversal of the chloride shift and subsequent dehydration by intracellular erythrocyte CA; in agnathans, mobilization occurs directly from the intraerythrocyte HCO_3^- pool. It is therefore the soluble CA within the erythrocyte that maintains the P_{CO_2} gradient that drives CO_2 diffusion out of the blood. This involves diffusion across two membrane barriers: the plasma membrane of the erythrocyte and the endothelial membrane of the gill. In other situations (e.g., the sarcolemma of muscle), membrane-associated CA facilitates CO_2 transport, but for agnathans and teleosts, there is no membrane-associated CA on the branchial endothelium (Henry et al., 1988, 1993), and there is no CA on the plasma membrane of the red cells in any of the three major groups of fish (Henry *et al.*, 1993). The high concentration of CA within the red cell maintains a large enough P_{CO_2} gradient to drive CO_2 diffusion across both membranes without the need for membrane-associated CA facilitated transport through the boundary layers. This could have eliminated the selection pressure for serosal membrane-associated CA in the gills of most fish. Elasmobranchs, however, do possess significant CA activity on the basolateral membrane of the gill (Swenson *et al.*, 1995, 1996; Henry *et al.*, 1997b), but its role in CO_2 transport and excretion is unclear. Selective inhibition of this pool of CA results in the slower clearance of an injected HCO_3^- load, but it does not result in an increase in arterial P_{CO_2} (Swenson *et al.*, 1995, 1996; Gilmour *et al.*, 1997), a response that is commonly associated with inhibition of red cell CA and the disruption of normal CO_2 excretion (Swenson and Maren, 1987; Henry *et al.*, 1988, 1995). It is possible that ECF CA plays a role in CO_2 excretion during periods of excess CO_2 loading (i.e., exercise) (Gilmour *et al.*, 1997), or it could have an entirely separate function, e.g., maintaining pH/P_{CO_2} equilibrium for ventilatory control (Henry *et al.*, 1997b).

The latter function may be an important factor in the selective pressure for the evolution of branchial CA in elasmobranchs. The most clearly defined function of vascular CA in vertebrates is to facilitate the equilibration of plasma CO_2 reactions during lung or gill transit. This minimizes, but does not necessarily abolish, the pH/P_{CO_2} disequilibrium in postcapillary blood (Bidani and Crandall, 1978; Gilmour *et al.*, 1997). Ventilation in most air-breathing vertebrates is driven primarily by blood pH/P_{CO_2} (e.g., Smatresk, 1990). Even the minor changes in blood P_{CO_2}/pH that arise from tidal breathing have been found to affect mammalian ventilation (Takahashi *et al.*, 1990, 1992). The action of vascular CA on the equilibration of blood CO_2 reactions during pulmonary capillary transit would reduce the background noise in the postcapillary P_{CO_2}/pH signal detected by peripheral chemoreceptors (Heming *et al.*, 1993). In water-breathing fishes, ventilation is driven primarily by blood PO_2 (Smatresk, 1990); as such, ventilatory control mechanisms might be less sensitive to postcapillary pH/P_{CO_2} disequilibria and therefore not as dependent on vascular CA activity as air-breathing vertebrates. This idea is supported by the fact that injection of CA inhibitors alters the magnitude and direction of the postcapillary pH/P_{CO_2} disequilibrium in teleosts (Gilmour *et al.*, 1994) but causes only transient changes in ventilation (apnea and recovery within 5 min postinjection) (Henry *et al.*, 1988).

The major exception to this pattern occurs in the elasmobranchs. This group displays a pronounced pH/P_{CO_2} ventilatory drive (Heisler *et al.*, 1988; Graham *et al.*, 1990; Wood *et al.*, 1990; Perry and Gilmour, 1996), and they are also the only known fish group to have significant branchial vasculature CA activity (Swenson et al., 1995, 1996; Henry et al., 1997b). It is possible that a pH/P_{CO_2} sensitive ventilatory drive evolved independently in elasmobranchs and that the presence of branchial CA was selected in response to this mechanism of ventilatory control. Both the evolution and function of branchial CA, and its relation to ventilation deserve further investigation in elasmobranchs.

Another group that deserves more systematic study is the air-breathing fishes. With the development of the lung as an air-breathing organ, it is possible that this group also independently evolved a pH/P_{CO_2} ventilatory drive, although existing evidence indicates that air-breathing frequency is not affected by hypercapnia (Smatresk and Cameron, 1982). It would still be interesting to examine the presence and potential function of pulmonary CA in these animals.

Transport of CO_2 through the branchial epithelium has been traditionally viewed as occurring by simple diffusion; the actual process may, in fact, be more complex. The branchial ICF compartment contains very high levels of CA activity (Henry et al., 1988, 1993); as a result, the chemical species of CO_2 are maintained in a virtual instantaneous equilibrium. Therefore, CA in the ICF boundary layer of the apical membrane may be important in facilitating CO_2 transport to the ambient medium. There is evidence indicating that CA is concentrated in the apical region of the respiratory epithelium (Conley and Mallatt, 1988; Rahim et al., 1988), and that the enzyme may also be localized to the extracellular surface of the apical cell membrane (Wright et al., 1986). However, definitive experimental evidence supporting a physiological role for apical CA is currently lacking.

In order to examine the rate of CO_2 hydration at the gill surface of trout, Wright et al. (1986) measured the downstream pH disequilibrium in expired water using a stopped-flow method. A consistent pH disequilibrium (0.07 unit decrease, indicative of continued CO_2 hydration in expired water) occurred when the CA inhibitor acetazolamide (Az, 1.6 mM) was added to the test water. Small and inconsistent changes in the stopped-flow pH (0.02–0.03 unit decreases in 11 of 29 trials; 0.02 unit increases in 2 of 29 trials; and no change in 16 of 29 trials) were recorded under control conditions. No pH disequilibrium was observed when bovine CA was added to the test water. The lack of a consistent downstream pH disequilibrium in control animals was taken as evidence for the presence of gill mucus CA, which was postulated as catalyzing the hydration of excreted CO_2 at the gill surface. However, the experimental design was highly unfavorable for detecting pH disequilibria. The inflow : volume ratio of the pH measurement chamber was 8–10 min^{-1}, which yields a 99% replacement time for the chamber of 0.5 to 2 min (see Fig. 2). Yet, given the exceptionally low nonbicarbonate buffer capacity (β_{nb}) of the test water (dechlorinated Vancouver tap water, 0.08 mM/pH unit), the half-time of the uncatalyzed hydration reaction in expired water ($T = 10$–$13°C$) was probably only several seconds (see Fig. 3). A major portion of any pH disequilibrium would be unobservable under such conditions (Fig. 4). In studies with the CA inhibitor, the inclusion of 1.6 mM Az (a weak base, $pK = 9.1$) would have increased the β_{nb} of the test water by more than an order of magnitude. This increase would be expected to increase the measured ΔpH independent of any effect on CA activity (Bidani and Heming, 1991; Heming and Bidani, 1992; see Fig. 4). Furthermore, gill surface pH is influenced by a complex of processes, including fluxes of titratable acids/bases, ammonia, and CO_2 (Playle and Wood, 1989).

Under conditions more favorable for measuring pH disequilibria: water $T = 8°C$ and β_{nb} of 0.51 mM/pH unit, yielding an uncatalyzed reaction half-time of probably 10–20 s; inflow : volume ratio of the pH chamber = 50^{-1} min providing 99% replacement of chamber volume in <20 s, significant pH disequilibria (0.10–0.11 unit decreases) were recorded in the expired water of both control trout and trout exposed to Az (0.44 mM) (Heming, 1986; Fig. 5). The postgill pH disequilibrium was abolished by adding bovine

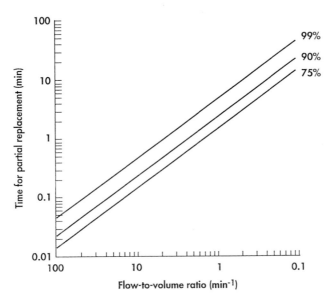

Fig. 2. Time required for partial molecular replacement of fluid in the pH electrode chamber, as a function of chamber volume and inflow rate. Adapted from Sprague (1973).

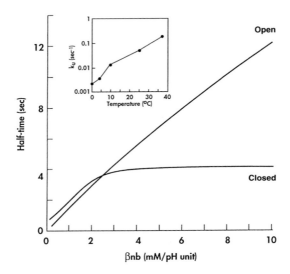

Fig. 3. Model predictions of half-time for pH change during uncatalyzed CO_2–HCO_3^-–H^+ equilibration at 37°C following a change in P_{CO_2}, as a function of nonbicarbonate buffering capacity (β_{nb}). Data for open system (i.e., total CO_2 content varies during equilibration) from Gray (1971) (P_{CO_2} = 35, "arrow to right symbol" 10 torr, initial pH 7.37). Data for closed system (i.e., total CO_2 content remains constant during equilibration) from Bidani and Heming (1991) P_{CO_2} = 35, "arrow to right symbol" 10 torr, initial pH 7.37). Half-time of the pH disequilibrium will be greater at lower temperatures (e.g., increased by about an order of magnitude at 15°C). Inset shows the effects of temperature on the uncatalyzed velocity constant (k_u) for unopposed CO_2 hydration. From Forster (1993).

CA to the test water. These data argue against the presence of CA activity on the gill surface.

V. AMMONIA AS A RESPIRATORY GAS

There are a number of similarities between the behavior of carbon dioxide and the behavior of ammonia in physiological solutions. Ammonia exists in gaseous (NH_3) and ionic (NH_4^+) forms. Its pK is between 9 and 10; therefore, at physiological pH, approximately 97% of the total ammonia is in the ionic (protonated) form in both the ECF and ICF. Ammonia gas, however, is the more mobile chemical species, having permeability values that are between 10 and 100 fold higher than those for NH_4^+ (Cameron and Heisler, 1985). Experimental evidence supports the idea that ammonia transport takes place primarily as NH_3 diffusion both across the sarcolemma of muscle (Wright *et al.*, 1988; Wang *et al.*, 1996) and across the branchial epithelium (Avella and Bornancin, 1989; Cameron and Heisler, 1983, 1985;

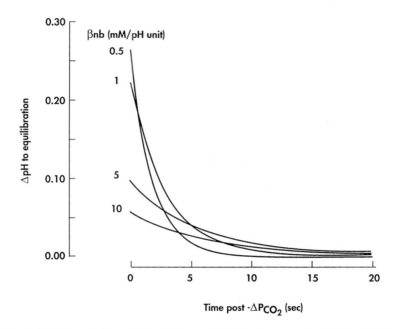

Fig. 4. Model predictions of the pH change (ΔpH) during uncatalyzed CO_2–HCO_3^-–H^+ equilibration at 37°C following a change in P_{CO_2} (ΔP_{CO_2}), showing the effects of β_{nb} (P_{CO_2} = 35 "arrow to right symbol" 10 torr, initial pH 7.37). The pH change will occur more slowly at lower temperatures (e.g., slowed by an order of magnitude at 15°C). The fraction of the overall ΔpH that can be measured in pH disequilibrium studies will be determined by the transit time for fluid to flow from the gas exchange site to disequilibria studies will be determined by the transit time for fluid to flow from the gas exchange site to the pH electrode, and the kinetics of the equilibration reaction. Adapted from Bidani and Heming (1991).

Wright and Wood, 1985; Claiborne and Evans, 1988; Evans and More, 1977; Evans *et al.,* 1989; Wilson *et al.,* 1994). As such, ammonia movements appear to parallel those for CO_2: mobilization of the diffusable form from a much larger and more stationary pool, and diffusion across a membrane down a relatively small partial pressure gradient (40–50 μtorr for NH_3) (Cameron and Heisler, 1983, 1985).

There is a growing body of evidence supporting a role for CA in NH_3 transport similar to that for CO_2. In general, acidification of the external boundary layer facilitates ammonia transport across membranes (Wright *et al.,* 1989; Wilson *et al.,* 1994). This is true in fish for both the apical membrane of the gill and for the sarcolemma (Wright *et al.,* 1989; Wilson *et al.,* 1994; Wang *et al.,* 1996). Ammonia diffusion across the sarcolemma is amiloride-insensitive, indicating that proton transport across the sarcolemma does not contribute to the ECF boundary layer acidification (Wang

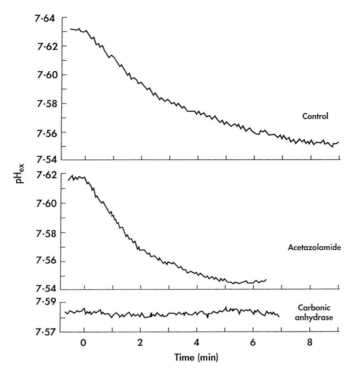

Fig. 5. Typical changes in the pH of expired water for a single rainbow trout at 8°C, after water flow past the pH electrode was stopped at time zero. The trout was exposed sequentially to (1) control test water, (2) test water containing bovine CA, and (3) after a 90-min recovery period in control water, test water containing acetazolamide. From Heming (1986).

et al., 1996). However, treatment of an isolated, perfused trout tail muscle preparation with quaternary ammonium sulfanilamide (QAS) or acetazolamide results in ammonia retention within the ICF and an increase in the PNH_3 gradient across the sarcolemma (Henry *et al.,* 1997a). This suggests that CA, associated with the extracellular surface of the sarcolemma, facilitates NH_3 transport through the generation of protons via the catalyzed hydration of CO_2; this results in ammonia trapping as NH_4^+ in the ECF boundary layer, thus maintaining the transsarcolemma PNH_3 gradient.

The acidification of the external boundary layer of fish gills was first suggested by Lloyd and Herbert (1960) as being an important factor in ammonia excretion, primarily as an explanation for the effects of water pH/CO_2 on the acute lethality of ammonia. Szumski *et al.* (1982) elaborated on this hypothesis and first suggested gill CA as facilitating the external acidification via the catalyzed hydration of excreted CO_2. But though acidification of the external gill surface boundary layer has been shown to be important in branchial ammonia excretion (Wright *et al.*, 1985; Wilson *et al.*, 1994), the role of CA in that process is less certain. It has been suggested that acidification of the external apical surface of the gill takes place by the catalyzed hydration of CO_2, with CA being present either on the membrane and/or in the mucus of the external boundary layer (Wright *et al.*, 1986; Rahim *et al.*, 1988; Wright *et al.*, 1989). There are some gaps in the evidence, however. Ammonia excretion at the gill is amiloride-sensitive, being reduced by about 30% (Wright *et al.*, 1989); that indicates a significant fraction of the protons come directly from apical Na^+/H^+ exchange. This reduction occurs despite an amiloride-stimulated increase in CO_2 excretion of about 70%. Presumably, an increase in CO_2 excretion, in the presence of CA, would provide an adequate H^+ supply for NH_3 protonation, especially since CO_2 excretion occurs at about 10 times the rate of ammonia excretion. The CA inhibitor acetazolamide (Az), when added to the perfusion medium of the gills (i.e., the blood), also reduces ammonia excretion (Wright *et al.*, 1989); however the inhibitor-treated blood was allowed to perfuse the gills for only 7 min before the measurement was made. Given the slow permeability of Az, this was not enough time for the drug to equilibrate across the gill epithelium and reach the external apical surface. Finally, acidification of the gill surface boundary layer via CO_2 hydration may not be dependent on the presence of CA activity. At the low β_{nb} of many aquatic environments (especially fresh water), the CO_2–HCO_3^-–H^+ system will equilibrate rapidly at the gill surface even in the absence of CA (see Fig. 3). The rate of uncatalyzed CO_2 hydration will depend on the β_{nb} of the mucus layer, and this has not yet been determined. At this point, direct experimental evidence for branchial CA playing a role in ammonia excretion remains lacking.

VI. PLASMA INHIBITORS OF CARBONIC ANHYDRASE

Less than 5 years after the discovery of CA, Booth (1938) found mammalian plasma contained a factor that inhibited the CA-catalyzed hydration/dehydration reactions. Initial studies with porcine serum suggested that the plasma inhibitor (pICA) was a globulinlike protein. In addition, pICAs were found to be widely distributed among mammalian species (see also Hill, 1986) (but absent in monkeys and humans); trace amounts were found in chicken plasma, but no inhibition was seen for either pigeons or ducks. After this pioneering study, the subject was largely ignored for almost three decades. Maetz (1956) was the first to find evidence for pICA in fish (perch, carp, and goldfish). Further, he demonstrated that pICA exhibited species specificity: perch and porcine pICA displayed minimal cross-reactivity with porcine and perch erythrocyte CA, respectively. Subsequent investigations have detected pICA in eels (Haswell *et al.,* 1983), trout (Heming and Watson, 1986), and a variety of other fish species (Henry *et al.,* 1997b). See Table 2 for a summary of the distribution of pICA in fish.

The available data suggest that a number of different plasma proteins are responsible for pICA function across broad phylogenetic lines. Fish pICA appears to be a relatively low MW protein: 10–30 kDa in eels (Haswell *et al.,* 1983) and 6.9 kDa in trout (Dimberg, 1994). Fish pICA demonstrate both species and isozyme specificity. Trout pICA appears to be more effective against gill than erythrocyte CA. In general, pICA in one species is more effective against CA from its own erythrocytes than it is against erythrocyte CA from other species, and the effectiveness appears to drop off with phylogenetic distance (Dimberg, 1994; Henry *et al.,* 1997b). No teleost pICA is active against elasmobranch erythrocyte CA or bovine CA II.

A different protein appears to be responsible for CA inhibition in mammalian plasma. Purified porcine pICA was found to be a 79-kDa monomeric glycoprotein that specifically and reversibly inhibits mammalian CA II with a 1 : 1 stoichiometry and a K_i of <0.4 nM (Roush and Fierke, 1992). Porcine pICA also shows isozyme and species specificity, being less effective against the CA I, III, and IV isozymes and having no effect against chicken erythrocyte CA II.

Plasma inhibitors have been detected, with some exceptions, in all fish species examined to date; dogfish shark, bowfin, and gravid salmonids lack pICA, and bullhead catfish have minimal detectable amounts (Table 2). Plasma CA activity has generally been measured in fishes that lack pICA, and although the source of that activity has yet to be definitively established, it is likely a result of the liberation of CA from one or more ICF pools. Any activity that causes red cell hemolysis will cause erythryocyte CA to be released into the plasma, including normal red cell aging, damage, and turnover, lysis from handling stress, and exercise. The plasma CA activity in dogfish sharks is believed to originate from natural, endogenous erythrocyte lysis (Henry *et al.,* 1997b). Further, cytosolic CA from somatic cells can leak into the circulation during exercise and in response to stress. For example, CA III, presumably eluted from hepatocytes and muscle fibers, is present in normal human serum (10–15 μg I^{-1}), and this increases dramatically after strenuous exercise and in association with certain diseases (Vaananen *et al.,* 1986; Syrjala *et al.,* 1990). Red and white muscle fibers of lower vertebrates, including fish, contain cytosolic CA III and II, respec-

Table II

Distribution of Plasma Inhibitors of CA (pICA) in Fishes: Effects of Blood Plasma on the Activity of CA from Lysed Erythrocytes

Species	Plasma inhibition	Reference
Squalus acanthias (spiny dogfish)	+ (H)	Henry *et al.*, 1997b
Amia calva (bowfin)	NS (E, H)	Heming and Watson, 1986
Anguilla anguilla (yellow eel)	− (H)	Haswell *et al.*, 1983
Anguilla rostrata (American eel)	− (H)	Henry *et al.*, 1997b
Oncorhynchus mykiss (rainbow trout)		
Nongravid	− (E, H)	Heming, 1984; Heming and Watson, 1986; Henry *et al.*, 1997b
Gravid	+ (E, H)	Heming, 1984
O. tshawytscha (chinook salmon)	− (H)	Henry *et al.*, 1997b
Salmo clarki (cutthroat trout)		
Nongravid	− (E, H)	Heming, 1984
Gravid	NS (E, H)	Heming, 1984
Carassius auratus (goldfish)	− (H)	Maetz, 1956
Cyprinus carpio (common carp)	− (H)	Maetz, 1956
Catostomus macrocheilus (largescale sucker)	− (E, H)	Heming, 1984
Ictalurus nebulosus (brown bullhead)	− (E)	Heming, 1984
	−/NS (H)	Henry *et al.*, 1997b
Perca fluviatilis (perch)	− (H)	Maetz, 1956
Xiphister atropurpureus (black prickleback)	− (E,H)	Heming, 1984
Sebastes melanops (black rockfish)	− (H)	Henry *et al.*, 1997b
Ophiodon elongatus (lingcod	− (H)	Henry *et al.*, 1997b
Platichthys stellatus (starry flounder)	− (H)	Henry *et al.*, 1997b

Note. +, significant augmentation of CA activity; −, significant inhibition of CA activity; NS, no significant change in CA activity. E, CA activity measured by acetazolamide-sensitive esterase activity; H, CA activity measured by the CO_2 hydration/dehydration reactions.

tively (Sanyal et al., 1982a,b; Scheid and Siffert, 1985), and this could elute into the circulation during periods of increased sarcolemmal permeability. Thus, in order to ascertain whether pICA is absent in a given animal, or simply overwhelmed by a flux of ICF CA into the circulation, it is necessary to measure the degree of plasma contamination by cytosolic contents (e.g., plasma hemoglobin for erythrocyte lysis, and plasma creatine kinase for myofibers).

Assessment of the species distribution of pICA is complicated further by potential temporal variations in inhibitor concentrations. As a plasma protein, it is possible that pICA is affected by any of the factors that influence plasma protein composition, which for fish include stock differences, seasonal temperature variations, starvation, and disease. For example, nongravid trout possess a potent pICA, but the protein is absent in gravid trout (Heming, 1984). This temporal variation in trout pICA activity correlates with the substantial variations in plasma protein pattern that occur during the course of gonadal development in trout (Borchard, 1978). Because of the scattered distribution of pICA across animal phyla, the evidence for multiple pICA proteins, and the potential temporal variations in pICA activity, it is difficult to speculate on the evolutionary significance and the reasons behind its selection.

The physiological significance of pICA is equally uncertain, and it is plausible that different pICA proteins subserve different functions. Booth (1938) originally proposed that pICA functioned to prevent the CA released during erythrocyte sequestration from short-circuiting intraerythrocyte CO_2–HCO_3^-–H^+ reactions, but his hypothesis was generally weakened by the failure to detect pICA in birds and primates. However, a modified version of Booth's idea has recently arisen for fish pICA, based on the discovery that intracellular pH (pH_i) of some teleost erythrocytes is regulated by a β-adrenergic-activated Na^+/H^+ exchange protein (NHE). Specifically, it has been suggested that maintenance of a plasma space devoid of CA activity is necessary to prevent the catalyzed dehydration of HCO_3^- from short-circuiting catecholamine-stimulated proton transport out of the red cell (Motais et al., 1989). Addition of bovine CA to teleost blood in vitro has been shown to reduce or prevent the catecholamine-induced pHi increase usually seen as a compensatory response to whole blood acidosis (Motais et al., 1989; Nikinmaa et al., 1990). Preservation of red cell pHi in the face of a blood acidosis (e.g., during hypoxia and/or exercise) is postulated as maintaining normal O_2 carrying capacity of blood (via the Bohr and Root effects), and the effect of CA infusion into the plasma is predicted to disrupt this ability. However, experiments involving iv injection of bovine CA II into trout that were subjected to either exhaustive exercise (Wood and Munger, 1994) or environmental hypoxia (water P_{O_2} = 30–35 torr) (Lessard *et al.*, 1995) failed to detect the expected reduction in blood O_2 content *in vivo*. Furthermore, the pH_i of erythrocytes, brain and muscle (red and white) of fish in the treatment groups were unaffected by CA infusion. Plasma CA injection also had no effect on the aerobic or anaerobic exercise performance of trout. The *in vivo* data do not support the *in vitro* blood experiments, and as a result the importance of pICA for O_2 transport in fish remains uncertain.

Among air-breathing vertebrates, the argument that pICA functions in maintaining a "CA-free" plasma space is weakened considerably by evidence that capillary endothelial cells of many organs (e.g., lungs) express membrane-bound CA IV on their extracellular luminal surface (Tashian, 1989; Heming and Bidani, 1991). This "vascular" enzyme participates directly in plasma CO_2 reactions, but it is present in low concentrations relative to those in the red cell and has only marginal effects on pulmonary CO_2 exchange (Bidani, 1991). Thus, pICA inhibition of this isozyme would have little effect on gas transport/exchange. Comparable vascular CA appears to be absent from the basal membrane of teleost gills (Henry *et al.*, 1988, 1993) and from the endothelial capillary wall in general (Currie *et al.*, 1995; Perry *et al.*, 1997). The only fish group found to have branchial membrane-associated CA is the elasmobranchs (Swenson *et al.*, 1995, 1996; Henry *et al.*, 1997b), but this group also appears to lack pICA. Evidence for extracellular CA has been found in isolated gas gland cells of the eel swim bladder (Pelster, 1995) and on the extracellular surface of the sarcolemma of white muscle in trout (Henry *et al.*, 1997a). It is doubtful that pICA has access to interstitial CA, so it is premature to conclude that fish pICA functions to regulate the activity of extracellular CA (intravascular or interstitial).

One possibility that has not been considered is that pICA proteins function in processes that are independent of gas exchange. In this regard, inhibition of plasma CA may be incidental to their primary function. In particular, the 79 kDa porcine pICA may serve as a scavenger/transporter of CA Zn. This pICA has 62% primary sequence homology to porcine transferrin (M. Wuebens, E. Rousch, and C. Fierke, unpublished data, personal communication). Transferrin is present in the plasma of all vertebrates and is an essential component of iron transport and metabolism. It is attractive to speculate that the transferrinlike pICA plays a comparable role in the transort and metabolism of zinc, the metal core of CA. This hypothesis is supported, albeit indirectly, by the studies of Applegren *et al.* (1989) and Ojteg and Wistrand (1994). In studies that monitored the clearance and translocation of ^{125}I labeled CA I, II, and III injected into plasma, the authors found that CA I and III were cleared from plasma with a half-time of 112 min and localized primarily to the renal cortex; CA

II was cleared more slowly ($t_{1/2}$ = 210 min) and localized preferentially to the liver. More important, the labeled CA II remaining in plasma at 150 min was present predominantly as a high MW complex of about 115 kDa. Given the MW of the CA II isozyme (~29 kDa), the authors concluded that the MW of the CA-binding factor was about 85 kDa, similar to the 79 kDa MW reported by Roush and Fierke (1979) for porcine pICA. Furthermore, the 115-kDa CA complex was reactive with antiserum against rat transferrin. The catalytic activity of the CA II complex was not measured, but rats are known to possess a potent pICA (Booth, 1938). Overall, these results suggest that the plasma CA-binding factor, responsible for localizing circulating CA II to the liver, could be pICA. Similar studies are warranted in fish to determine the fate of CA and the associated Zn that is released into the circulation.

VII. ANALYTICAL TECHNIQUES

A. CA Assays

There are numerous assay methods to measure CA activity, but all are based on following the production or consumption of one of the chemical species involved in the hydration/dehydration reaction. This can be accomplished through a variety of approaches, but the most commonly used CA assays can be grouped as follows: electrometric, spectrophotometric, manometric, and isotopic. Of these, the most widespread type of assay is electrometric, involving the measurement of pH as an indicator of CA activity. This approach has the advantage of being simple and straightforward. There are two categories of pH-based CA assays: the delta pH assay and the pH stat assay. The delta pH assay, based on the method of Stadie and O'Brien (1933), was shown to be simpler and more reliable than other techniques used at the time (Wilbur and Anderson, 1948). In its original form, buffer and CO_2-saturated water were simultaneously forced into the reaction vessel from hand-operated syringes; no other mixing was used. The rate was determined by measuring the time it took for a pH change (from 8.0 to 6.3) to occur (up to 120 sec for the uncatalyzed rate). The assay was run at 4°C by packing the vessel in ice. This assay had two drawbacks. First, both the hydration and dehydration reactions are highly pH sensitive, with pK_a values near 7.0 (the reaction mechanism, in both directions, depends on the reversible dissociation of a water molecule at the active site). The pH-dependent activity profile is sigmoid; therefore, changes in pH between 6 and 8 significantly alter the CA activity (Roughton and Booth, 1946a; Kernohan, 1964; Coleman, 1980). As a result, true CA activity is underestimated. Second, the reaction is being measured only at an arbitrary end point, preventing an accurate measure of initial velocity, and the rate can be expressed only in semiquantitative "enzyme units" (Wilbur and Anderson, 1948).

In its current form, the delta pH assay developed by Davis (1958, 1959) requires instrumentation commonly available: a pH meter, rapid sensing electrodes, recorder, reaction vessel, and constant temperature circulating water bath. A modification of the method is described by Henry (1991). Briefly, the reaction is started by the addition of a known volume of CO_2-saturated water to a buffer containing CA. The reaction medium is rapidly mixed, and the drop in pH is measured using pH electrodes that have a fast response time (100 ms or less). Because the hydration reaction occurs so rapidly, the assay must be performed at low temperatures; this necessitates having a thermostated reaction vessel. The stoichiometry of CO_2 hydration to H^+ production is $1:1$, so the slope of the pH change represents the rate of the catalyzed reaction. Reliable rates can be obtained from small changes in pH (0.10–0.25 pH unit), and the assay can be calibrated by the addition of increasing volumes of a known concentration of acid. This assay can be used for the quantitative measurement of CA activity in cell or tissue homogenates, and it can be used in kinetic studies of the enzyme as well. Although the delta pH assay is used most commonly to measure the catalyzed rate of CO_2 hydration, it can also be adapted to measure bicarbonate dehydration (Maren and Couto, 1979). The primary alteration is in the reaction vessel, which in this case is fitted with a sintered glass disk through which nitrogen gas is vigorously bubbled. The reaction is started by the addition of $NaHCO_3$ to a buffer containing CA, and an increase in pH is followed as protons are consumed. The N_2 stream carries CO_2 gas out of solution, preventing the assay from being limited by the diffusion of CO_2 out of solution and the resultant product inhibition by proton buildup.

The second electrometric CA assay, the pH stat assay, is based on the automated addition of either acid or base to keep the pH of the reaction medium constant during the dehydration or hydration reactions, respectively. This method was developed around commercially available autotitrators (Liebman *et al.*, 1961; Hansen and Magid, 1966; McIntosh, 1968; Henry, 1991). For measuring hydration, CO_2 gas is bubbled through the buffer in the reaction vessel, and the reaction is started by the addition of CA. The rate of addition of a known concentration of base (e.g., NaOH) gives a direct measure of the rate of catalysis. To initiate the dehydration reaction, $NaHCO_3$ is added to a buffer containing CA, and the rate of addition of a known concentration of acid (e.g., HCl) gives the rate of catalysis. As with measuring dehydration with the delta pH method, N_2 must be vigorously bubbled through the reaction mixture in order to strip CO_2 out of solution and prevent the diffusion of CO_2 from the liquid to the gas phase from being the rate-limiting step in the assay (see discussion of manometric assays that follows). Henry and Cameron (1982) found that the rate of N_2 gas flow had to be greater than 300 ml min^{-1} for the measured rate to be independent of diffusion limitations. The pH stat assay has two advantages over the delta pH method: the rates of both the hydration and dehydration reactions can be measured at constant pH, and there is no need for external calibration. One limitation, however, is that samples containing high concentrations of protein are very difficult to assay because the bubbling of either CO_2 or N_2 causes excessive foaming. Also, neither the delta pH assay nor the pH stat assay can be used to measure CA activity of intact cells.

Spectrophotmetric CA assays are also based on measuring the change in pH that occurs during hydration/dehydration, but they do so indirectly by following the change in absorbance of a pH-sensitive dye. This approach includes the following types of assays: indicator dye assay, rapid flow assay, and enzyme-linked assay. The original indicator dye assay, based on the method of Philpott and Philpott (1936), simply involved measuring the amount of time required for a color change in pH-sensitive dye (bromthymol blue, phenol red, or indophenol). This method suffered from a variety of drawbacks, including inhibitory effects caused by the indicator dye and by the large changes in pH. Improvements in this method were made that partially overcame these difficulties, such as directly measuring the change in absorbance of a Veronal buffer at 276 nm (Nyman, 1963), and reducing the magnitude of the pH change by using a very sensitive photometric stop-flow apparatus (Kernohan et al., 1963). Despite these changes, however, this approach has been largely abandoned.

Another spectrophotometric assay is based on the ability of CA to catalyze other reactions, particularly the hydrolysis of esters (Verpoorte et al., 1967). Briefly, the change in absorbance at 348 nm is monitored as p-nitrophenyl acetate is hydrolyzed by CA. Because the rate of esterase activity of CA is orders of magnitude slower than that of the CO_2 reactions, this assay can be run at ambient temperatures. It can also be used for kinetic studies, but it will greatly underestimate the actual CA activity.

Manometric assays measure the increase in gas pressure that results from the dehydration of HCO_3^- and the subsequent transfer of CO_2 gas from the liquid to the gas phase in the reaction vessel. First developed by Meldrum and Roughton (1933), this method was used in the initial discovery of CA. Numerous problems and errors associated with this approach require correction, so though the assay is technically simple, it is complex and cumbersome to use in a quantitatively accurate manner. The most serious problem, among many, is probably the fact that the rate-limiting step in the assay is not the catalyzed generation of CO_2 but rather the diffusion of CO_2 out of the reaction buffer. Roughton and Booth (1946b) outlined extensive procedures used in correcting for diffusion limitations, but these have hardly ever been employed by other researchers. The problems associated with diffusion limitations are true for any manometric assay, and it is probably the main reason that these assays have been abandoned as inaccurate.

A good example of the difficulty of the manometric class of assays was the inability of Haswell and Randall (1976) to detect CA activity in whole blood of rainbow trout using the original manometric boat assay. Their results led them to conclude that intraerythrocyte CA did not have access to plasma bicarbonate, that the CA inhibitor in trout plasma acted to prevent Cl^-/HCO_3^- exchange, that erythrocyte CA did not play a role in systemic CO_2 excretion, and that plasma bicarbonate was mobilized to CO_2 for excretion by CA in the gill. This was shown to be incorrect by a number of subsequent studies (Obaid et al., 1978; Cameron, 1978; Perry et al., 1982; Swenson and Maren, 1987; Henry et al., 1988). It is instructive, however, in illustrating how results obtained from a flawed assay can skew the interpretation of kinetic processes in the wrong direction.

A recent modification of the traditional manometric assay, which measures the accumulation of ^{14}C-labeled CO_2 in air generated from labeled HCO_3^- in solution (Wood and Perry, 1991), has been used in a variety of studies on CA activity and CO_2 production of red cells and plasma of fish. The rate-limiting step in this procedure also appears to be the diffusion of CO_2 out of solution. The accumulation in the assay trap, of labeled CO_2 produced by trout red cells, is linear over a 20-min time course of a single measurement (Fig. 2A in Wood and Perry, 1991). The actual production of CO_2 from bicarbonate takes place much more quickly: the chloride shift in fish red cells is very rapid, having a half-time on the order of a couple hundred milliseconds (Cameron, 1978; Obaid et al., 1978), and fish red cells have very high concentrations of CA (Henry et al., 1988, 1993), making the catalyzed conversion of HCO_3^- to CO_2 instantaneous. So the process of converting extracellular HCO_3^- to CO_2 is probably complete within seconds, but it takes more than 20 min for the full amount of CO_2 to diffuse out of the assay buffer and be collected in the assay trap. Other assays used to measure CO_2/HCO_3^- kinetics in red cells come to completion (equilibrium) in seconds (Cameron, 1978; Obaid et al., 1978). The high solubility of CO_2 in solution, plus the inefficiency of removing the CO_2 from solution via gentle shaking, limits the rate of CO_2 accumulation in the trap (and hence the measurement of the rate of CO_2 production) to the rate of diffusion of CO_2 from the liquid to the gas phase of the assay, a process that takes minutes. As a result, the actual rate of CO_2 production is underestimated, and the error becomes progressively larger as the actual rate of CO_2 excretion increases, ultimately causing the largest error in the measurement of whole blood. The usefulness of this assay in measuring the rate of CO_2 production by red cells, in determining the factors that affect red cell CO_2 production, and in comparing relative rates of CO_2 production in red cells among different species is doubtful.

All the aforementioned assays are useful only in measuring the activity of CA in cell-free homogenates; only one technique, the only true isotopically based assay, has been developed that will accurately measure CA activity within suspensions of intact cells. This is a stable isotope assay based on the rate of transfer of ^{18}O from HCO_3^- to H_2O (Itada and Forster, 1977; Forster and Itada, 1980). The high concentration of water (55 M) acts as a sink for the ^{18}O, causing a slow decay in the labeled CO_2. The disppearance of ^{18}O-labeled CO_2 is monitored by a mass spectrometer and can be converted to a measurement of the rate of CA activity. The assay can be used for intact cells or free enzyme, and it can be used at ambient temperature. The single major limitation is the requirement for highly specialized instrumentation.

B. CA Inhibitors

The most powerful tool in the study of CA function is the use of various sulfonamide CA inhibitors. These compounds, for the most part, have a high affinity for CA and inhibit enzyme activity at very low concentrations. For example, acetazolamide has a dissociation constant (K_i) of about 10 nM for CA II (Maren, 1977), and many of the other commonly used sulfonamide inhibitors (e.g., methoxzolamide, benzolamide, ethoxzolamide) have K_i values in the range of 1–50 nM. For *in vivo* work, it is generally considered that using an inhibitor concentration of 1000 times the K_i (e.g., a concentration of 100 μM for Az) will produce 99.99% inhibition of CA and cause the full physiological effect to be observed (Maren, 1967b, 1977). However, it is also a good idea to use multiple concentrations of the same inhibitor to generate a dose–response curve for the physiological effect being measured or to use multiple inhibitors to generate the same effect.

Because carbonic anhydrase is distributed across a wide variety of organ systems and tissue and cell types, and because it plays a role in so many different physiological and biochemical processes, it is difficult sometimes to interpret the results of CA inhibition studies without surgically or pharmacologically isolating the target of interest. One practical approach is to use inhibitors that are specific to different fluid compartments or CA isozymes based on the inhibitors' permeability or affinity for CA. Sulfonamides (e.g., prontosil) bound to various-sized dextran molecules have been used in isolated organ preparations to differentiate the roles of intracellular, extracellular interstitial, and extracellular intravascular CA (cf. Heming *et al.*, 1986). Similarly, a quarternary ammonium sulfanilamide complex (QAS) can be used to distinguish the roles of intracellular versus extracellular CA (Henry, 1987). In fish, QAS was used to study the role of a putative CA IV isozyme on the extracellular surface of trout white muscle (Henry *et al.*, 1997a), and a polyoxyethylene–aminobenzolamide was used to study the role of CA IV in dogfish gills (Swenson *et al.*, 1995, 1996). Conversely, low concentrations of certain sulfonamides of low permeability (i.e., benzolamide) can also be used to specifically block ECF CA activity (e.g., Gilmour *et al.*, 1997). Some inhibitors have been used as tissue-specific probes of CA activity. Swenson and Maren (1987) used benzolamide to specifically inhibit gill CA activity and methazolamide to inhibit both gill and red cell CA activity. Henry *et al.* (1988), on the other hand, used both acetazolamide and benzolamide to inhibit red cell CA activity. Regardless of which inhibitors are used and for which purpose, it is important to independently document the effectiveness of the inhibitor on the target tissue. Swenson and Maren (1987) measured the inhibitor concentrations in both gills and red cells and calculated the fraction of inhibition achieved in each, whereas Henry *et al.* (1988, 1995) measured the CA activity of the target tissue to ensure that inhibition was achieved.

C. pH Disequilibria

As CO_2 is either loaded into the blood in the systemic capillaries or unloaded at the gas exchange organ, the chemical equilibrium among blood CO_2, H^+, and HCO_3^- is temporarily upset. The rate at which the CO_2 system reequilibrates is determined by the kinetics of the hydration/dehydration reactions in plasma and erythrocytes and the rates of acid–base exchange across the erythrocyte membrane (Bidani and Crandall, 1978). A variety of stopped-flow and continuous flow techniques have been developed to monitor the changes in extracellular pH during the reequilibration process (Crandall and O'Brasky, 1978; Bidani *et al.*, 1978; Crandall and Bidani, 1981; Bidani *et al.*, 1983) and to study the contributions of CA and anion exchange to pulmonary CO_2 exchange and pH regulation in mammals (Bidani *et al.*, 1978; Crandall *et al.*, 1981). Similar approaches have recently been applied to study branchial CO_2 exchange and intravascular pH disequilibria in fish blood (Gilmour *et al.*, 1994, 1997; Gilmour and Perry, 1994) and in expired water downstream from the gill (Heming, 1986).

The basis of these techniques is measurement of the pH at different time points during reequilibration, but interpretation of the pH difference (ΔpH) of mixed phase solutions (e.g., whole blood) is complex and cannot be accomplished without consideration of the CO_2 reactions in each phase and the acid–base fluxes between the two phases. The ΔpH of whole blood cannot be used alone to interpret *in situ* CA activity at the site of gas exchange (Bidani and Crandall, 1978; Crandall *et al.*, 1981). Chapter 9 of this volume (Gilmour, 1998) is a more detailed discussion of the pH/CO_2 disequilibria in fish blood; the remainder of this section focuses on single phase solutions (i.e., physiological saline or expired water), which are reflective of the CO_2 reactions at the gas exchange site. Failure to detect a pH disequilibrium indicates the presence of CA, and detection of a ΔpH indicates that CA is absent (or limiting). Experimental manipulation of the measured ΔpH with exogenous CA and/or inhibitors can be employed to quantify the *in situ* CA activity at the gas exchange site (cf. Heming and Bidani, 1991).

There are several technical and interpretive issues to consider when using pH disequilibria methods. The sensitivity for detecting pH disequilibria (and, hence, *in situ* CA activity) depends on the kinetics of the reequilibration process and the physical characteristics of the experimental setup, particularly the volume of the pH electrode chamber and the rate of fluid flow through the chamber. Figure 2 shows the relationship between chamber volume, flow rate, and the time required for partial molecular replacement of the chamber fluid. As a minimum guideline, 99% replacement of the chamber fluid should occur within one half-time of the uncatalyzed equilibration reaction (e.g., for a half-time of 60 s, chamber flow rate should be 5 times the chamber volume; and for a half-time of 6 s, it should be 90 times).

The speed at which the uncatalyzed CO_2–HCO_3^-–H^+ reactions approach equilibrium after a change in P_{CO_2} depends on the rates of hydration versus dehydration, the dissociation constant for CO_2, and the nonbicarbonate buffer capacity (β_{nb}) of the system. Models of an open system in which total CO_2 varies (i.e., at a gas exchange site) (Gray, 1971), and a serial arrangement of open and closed systems (i.e., fluid flow through a gas exchange site then into closed blood vessels such as arteries) (Bidani and Heming, 1991) both predict that β_{nb} will markedly affect equilibration rate (Fig. 3). This has been verified experimentally (Heming and Bidani, 1992); equilibration occurs very rapidly at low β_{nb}, even in the absence of CA (e.g., at 15°C and a β_{nb} = 5 mM/pH unit $t_{1/2}$ = 40–60 s, but for β_{nb} = 0.5, $t_{1/2}$ = 8–13 s). The consequence of this effect is that the addition of CA to systems of low β_{nb} (e.g., freshwater flowing over the gill surface) will have little effect on the measured ΔpH (the uncatalyzed and catalyzed reactions both occur so rapidly that it is difficult to distinguish between them) and adding CA inhibitors may increase the ΔpH by increasing the buffering capacity of the system.

Finally, the transit time for fluid flow from the gas exchange site to the pH electrode is also a determinant of the setup's sensitivity. Sensitivity decreases with increasing transit time as more of the overall pH disequilibrium is missed (Fig. 4); i.e., the observable ΔpH reflects the difference between the overall pH disequilibrium and the portion of the reequilibration reaction that occurs upstream of the electrode and so is missed. In this regard, an interesting effect is created by the actions of β_{nb} on the dynamics and kinetics of the pH disequilibrium. Increments in β_{nb} reduce the overall change in pH but increase the time required for reequilibration such that less of the overall disequilibrium is missed during fluid transit to the electrode (Fig. 4). For this reason, it is often possible to increase the sensitivity of measurement (the observable part of the disequilibrium) by increasing the experimental β_{nb}. However, one must be cautious because increases in β_{nb} do not necessarily lead to a decrement in the overall magnitude of the measured pH change (Bidani and Heming, 1991; Heming and Bidani, 1992).

VIII. SUMMARY

Carbonic anhydrase is a central enzyme in the transport and excretion of CO_2 at both the cellular and systemic levels. The enzyme functions in maintaining a near instantaneous equilibrium between the chemical species involved in the CO_2 hydration/dehydration reactions. The wide distribution of CA allows these reactions to proceed at the catalyzed rate in a number of separate fluid compartments within the animal, thus ensuring that CO_2 transport will not be limited at any one critical point. The transport and excretion of ammonia parallel that of CO_2 in many respects. Although CA appears to have been selected as primarily an enzyme of facilitated transport, there is some indication that it might also play a role in ventilatory control in some groups of fishes. The evolution of CA function continues to be one of the most fertile areas of study of this enzyme.

REFERENCES

Appelgren, L.-E., Odlind, B., and Wistrand, P. J. (1989). Tissue distribution of ^{125}I-labelled carbonic anhydrase isozymes I, II and III in the rat. *Acta Physiol. Scand.* **137,** 449–456.
Avella, M., and Bornancin, M. (1989). A new analysis of ammonia and sodium transport through the gills of the freshwater rainbow trout (*Salmo gairdneri*). *J. Exp. Biol.* **142,** 155–175.
Bauer, C., Gros, G., and Bartels, H., eds. (1980). "Biophysics and Physiology of Carbon Dioxide," 453 pp. Springer-Verlag, New York.
Bergenhem, N., and Carlsson, U. (1990). Partial amino acid sequence of erythrocyte carbonic anhydrase from tiger shark. *Comp. Biochem. Physiol. B* **95,** 205–213.

Bidani, A. (1991). Analysis of abnormalities of capillary CO_2 exchange in vivo. *J. Appl. Physiol.* **70,** 1686–1699.

Bidani, A., and Crandall, E. D. (1978). Slow postcapillary changes in blood pH in vivo: Titration with acetazolamide. *J. Appl. Physiol.* **45,** 565–573.

Bidani, A., Crandall, E. D., and Forster, R. E. (1978). Analysis of post-capillary changes in blood pH in vivo after gas exchange. *J. Appl. Physiol.* **44,** 770–781.

Bidani, A., and Heming, T. A. (1991). Effects of perfusate buffer capacity on capillary CO_2–HCO_3^-–H^+ reactions: Theory. *J. Appl. Physiol.* **71,** 1460–1468.

Bidani, A., Matthew, S. J., and Crandall, E. D. (1983). Pulmonary vascular carbonic anhydrase activity. *J. Appl. Physiol.* **55,** 75–83.

Booth, V. H. (1938). The carbonic anhydrase inhibitor in serum. *J. Physiol.* **91,** 474–489.

Borchard, B. (1978). Studies on the rainbow trout (*Salmo gairdneri* Richardson). 1. Correlation between gonadal development and serum protein pattern. *Annls. Biol. Anim. Biochim. Biophys.* **18,** 1027–1034.

Brinkman, R., Margaria, R., Meldrum, N. U., and Roughton, F. J. W. (1932). The CO_2 catalyst present in blood. *J. Physiol.* **75,** 3P–4P.

Burggren, W., and Haswell, M. S. (1979). Aerial CO_2 excretion in the obligate air breathing fish *Trichogaster trichopterus:* A role for carbonic anhydrase. *J. Exp. Biol.* **82,** 215–225.

Cameron, J. N. (1978). Chloride shift in blood. *J. Exp. Zool.* **206,** 289–295.

Cameron, J. N. (1989). "The Respiratory Physiology of Animals," 353 pp. Oxford Univ. Press, New York.

Cameron, J. N., and Heilser, N. (1983). Studies of ammonia in the rainbow trout: Physicochemical parameters, acid–base behavior and respiratory clearance. *J. Exp. Biol.* **105,** 107–125.

Cameron, J. N., and Heisler, N. (1985). Ammonia transfer across fish gills: A review. *In* "Proceedings in Life Sciences: Circulation, Respiration, and Metabolism" (R. Gilles, ed.), pp. 91–100. Springer-Verlag, Heidelberg.

Carlsson, U., Kjellstrom, B., and Antonsson, B. (1980). Purification and properties of cyclostome carbonic anhydrase from erythrocytes of hagfish. *Biochim. Biophys. Acta* **612,** 160–170.

Chen, J. C. T., and Chesler, M. (1992). pH transients evoked by excitatory synaptic transmission are increased by inhibition of extracellular carbonic anhydrase. *Proc. Natl. Acad. Sci. USA* **89,** 7786–7790.

Claiborne, J. B., and Evans, D. H. (1988). Ammonia and acid–base balance during high ammonia exposure in a marine teleost (*Myoxocephalus octodecimspinosus*). *J. Exp. Biol.* **140,** 89–105.

Coleman, J. E. (1980). Current concepts of the mechanism of action of carbonic anhydrase. *In* "Biophysics and Physiology of Carbon Dioxide" (C. Bauer, G. Gros, and H. Bartels, eds.), pp 133–150. Springer-Verlag, New York.

Conley, D. M., and Malalatt, J. (1988). Histochemical localization of Na^+-K^+ ATPase and carbonic anhydrase activity in gills of 17 fish species. *Can. J. Zool.* **66,** 2398–2405.

Crandall, E. D., and O'Brasky, J. E. (1978). Direct evidence for participation of rat lung carbonic anhydrase in CO_2 reactions. *J. Clin. Invest.* **62,** 618–622.

Crandall, E. D., and A. Bidani, A. (1981). Effects of red blood cell HCO_3^-/Cl^- exchange kinetics on lung CO_2 transfer: Theory. *J. Appl. Physiol.* **50,** 265–271.

Crandall, E. D., Mathew, S. J., Fleischer, R. S., Winter, H. I., and Bidani, A. (1981). Effects of inhibition of RBC HCO_3^-/Cl^- exchange on CO_2 excretion and downstream pH disequilibrium in isolated rat lungs. *J. Clin. Invest.* **68,** 853–862.

Currie, S., Kieffer, J. D., and Tufts, B. L. (1995). The effects of blood CO_2 reaction rates in CO_2 removal from muscle in exercised trout. *Respir. Physiol.* **100,** 261–269.

Davis, R. P. (1958). The kinetics of the mechanism of the reaction of human carbonic anhydrase. I. Basic mechanism and the effect of electrolytes on enzyme activity. *J. Amer. Chem. Soc.* **80,** 5209–5214.

Davis, R. P. (1959). The kinetics of the reaction of human erythrocyte carbonic anhydrase. II. The effect of sulfanilamide, sodium sulfide and various chelating agents. *J. Amer. Chem. Soc.* **81,** 5674–5678.

Dimberg, K. (1994). The carbonic anhydrase inhibitor in trout plasma: Purification and its effect on carbonic anhydrase activity and the Root effect. *Fish Physiol. Biochem.* **12,** 381–386.

Dodgson, S. J., Tashian, R. E., Gros, G., and Carter, N. D. (1991). "The Carbonic Anhydrases: Cellular Physiology and Molecular Genetics," 379 pp. Plenum, New York.

Edsall, J. T. (1968). Carbon dioxide, carbonic acid, and bicarbonate ion: Physical properties and kinetics of interconversion. *In* "CO_2: Chemical Biochemical, and Physiological Aspects" (R. E. Forster, J. T. Edsall, A. B. Otis, and F. J. W. Roughton, eds.), pp. 15–28. NASA SP #188, Washington, D.C.

Effros, R. M., Mason, G., and Silverman, P. (1981). Role of perfusion and diffusion in $^{14}CO_2$ exchange in the rabbit lung. *J. Appl. Physiol.* **51,** 1136–1144.

Ellory, J. C., Wolowyk, M. W., and Young, J. D. (1987). Hagfish (*Eptatretus stouti*) erythrocytes show minimal chloride transport activity. *J. Exp. Biol.* **129,** 377–383.

Evans, D. H., and More, K. J. (1988). Modes of ammonia transport across the gill epithelium of the dogfish pup (*Squalus acanthias*). *J. Exp. Biol.* **138,** 375–397.

Evans, D. H., More, K. J., and Robbins, S. L. (1989). Modes of ammonia transport across the gill epithelium of the marine teleost fish *Opsanus beta*. *J. Exp. Biol.* **144,** 339–356.

Ferguson, R. A., Sehdev, H., Bagatto, B., and Tufts, B. L. (1992). In vitro interactions between oxygen and carbon dioxide transport in the blood of the sea lamprey (*Petromyzon marinus*). *J. Exp. Biol.* **173,** 25–41.

Forster, R. E. (1993). Carbonic anhydrase in the carotid body. *In* "Neurobiology and Cell Physiology of Chemoreception," (P. G. Data *et al.,* eds.), pp. 137–147. Plenum Press, New York.

Forster, R. E., and Itada, N. (1980). Carbonic anhydrase activity in intact red cells as measured by means of ^{18}O exchange between CO_2 and water. *In* "Biophysics and Physiology of Carbon Dioxide" (C. Bauer, G. Gros, and H. Bartels, eds.), pp. 177–183. Springer-Verlag, New York.

Gilmour, K. M. (1998). Causes and consequences of acid–base disequilibria. *In* "Fish Physiology: Fish Haemoglobin and Respiration" (S. F. Perry and B. L. Tufts, eds.), Vol. 17. Academic Press, New York.

Gilmour, K. M., Randall, D. J., and Perry, S. F. (1994). Acid–base disequilibrium in the arterial blood of rainbow trout. *Respir. Physiol.* **96,** 259–272.

Gilmour, K. M., and Perry, S. F. (1994). The effects of hypoxia, hyperoxia or hypercapnia on the acid–base disequilibrium in the arterial blood of rainbow trout. *J. Exp. Biol.* **192,** 269–284.

Gilmour, K. M., Henry, R. P., Wood, C. M., and Perry, S. F. (1997). Extracellular carbonic anhydrase and an acid–base disequilibrium in the blood of the dogfish, *Squalus acanthias, J. Exp. Biol.* **200,** 173–183.

Graham, J. B. (1994). An evolutionary perspective for bimodal respiration: A biological synthesis of fish air breathing. *Amer. Zool.* **24,** 229–237.

Graham, M. S., Turner, J. D., and Wood, C. M. (1990). Control of ventilation in the hypercapnic skate *Raja ocella*. I. Blood and extracellular fluid. *Respir. Physiol.* **80,** 259–277.

Gray, B. A. (1971). The rate of approach to equilibrium in uncatalyzed CO_2 hydration reactions: The theoretical effect of buffering capacity. *Respir. Physiol.* **11,** 223–234.

Gros, G., and Dodgson, S. J. (1988). Velocity of CO_2 exchange in muscle and liver. *Ann. Rev. Physiol.* **50**, 669–694.

Gutknecht, J., Bisson, M. A., and Tosteson, F. C. (1977). Diffusion of carbon dioxide through lipid bilayer membranes: Effects of carbonic anhydrase, bicarbonate and unstirred layers. *J. Gen. Physiol.* **69**, 779–794.

Hall, G. E., and Schraer, R. (1983). Characterization of high activity carbonic anhydrase isozyme purified from erythrocytes of *Salmo gairdneri. Comp. Biochem. Physiol. B* **75**, 81–92.

Hansen, P., and Magid, E. (1966). Studies on a method of measuring carbonic anhydrase activity. *Scand. J. Clin. and Lab. Invest.* **18**, 21–32.

Haswell, M. S. (1977). Carbonic anhydrase in flounder erythrocytes. *Comp. Biochem. Physiol. A* **56**, 281–282.

Haswell, M. S., and Randall, D. J. (1976). Carbonic anhydrase inhibitor in trout plasma. *Respir. Physiol.* **28**, 17–27.

Haswell, M. S., Raffin, J.-P., and LeRay, C. (1983). An investigation of the carbonic anhydrase inhibitor in eel plasma. *Comp. Biochem. Physiol. A* **74**, 175–177.

Heisler, N., Toews, D. P., and Holeton, G. F. (1988). Regulation of ventilation and acid–base status in the elasmobranch *Scyliorhinus stellaris* during hyperoxia induced hypercapnia. *Respir. Physiol.* **71**, 227–246.

Heming, T. A. (1984). The role of fish erythrocytes in transport and excretion of carbon dioxide. Ph.D. thesis. Univ. British Columbia, Vancouver, B.C.

Heming, T. A. (1986). Carbon dioxide excretion and ammonia toxicity in fishes: Is there a relationship? *In* "Problems of Aquatic Toxicology, Biotesting, and Water Quality Management," pp. 84–94. U.S. Environmental Protection Agency, Athens, Ga. EPA/600/9-86/024.

Heming, T. A., and Watson, T. A. (1986). Activity and inhibition of carbonic anhydrase in *Amia calva,* a biomodal-breathing holostean fish. *J. Fish. Biol.* **28**, 385–392.

Heming, T. A., and Bidani, A. (1991). Kinetic properties of pulmonary carbonic anhydrase. *In* "Carbonic Anhydrase, from Biochemistry and Genetics to Physiology and Clinical Medicine" (F. Botre, G. Gros, and B. T. Storey, eds.), pp. 379–392. Verlag-Chemie, Heidelburg.

Heming, T. A., and Bidani, A. (1992). Influence of proton availability on intracapillary CO_2–HCO_3^-–H^+ reactions in isolated art lungs. *J. Appl. Physiol.* **72**, 2140–2148.

Heming, T. A., Geers, C., Gros, G., Bidani, A., and Crandall, E. D. (1986). Effects of dextran-bound inhibitors on carbonic anhydrase activity in isolated rat lung. *J. Appl. Physiol.* **61**, 1849–1856.

Heming, T. A., Vanoye, C. G., Stabenau, E. K., Boush, E. D., Fierke, C. A., and Bidani, A. (1993). Inhibitor sensitivity of pulmonary vascular carbonic anhydrase. *J. Appl. Physiol.* **75**, 1642–1649.

Henry, R. P. (1984). The role of carbonic anhydrase in blood ion and acid–base regulation. *Amer. Zool.* **24**, 241–251.

Henry, R. P. (1987). Quaternary ammonium sulfanilamide: A membrane-impermeant carbonic anhydrase inhibitor. *Am. J. Physiol.* **252**, R959–R965.

Henry, R. P. (1988). Multiple functions of crustacean gill carbonic anhydrase. *J. Exp. Zool.* **248**, 19–24.

Henry, R. P. (1991). Techniques for measuring carbonic anhydrase activity in vitro and in vivo: The electrometric delta pH method and the pH stat method. *In* "The Carbonic Anhydrases: Cellular Physiology and Molecular Genetics" (S. J. Dodgson, ed.), pp. 119–125. Plenum, New York.

Henry, R. P. (1996). Multiple roles of carbonic anhydrase in cellular transport and metabolism. *Ann. Rev. Physiol.* **58**, 523–538.

Henry, R. P., and Cameron, J. N. (1982). The distribution and partial characterization of carbonic anhydrase in selected aquatic and terrestrial decapod crustaceans. *J. Exp. Zool.* **221**, 309–321.

Henry, R. P., Smatresk, N. J., and Cameron, J. N. (1988). The distribution of branchial carbonic anhydrase and the effects of gill and erythrocytic carbonic anhydrase inhibition in the channel catfish *Ictalurus punctatus. J. Exp. Biol.* **134**, 201–218.

Henry, R. P., Tufts, B. L., and Boutilier, R. G. (1993). The distribution of carbonic anhydrase type I and II isozymes in lamprey and trout: Possible co-evolution with erythrocyte chloride/bicarbonate exchange. *J. Comp. Physiol. B* **163**, 380–388.

Henry, R. P., Boutilier, R. G., and Tufts, B. L. (1995). Effects of carbonic anhydrase inhibition on the acid–base status in lamprey and trout. *Respir. Physiol.* **99**, 241–248.

Henry, R. P., Wang, Y., and Wood, C. M. (1997a). Carbonic anhydrase facilitates CO_2 and NH_3 transport across the sarcolemma of trout white muscle. *Amer. J. Physiol.* (In press).

Henry, R. P., Gilmour, K. M., Wood, C. M., and Perry, S. F. (1997b). Extracellular carbonic anhydrase activity and carbonic anhydrase inhibitors in the circulatory system of fish. *Physiol. Zool.* (In press).

Hill, E. P. (1986). Inhibition of carbonic anhydrase by plasma of dogs and rabbits. *J. Appl. Physiol.* **60**, 191–197.

Hjelmeland, K., Christie, M., and Raa, J. (1983). Skin mucus protease from rainbow trout, *Salmo gairdnerii* Richardson, and its biological significance. *J. Fish Bowl.* **23**, 13–22.

Hoar, W. S., and Randall, D. J., eds. (1984a). "Fish Physiology, vol XA, Anatomy, Gas Transfer, and Acid–Base Regulation." Academic Press, New York, 456 pp.

Hoar, W. S., and Randall, D. J., eds. (1984b). "Fish Physiology, vol XB, Ion and Water Transfer." Academic Press, New York, 416 pp.

Houston, A. H., and Mearow, K. M. (1979). Temperature-related changes in the erythrocytic carbonic anhydrase (acetazolamide-sensitive esterase) activity of goldfish, *Carassius auratus. J. Exp. Biol.* **78**, 255–264.

Itada, N., and Forster, R. E. (1977). Carbonic anhydrase activity in intact red blood cells measured with ^{18}O exchange. *J. Biol. Chem.* **252**, 3881–3890.

Kernohan, J. C. (1964). The activity of bovine carbonic anhydrase in imidazole buffers. *Biochim. Biophys. Acta* **81**, 346–356.

Kernohan, J. C., Forrest, W. W., and Roughton, F. J. W. (1963). The activity of concentrated solutions of carbonic anhydrase. *Biochim. Biophys. Acta* **67**, 31–41.

Kifor, G., Toon, M. R., Janoshazi, A., and Solomon, A. K. (1993). Interaction between red cell membrane band-3 and cytosolic carbonic anhydrase. *J. Membrane Biol.* **134**, 169–179.

Kim, J.-S., Gay, C. V., and Schraer, R. (1983). Purification and properties of carbonic anhydrase from salmon erythrocytes. *Comp. Biochem. Physiol. B* **76**, 523–527.

Kirshner, L. B. (1979). Control mechanisms in crustaceans and fishes. *In* "Mechanisms of Osmoregulation in Animals: Maintenance of Cell Volume" (R. Gilles, ed.), pp. 152–222. Wiley, New York.

Klocke, R. A. (1978). Catalysis of CO_2 reactions by lung carbonic anhydrase. *J. Appl. Physiol.* **44**, 882–888.

Lacy, E. R. (1983). Histochemical and biochemical studies of carbonic anhydrase activity in the opercular epithelium of the euryhaline teleost, *Fundulus heteroclitus. Am. J. Anatomy* **166**, 19–39.

Leibman, K. C., Alford, D., and Boudet, R. A. (1961). Nature of the inhibition of carbonic anhydrase by acetazolamide and benzthiazide. *J. Pharmacol. Exptl. Therap.* **131**, 271–274.

Lessard, J., Val, A. L., Aota, S., and Randall, D. J. (1995). Why is there no carbonic anhydrase activity available to fish plasma? *J. Exp. Biol.* **198**, 31–38.

Lindskog, S. (1980). Rate-limiting steps in the catalytic action of carbonic anhydrase. *In* "Biophysics and Physiology of Carbon Dioxide" (C. Bauer, G. Gros, and H. Bartels, eds.), pp. 230–237. Springer-Verlag, New York.

Lloyd, R., and Herbert, D. W. M. (1960). The influence of carbon dioxide on the toxicity of un-ionized ammonia to rainbow trout (*Salmo gairdnerii* Richardson). *Ann. Appl. Biol.* **48**, 399–404.

Maetz, J. (1956). Le dosage de l'anhydrase carbonique. Etude de quelques substances inhibitrices et activatrices. *Bull. Soc. Chim. Biol.* **38**, 447–474.

Maren, T. H. (1967a). Carbonic anhydrase in the animal kingdom. *Fed. Proc.* **26**, 1097–1103.

Maren, T. H. (1967b). Carbonic anhydrase: Chemistry, physiology and inhibition. *Physiol. Rev.* **47**, 595–781.

Maren, T. H. (1977). Use of inhibitors in physiological studies of carbonic anhydrase. *Am. J. Physiol.* **232**, F291–F297.

Maren, T. H., and Couto, E. O. (1979). The nature of anion inhibition of human red cell carbonic anhydrases. *Arch. Biochem. Biophys.* **196**, 501–510.

Maren, T. H., Friedland, B. R., and Rittmaster, R. S. (1980). Kinetic properties of primitive vertebrate carbonic anhydrases. *Comp. Biochem. Physiol. B* **67**, 69–74.

Maynard, J. R., and Coleman, J. E. (1971). Elasmobranch carbonic anhydrase: Purification and properties of the enzyme from two species of shark. *J. Biol. Chem.* **246**, 4455–4464.

McIntosh, J. E. A. (1968). Assay of carbonic anhydrase by titration at constant pH. *Biochem. J.* **109**, 203–207.

Meldrum, N. U., and Roughton, F. J. W. (1933). Carbonic anhydrase: Its preparation and properties. *J. Physiol.* (London) **80**, 113–142.

Motais, R., Fievet, B., Garcia-Romeu, F., and Thomas, S. (1989). Na^+-H^+ exchange and pH regulation in red blood cells: Role of uncatalyzed H_2CO_3 dehydration. *Am. J. Physiol.* **256**, C728–C735.

Nikinmaa, M. (1992). Membrane transport and control of hemoglobin–oxygen affinity in nucleated erythrocytes. *Physiol. Rev.* **72**, 301–321.

Nikinmaa, M., and Raillo, E. (1987). Anion movements across lamprey (*Lampetra fluviatilis*) red cell membranes. *Biochem. Biophys. Acta* **899**, 134–136.

Nikinmaa, M., Tithonen, K., ,and Paajaste, M. (1990). Adrenergic control of red cell pH in samonid fish: Roles of the sodium/proton exchange, Jacobs–Stewart cycle and membrane potential. *J. Exp. Biol.* **154**, 257–271.

Nyman, P. O. (1963). A spectrophotometric method for the assay of carbonic anhydrase activity. *Acta Chem. Scan.* **17**, 429–435.

Obaid, A. L., Critz, A. M., and Crandall, E. D. (1978). Kinetics of bicarbonate/chloride exchange in dogfish erythrocytes. *Amer. J. Physiol.* **237**, R132–R138.

Ojteg, G., and Wistrand, P. J. (1994). Renal handling and plasma elimination kinetics of carbonic anhydrase isoenzymes I, II and III in the rat. *Acta Physiol. Scand.* **151**, 531–539.

Paranawithans, S. R., Tu, K., Laipis, P. J., and Silverman, D. N. (1990). Enhancement of the catalytic activity of carbonic anhydrase III by phosphates. *J. Biol. Chem.* **265**, 22270–22274.

Parkes, J. L., and Coleman, P. S. (1989). Enhancement of carbonic anhydrase activity by erythrocyte membranes. *Arch. Biochem. Biophys.* **275**, 459–468.

Pelster, B. (1995). Mechanisms of acid release in isolated gas gland cells of the European eel *Anguilla anguilla*. *Amer. J. Physiol.* **269**, R793–R799.

Perry, S. F., Davie, P. S., Daxboeck, C., and Randall, D. J. (1982). A comparison of CO_2 excretion in a spontaneously ventilating blood-perfused trout preparation and saline-perfused gill preparations: Contribution of the branchial epithelium and red blood cell. *J. Exp. Biol.* **101**, 47–60.

Perry, S. F., and Laurent, P. (1990). The role of carbonic anhydrase in carbon dioxide excretion, acid–base balance and ionic regulation in aquatic gill breathers. *In* "Comparative Physiology, Animal Nutrition and Transport Processes 2, Transport Respiration and Excretion: Comparative and Environmental Aspects" (J. P. Truchot and B. Lahlou, eds.), Vol. 2, pp. 39–57. S. Karger, Basel, Switzerland.

Perry, S. F., and Gilmour, K. M. (1996). Consequences of catecholamine release on ventilation and blood oxygen transport during hypoxia and hypercapnia in an elasmobranch (*Squalus acanthias*) and a teleost (*Oncorhynchus mykiss*). *J. Exp. Biol.* **199**, 2105–2118.

Perry, S. F., Brauner, C. J., Tufts, B., and Gilmour, K. M. (1997). Acid–base disequilibrium in the venous blood of rainbow trout (*Oncorhynchus mykiss*). *Exp. Biol. Online* **2**, 1–11.

Philpott, F. J., and Philpott, J. St. L. (1936). A modified colorimetric estimation of carbonic anhydrase. *Biochem. J.* **30**, 2191–2193.

Playle, R. C., and Wood, C. M. (1989). Water chemistry changes in the gill micro-environment of rainbow trout: Experimental observations and theory. *J. Comp. Physiol. B* **159**, 527–537.

Rahim, S. M., Delaunoy, J.-P., and Laurent, P. (1988). Identification and immunocytochemical localization of two different carbonic anhydrase isoenzymes in teleostean fish erythrocytes and gill epithelia. *Histochem.* **89**, 451–459.

Rahn, H., and Howell, B. J. (1976). Bimodal gas exchange. *In* "Respiration of Amphibious Vertebrates" (G. M. Hughes, ed.), pp. 271–285. Academic Press, New York.

Randall, R. F., and Maren, T. H. (1972). Absence of carbonic anhydrase in red cell membranes. *Biochim. Biophys. Acta* **268**, 730–732.

Randall, D. J., Burggren, W. W., Farrell, A. P., and Haswell, M. S. (1982). "The Evolution of Air Breathing in Vertebrates." Cambridge Univ. Press, Cambridge. 242 pp.

Roughton, F. J. W. (1935). Recent work on carbon dioxide transport by the blood. *Physiol. Rev.* **15**, 263–314.

Roughton, F. J. W., and Booth, V. H. (1946a). The effect of substrate concentration, pH and other factors upon the activity of carbonic anhydrase. *Biochem. J.* **40**, 319–330.

Roughton, F. J. W., and Booth, V. H. (1946b). The manometric determination of the activity of carbonic anhydrase under varied conditions. *Biochem. J.* **40**, 309–319.

Roush, E. D., and Fierke, C. A. (1992). Purification and characterization of a carbonic anhydrase II inhibitor from porcine plasma. *Biochem.* **31**, 12536–12542.

Sanyal, G., Swenson, E. R., and Maren, T. H. (1982a). The isolation of carbonic anhydrase from the muscle of *Squalus acanthias* and *Scomber scombrus:* Inhibition studies. *Bull. Mt. Desert Isl. Biol. Lab.* **24**, 66–68.

Sanyal, G., Sweson, E. R., Pessah, N. I., and Maren, T. H. (1982b). The carbon dioxide hydration activity of skeletal muscle carbonic anhydrase. *Mol. Pharmacol.* **22**, 211–220.

Sanyal, G., Pessah, N. I., Swenson, E. R., and Maren, T. H. (1982c). The carbon dioxide hydration activity of purified teleost red cell carbonic anhydrase. Inhibition by sulfonamides and anions. *Comp. Biochem. Physiol. B* **73**, 937–944.

Scheid, P., and Siffert, W. (1985). Effects of inhibiting carbonic anhydrase on isometric contraction of frog skeletal muscle. *J. Physiol.* **361**, 91–101.

Silverman, D. N. (1991). The catalytic mechanism of carbonic anhydrase. *Can. J. Bot.* **69**, 1070–1078.

Smatresk, N. J. (1990). Chemoreceptor modulation of endogenous respiratory rhythms in vertebrates. *Am. J. Physiol.* **259**, R887–R897.

Smatresk, N. J., and Cameron, J. N. (1982). Respiration and acid–base physiology of the spotted gar, a bimodal breather. II. Responses to temperature changes and hypercapnia. *J. Exp. Biol.* **96**, 281–293.

Smeda, J. S., and Houston, A. H. (1979). Carbonic anhydrase (acetazolamide-sensitive esterase) activity in the red blood cells of thermally-acclimated rainbow trout, *Salmo gairdneri*. *Comp. Biochem. Physiol. A* **62**, 719–723.

Smith, A. C., and Ramos, F. (1976). Occult hemoglobin in fish skin mucus as an indicator of early stress. *J. Fish. Biol.* **9**, 537–541.

Sprague, J. B. (1973). The ABC's of pollutant bioassay using fish. *In* "Biological Methods for the Assessment of Water Quality" (J. Cairns, Jr. and K. L. Dickson, eds.), pp. 6–30. Am. Soc. Testing Mat., Philadelphia.

Stabenau, E. K., Bidani, A., and Heming, T. A. (1996). Physiological characterization of pulmonary carbonic anhydrase in the turtle. *Respir. Physiol.* **104**, 187–196.

Stadie, W. C., and O'Brien, H. (1933). The catalysis of the hydration of carbon dioxide and the dehydration of carbonic acid by an enzyme isolated from red blood cells. *J. Biol. Chem.* **103**, 521–529.

Swenson, E. R., and Maren, T. H. (1987). Roles of gill and red cell carbonic anhydrase in elasmobranch HCO_3^- and CO_2 excretion. *Amer. J. Physiol.* **253**, R450–R458.

Swenson, E. R., Lippincott, L., and Martin, T. H. (1995). Effect of gill membrane-bound carbonic anhydrase inhibition on branchial bicarbonate excretion in the dogfish shark, *Squalus acanthias. Bull. Mt. Desert, Isl. Biol. Lab.* **34**, 94–95.

Swenson, E. R., Taschner, B. C., and Maren, T. H. (1996). Effect of membrane-bound carbonic anhydrase (CA) inhibition on bicarbonate excretion in the shark, *Squalus acanthias. Bull. Mt. Desert. Isl. Biol. Lab.* **35**, 35.

Syrjhala, H., Vuori, J., Huttunen, K., and Vaananen, H. K. (1990). Carbonic anhydrase III as a serum marker for diagnosis of rhabdomyolysis. *Clin. Chem.* **36**, 696.

Szumski, D. S., Barton, D. A., Putnam, H. D., and Polta, R. C. (1982). Evaluation of EPA un-ionized ammonia toxicity criteria. *J. Water Pollut. Control Red.* **54**, 281–291.

Takahashi, E., Menon, A. S., Kato, H., Slutsky, A. S., and Phillipson, E. A. (1990). Control of expiratory duration by arterial CO_2 oscillations in vagotomized dogs. *Respir. Physiol.* **79**, 45–56.

Takahashi, E., Tejima, K., and Yamakoshi, Y.-I. (1992). Entrainment of respiratory rhythm to respiratory oscillations of arterial P_{CO_2} in vagotomized dogs. *J. Appl. Physiol.* **73**, 1052–1057.

Tang, Y., H. Lin, and Randall, D. J. (1992). Compartmental distribution of carbon dioxide and ammonia in rainbow trout at rest and following exercise and the effect of bicarbonate infusion. *J. Exp. Biol.* **169**, 235–249.

Tashian, R. E. (1989). The carbonic anhydrases: Widening perspectives on their evolution, expression and function. *BioEssays* **10**, 186–192.

Tashian, R. E., and Hewett-Emmett, D., eds. (1984). Biology and chemistry of the carbonic anhydrases. *N.Y. Acad. Sci.* Vol. 429. 640 pp.

Tu, C. K., Silverman, D. N., Forsman, C., Jonsson, B. H., and Lindskog, S. (1989). Role of the histidine 64 in the catalytic mechanism of human carbonic anhydrase II studied with a site-specific mutant. *Biochemistry* **28**, 7913–7918.

Tufts, B. L., and Boutilier, R. G. (1989). The absence of rapid chloride-bicarbonate exchange in lamprey erythrocytes: Implications for CO_2 transport and ion distributions between plasma and erythrocytes in the blood of *Petromyzon marinus. J. Exp. Biol.* **144**, 565–576.

Tufts, B. L., and Boutilier, R. G. (1990a). CO_2 transport and ion distributions between plasma and erythrocytes in the blood of a primitive vertebrate. *Myxine glutinosa. Exp. Biol.* **48**, 341–347.

Tufts, B. L., and Boutillier, R. G. (1990b). CO_2 transport in agnoathan blood: Evidence of erythrocyte Cl_-/HCO_3^- limitations. *Respir. Physiol.* **80**, 335–348.

Tufts, B. L., Bagatto, B., and Cameron, B. (1992). In vivo analysis of gas transport in arterial and venous blood of the sea lamprey *Petromyzon marinus. J. Exp. Biol.* **169**, 105–119.

Tufts, B. L., and Perry, S. F. (1998). Blood carbon dioxide transport. *In* "Fish Physiology, Haemoglobin and Respiration" (B. L. Tufts and S. F. Perry, eds.), Vol. 17. Academic Press, New York.

Vaananen, H. K., Leppilampi, M., Vuori, J., and Takala, T. E. S. (1986). Liberation of muscle carbonic anhydrase in serum during extensive exercise. *J. Appl. Physiol.* **61,** 561–564.

Verpoorte, J. A., Mehta, S., and Edsall, J. T. (1967). Esterase activities of human carbonic anhydrase B and C. *J. Biol. Chem.* **242,** 4221–4229.

Walsh, P. J., and Henry, R. P. (1991). Carbon dioxide and ammonia metabolism and exchange. *In* "Biochemistry and Molecular Biology of Fishes" (P. W. Hochachka and T. P. Mommsen, eds.), Vol. 1, pp. 181–207. Elsevier, New York.

Wang, Y., Heigenhauser, G. J. F., and Wood, C. M. (1996). Ammonia transport and distribution after exercise across white muscle cell membranes in rainbow trout: A perfusion study. *Amer. J. Physiol. Regulatory* **40,** R738–R745.

Wilbur, K. M., and Anderson, N. G. (1948). Electrometric and colorimetric determination of carbonic anhydrase. *J. Biol. Chem.* **176,** 147–154.

Wilson, R. W., Wright, P. M., Munger, S., and Wood, C. M. (1994). Ammonia excretion is freshwater rainbow trout (*Oncorhynchus mykiss*) and the importance of gill boundary layer acidification: Lack of evidence for Na^+/NH_4^+ exchange. *J. Exp. Biol.* **191,** 37–58.

Wood, C. M., and Perry, S. F. (1991). A new in vitro assay for carbon dioxide excretion by trout red blood cells: Effects of catecholamines. *J. Exp. Biol.* **157,** 349–366.

Wood, C. M., and Munger, R. S. (1994). Carbonic anhydrase injection provides evidence for the role of blood acid–base status in stimulating ventilation after exhaustive exercise in rainbow trout. *J. Exp. Biol.* **194,** 225–253.

Wood, C. M., Turner, J. D., Munger, R. S., and Graham, M. S. (1990). Control of ventilation in the hypercapnic skate *Raja ocella.* II. Cerebrospinal fluid and intracellular pH in the brain and other tissues. *Respir. Physiol.* **80,** 277–298.

Wood, C. M., Perry, S. F., Walsh, P. J., and Thomas, S. (1994). HCO_3^- dehydration by the blood of an elasmobranch in the absence of a Haldane effect. *Respir. Physiol.* **98,** 319–337.

Wright, P., Heming, T., and Randall, D. J. (1986). Downstream pH changes in water flowing over the gills of rainbow trout. *J. Exp. Biol.* **126,** 499–512.

Wright, P. A., and Wood, C. M. (1985). An analysis of branchial ammonia excretion in the freshwater rainbow trout: Effects of environmental pH change and sodium uptake blockade. *J. Exp. Biol.* **114,** 329–353.

Wright, P. A., Randall, D. J., and Wood, C. M. (1988). The distribution of ammonia and H^+ between tissue compartments in lemon sole (*Parophrys vetulus*) at rest, during hypercapnia, and following exercise. *J. Exp. Biol.* **136,** 149–175.

Wright, P. A., Randall, D. J., and Perry, S. F. (1989). Fish gill water boundary layer: A site of linkage between carbon dioxide and ammonia excretion. *J. Comp. Physiol. B* **158,** 627–635.

4

THE PHYSIOLOGY OF THE ROOT EFFECT

BERND PELSTER
DAVID RANDALL

I. Introduction
   A. Definition
   B. Measurement of the Root Effect
II. Occurrence of the Root Effect
   A. Interspecific Distribution
   B. Intraspecific Differentiation between Multiple Hemoglobins
III. Characterization of the Root Effect
   A. H$^+$ and CO$_2$ Root Effects
   B. Influence of Allosteric Ligands
   C. Molecular Mechanisms
   D. Kinetics of the Root Effect
IV. Root Effect and Swim Bladder Function
   A. The Swim Bladder as a Hydrostatic Organ
   B. Generation of High Oxygen Partial Pressures
V. Oxygen-Concentrating Mechanisms in the Fish Eye
   A. Circulation to the Fish Eye
   B. Oxygen Supply to the Fish Eye
VI. Adrenergic Effects
   A. Catecholamines and Erythrocytic pH
   B. The Root and Haldane Effects
      References

I. INTRODUCTION

A. Definition

In almost all species, hemoglobin oxygen affinity is markedly modified by protons, a phenomenon known as the Bohr effect. Analyzing the oxygen-binding characteristics of toadfish, sea robin, and mackerel blood, Root (1931) observed that in the presence of CO$_2$ the oxygen dissociation curve of the hemoglobin became nearly asymptotic with respect to the abscissa

113

Fish Physiology, Volume 17:
FISH RESPIRATION

before saturation was complete. Thus, even at high P_{O_2}, the oxygen content of the blood was markedly reduced in the presence of increasing amounts of CO_2. The same effect was obtained by acid addition to the blood (Root and Irving, 1943). Scholander and Van Dam (1954) termed the reduction in hemoglobin oxygen-carrying capacity caused by a decrease in blood pH the Root effect, as distinct from the Bohr effect, which describes the change in hemoglobin oxygen affinity associated with a change in blood pH (Δlog P_{50}/ΔpH) (Fig. 1). The distinction between the Bohr and Root effects found support in the observation that the oxygen-carrying capacity of the hemoglobin of some fish species was so drastically diminished at low pH that oxygen pressures of 140 atmospheres were unable to fully saturate the blood (Scholander and Van Dam, 1954). It is generally accepted that, at low pH, Root effect hemoglobins are locked in the deoxygenated conformational state and exhibit a depression in cooperativity of oxygen binding (Brittain, 1987; Riggs, 1988; Mylvaganam *et al.*, 1996; Perutz, 1996).

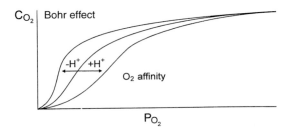

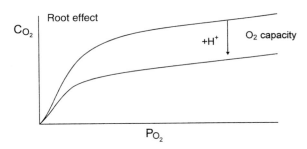

Fig. 1. A change in proton concentration induces a modification of the hemoglobin–oxygen affinity (Bohr effect) in most vertebrates, whereas in fish blood an increase in proton concentration may also induce a decrease in the hemoglobin oxygen-carrying capacity (Root effect). Although the molecular basis for the Bohr effect is a stabilization of the T state of the hemoglobin due to proton binding, recent results indicate that the Root effect is due to a destabilization of the R state.

B. Measurement of the Root Effect

Different techniques have been adopted to assess the Root effect. One strategy is to measure the changing oxygen content of the blood in the face of increasing P_{CO_2} at constant high P_{O_2}, where the oxygen content can be measured polarographically or gasometrically. Gasometric methods have the advantage that changes in oxygen-carrying capacity in undiluted whole blood can be measured. The optical method makes use of the different light absorption characteristics of oxygenated and deoxygenated hemoglobin. In the presence of the Root effect, an increase in the fraction of deoxygenated hemoglobin with decreasing pH is reflected by decreases in the absorbance differences between the oxygenated and deoxygenated samples, compared to those at alkaline pH values, where the Root effect is absent. This method requires diluted hemoglobin solutions and thus needs much smaller sample volumes compared to the gasometric techniques. It also allows for the suspension of the hemoglobin in a chemically well-defined medium.

Information on the Root effect can also be obtained from the kinetics of gas binding and release, using stopped-flow spectrometry or flash photolysis of carboxyhemoglobin (Noble *et al.*, 1986, 1970; Tan and Noble, 1973; Parkhurst *et al.*, 1983; Brittain, 1985; cf. Riggs, 1979).

II. OCCURRENCE OF THE ROOT EFFECT

A. Interspecific Distribution

The Root effect is generally considered to be an exclusive property of fish hemoglobins; however, a few studies report a reduction in oxygen-carrying capacity with decreasing pH in mollusc hemocyanins (Bridges *et al.*, 1984, Miller and Mangum, 1988). A reversed Root effect was found in *Buccinum undatum,* where the hemocyanin-bound oxygen concentration increased with increasing P_{CO_2} (Brix, 1982).

Among the fishes, the Root effect is almost exclusively restricted to teleosts (Baines, 1975; Farmer *et al.*, 1979; Wilhelm and Reichl, 1981; Ingermann and Terwillinger, 1982; Di Prisco, 1988; Noble *et al.*, 1986; Dafré and Wilhelm, 1989), although a few accounts among the elasmobranchs have been reported (Farmer *et al.*, 1979; Dafré and Wilhelm, 1989; cf. Pelster and Weber, 1991). Farmer *et al.* (1979) suggested that the Root effect is more pronounced in primitive orders, although its magnitude varies greatly. In the suborder Gymnotoidei (Cypriniformes), for example, it is pronounced in some families and completely absent in others (see, Pelster and Weber, 1991).

The adaptational value and physiological importance of the Root effect are of great interest. Several attempts to correlate the presence of the Root effect with special environmental conditions, however, have provided little in the way of conclusive answers. No correlation was found between the presence of a Root effect and the ability to breath air, or between the presence of a Root effect and the depth of occurrence (Farmer *et al.*, 1979; Powers *et al.*, 1979; Nobel *et al.*, 1986). A strong Root effect, however, seems to be typical in very active fish that are highly dependent on vision (Wittenberg and Haedrich, 1974; Dafré and Wilhelm, 1989).

A clear correlation exists between presence of the Root effect and the presence of a gas-filled swim bladder or of a choroid rete in the fish eye (Wittenberg and Haedrich, 1974; Farmer *et al.*, 1979; Ingermann and Terwillinger, 1982; Dafré and Wilhelm, 1989). In evolutionary terms, all these studies favor the hypotheses that the Root effect is originally associated with the presence of a choroid rete in the fish eye. As will be discussed in this review, the Root effect indeed is crucial for the functioning of the swim bladder and for the oxygen supply to the retina. In elasmobranchs, neither a swim bladder nor a choroid rete is present. The physiological importance of the Root effect in elasmobranchs therefore remains obscure. Nevertheless, the possibility exists that the Root shift serves to enhance oxygen delivery to at least some tissues in these animals.

B. Intraspecific Differentiation between Multiple Hemoglobins

Many fish species possess multiple hemoglobins, which may differ in structure as well as in function (cf. Brunori, 1975; Weber, 1990), and appear to be equally distributed between different erythrocytes (Brunori *et al.*, 1974). The complement and concentrations of these multiple hemoglobins vary during development. In salmonid hemoglobins, for example, anodic components constitute 95% of the hemoglobin in freshwater salmon fry blood but only around 55% in adult seawater coho salmon (Giles and Randall, 1980; Sauer and Harrington, 1988). Electrophoretically cathodic hemoglobins usually exhibit high oxygen affinities and small Bohr effects, whereas the anodic ones have lower affinities and larger sensitivities to pH (Gillen and Riggs, 1973; Weber *et al.*, 1976a,b). The anodic components show strong Root effect (Itada *et al.*, 1970; Binotti *et al.*, 1971; Pelster and Weber 1990), which may be potentiated by organic phosphates, whereas the cathodic ones completely lack this phenomenon, even in the presence of saturating concentrations of organic phosphates (Pelster and Weber, 1990). Thus salmonid fry, for example, in freshwater have blood oxygen dissociation curves that are much more sensitive to changes in pH, organic phosphates, and temperature than the adult coho salmon blood.

III. CHARACTERIZATION OF THE ROOT EFFECT

A. H^+ and CO_2 Root Effects

Metabolic activity of tissues produces and releases either CO_2 or lactic acid in aerobic or anaerobic energy metabolism, respectively, and both metabolites will acidify the blood. Whereas CO_2 readily crosses cell membranes due to its high diffusivity, protons typically require special transport systems for passage through membranes. Already in 1931, Root considered possible differences in the effect of CO_2 and fixed acid (lactic acid) on the Root effect, and reported a greater effect of lactic acid. In a later reanalysis (Root and Irving, 1943), the effects of CO_2 and lactic acid addition were reported to be similar in the marine fish *Tautoga onitis*. This finding has been confirmed for eel whole blood. In eel hemolysate, however, the CO_2 Root effect was more pronounced than the fixed acid Root effect (Bridges *et al.,* 1983). This slight CO_2 effect, observed at 25°C, but hardly detectable at 15°C, was attributed to carbamino formation and assumed to be masked in whole blood either by a different intracellular pH/plasma pH relation or by the action of organic phosphates (Bridges *et al.,* 1983).

B. Influence of Allosteric Ligands

The major cofactors modifying hemoglobin oxygen affinity by allosteric interaction in fish blood are nucleoside triphosphates (NTP, mainly ATP and GTP; cf. Weber *et al.,* 1976a). The Root effect of red cell hemolysates from several freshwater fish species is increased by organic phosphate addition. In stripped hemolysates of carp and eel, a Root effect is observed only in the presence of organic phosphates, even at pH values as low as 6.1 and 7.0 (Vaccaro Torracca *et al.,* 1977; Pelster and Weber, 1990). The magnitude of the phosphate effect depends on pH. Below pH 6.8, physiological NTP/tetrameric Hb ratios of 1–2 induce an almost maximal effect (Pelster and Weber, 1990), which indicates a saturation of the available phosphate-binding sites of the hemoglobin. GTP, which forms an extra hydrogen bond with the hemoglobin compared to ATP (Gronenborn *et al.,* 1984), is more effective than ATP in depressing oxygen affinity (Weber *et al.,* 1976a). Accordingly, GTP augments the Root effect more than ATP at a given NTP/Hb_4 ratio (Pelster and Weber, 1990). As with the Bohr effect, chloride concentrations enhance the Root effect. In the physiological concentration range, however, the influences of chloride and lactate seem to be negligible (Pelster and Weber, 1990).

C. Molecular Mechanisms

Apart from the cyclostome fish, where the hemoglobin consists of identical monomers when oxygenated but aggregates to oligomers upon deoxygenation, the hemoglobins of other vertebrates are tetrameric, consisting of two α and two β chains (see Chapter 1; (Weber)). Oxygenation is associated with breakage of inter- and intrachain salt bridges, which allow the molecule to shift from a "tense" state (which has a low affinity of oxygen and high affinity for other ligands) to a "relaxed" state (with opposite affinities). This shift is basic to cooperativity in oxygen binding between the hemes. All allosteric ligands, like organic phosphates and chloride, that reduce the oxygen affinity of the hemoglobin bind to the deoxygenated T state and thus stabilize this conformational state, impeding the T–R transition. Consequently, the molecular explanation for the Root effect was thought to be a replacement of one or more amino acid residues of the hemoglobin, which allow additional binding of protons, locking the hemoglobin in the deoxygenated T state (Perutz and Brunori, 1982; Parkhurst *et al.*, 1983).

Recent studies on the Root hemoglobin of the spot *Leiostomus xanthurus* suggest, however, that the molecular basis of the Root effect is not a stabilization of the T state, but rather a destabilization of the R state (Mylvaganam *et al.*, 1996). Crystallographic analysis of the R state of this hemoglobin revealed that seven amino acids, critical to the presence of the Root effect, in conjunction with four additional amino acids that are present in most hemoglobins, form positive-charge clusters in the $\beta_1-\beta_2$ interphase of the oxygenated tetramer. At low pH, within these clusters, protonation of histidine 147 at the terminal of the β chain destabilizes the R state and provokes a release of oxygen (see also Perutz, 1996).

D. Kinetics of the Root Effect

The kinetics of the Root effect are more complex than the kinetics of oxygen binding in red cells. Although the kinetics of oxygen binding comprise the chemical reaction and the diffusion of oxygen, the kinetics of the Root effect also require a third component, the diffusion of acid, as illustrated in Fig. 2.

The time course of the hemoglobin reaction with O_2 is almost pH independent in the cathodal hemoglobin components from the trout *Salmo*

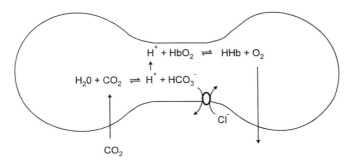

Fig. 2. Sequence of diffusional processes and chemical reactions involved in the release of oxygen from the hemoglobin via the Root effect (CA, carbonic anhydrase; Hb, hemoglobin). From Pelster *et al.* (1992), with permission.

irideus that lack a Root effect. The oxygen-binding kinetics of the anodic hemoglobins, however, which exhibit a Root effect, are biphasic, showing a slow and a fast phase. In species with a strong Root effect, the difference between the two rates may be up to 40-fold at low pH, while it is usually abolished at high pH by a dominance of the slow phase (Giardina *et al.*, 1973; Riggs, 1979). Lowering the pH from 7.5 to 7.0 causes a more than 5-fold increase in the dissociation rate of oxygenated Hb IV. In the marine teleost *Leiostomus xanthurus*, which has a single Root effect hemoglobin component, a pH decrease from 7 to 6 drastically raises the oxygen dissociation constant k (from about 50 to 340 s^{-1}) and decreases the oxygen combination velocity constant k' by about half (Bonaventura *et al.*, 1976).

Such biphasicity and the wavelength dependence of the time course of ligand binding suggest differential effects of pH on the hemes of the α and β chains of the hemoglobin molecules (Bonaventura *et al.*, 1976). The α and β subunits may represent the two roughly equal populations of heme groups with markedly different ligand-binding affinities observed in deep-sea fish species with swim bladders (Noble *et al.*, 1986). Heterogeneity in the oxygen affinities of the heme groups, moreover, may account for the negative cooperativity observed at low pH in Root effect hemoglobins (cf. Noble *et al.*, 1975).

The diffusion of oxygen into or out of the red cell is much slower than the chemical reaction with hemoglobin and represents the rate-determining step in the kinetics of red cell oxygenation (Heidelberger and Reeves, 1990; Pelster *et al.*, 1992). A detailed analysis of the kinetics in eel blood using a unicellular thin layer of red cells revealed a time course of the Root effect that was much slower than the oxygenation/deoxygenation reaction (Pelster *et al.*, 1992). There was no difference between the Root-on and Root-off reactions. Half times for the Root-on and Root-off reaction in eel blood were about 45–65 ms (Pelster *et al.,* 1992). In these experiments, the acidification was achieved by an increase in P_{CO_2}. With increasing P_{CO_2}, HCO_3^- and proton concentrations of the red cell will rapidly increase due to carbonic anhydrase activity. Protons will bind to the hemoglobin, causing a transition to the deoxygenated state and oxygen is released irrespective of the high P_{O_2}. Although CO_2 readily diffuses through membranes, acidification of the blood by fixed acid, i.e., the release of lactic acid produced by anaerobic metabolism, requires additional ion transport mechanisms within cell membranes, which typically have a low proton permeability. During alkalinization, all these reactions are reversed. The decrease in red cell proton concentration results in a release of protons bound to the hemoglobin, and the oxygenated state of the pigment can be reestablished.

Given this rather complex sequence of reactions and the kinetic properties determined *in vitro,* the important question now is whether the Root effect can reach an equilibrium state during passage of the swim bladder tissue or of the eye. Circulation times of the various sections of the swim bladder circulatory system have not yet been determined, but based on blood flow measurements and morphometric data of the rete mirabile of the European eel, Pelster *et al.* (1992) estimated that blood passage through the arterial section of the rete mirabile would take more than 1 s. Given the larger diameter of the venous capillaries, transit times for venous vessels of the rete should be even longer. The authors conclude that due to the reaction half-times of about 45 to 65 ms for the Root kinetics, the Root effect should be close to equilibrium after passage of the rete mirabile. For the fish eye, no information on blood flow is available.

IV. ROOT EFFECT AND SWIM BLADDER FUNCTION

A. The Swim Bladder as a Hydrostatic Organ

In many fish species, a gas-filled swim bladder serves as a hydrostatic organ to achieve neutral buoyancy. The swim bladder is equipped with a flexible wall, which means that pressure and volume change, according to Boyle's law, with changes in hydrostatic pressure. Hydrostatic pressure increases by 1 atm for every 10 m of water depth. In order to keep the swim bladder volume constant and to retain a status of neutral buoyancy during vertical migrations, fish must either reabsorb gas from the swim bladder while ascending or deposit gas while descending. To accomplish both tasks, the swim bladder typically consists of a resorbing and a secretory section (Fig. 3). The resorbing section of the bladder can be the oval, which can be closed off from the rest of the bladder by muscular activity, or even a special separate chamber. Within the resorbing section of the swim bladder, gas is reabsorbed as in an alveolar lung by diffusion along partial pressure gradients. In the swim bladder gas, partial pressures are higher than in arterial blood, because the partial pressure of gases dissolved in water, and thus in arterial blood, hardly increases with water depth (Enns *et al.*, 1967).

Deposition of gas is necessary, even in fish staying at the same depth, to compensate for diffusional loss of gas out of the swim bladder, as well as to keep the swim bladder volume constant to the face of increasing hydrostatic pressure when descending. Gas deposition is also a diffusional process with gases diffusing from the swim bladder capillaries into the swim bladder lumen. Depending on water depth, hydrostatic pressure in the swim bladder may be as high as several hundred atmospheres, whereas the partial pressure of gases in arterial blood hardly increases with water depth and hardly ever exceeds 1 atm, even at great depth. To establish a diffusion gradient toward the swim bladder lumen, very high gas partial pressures— exceeding the partial pressures in the swim bladder—must be established in swim bladder blood.

B. Generation of High Oxygen Partial Pressures

In the swim bladder, high partial pressures of gas are generated in two steps: an initial increase in partial pressure and a subsequent countercurrent multiplication (Kuhn *et al.*, 1963). The initial increase in partial pressure

Physoclist fishes (perch)

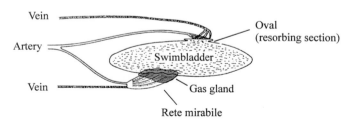

Fig. 3. Swim bladder morphology of a physoclist fish, showing the perch as a typical example. Adapted from Denton (1961). Characteristical features involved in the deposition of gases are a rete mirabile and the gas gland. For gas reabsorption, a special section of the swim bladder is responsible (oval), which has its own blood supply, bypassing the rete mirabile.

is achieved by a decrease in physical gas solubility and/or a release of gas molecules from a chemical binding site; for oxygen, this is achieved by liberation from the hemoglobin via the Root effect.

To initiate the Root effect, blood must be acidified during passage through the swim bladder, and this is achieved by the metabolic and secretory activity of gas gland cells in the swim bladder epithelium. Gas gland cells are cuboidal or cylindrical with a size ranging from $10-25$ μm to giant cells of $50-100$ μm. Gas gland cells may be lumped together, forming a compact gas gland as in cod, or spread over the whole swim bladder epithelium as in the eel. They are polar with extensive basal membrane foldings, typical for secretory cells. In contrast to other secretory cells, however, these infoldings are not associated with large numbers of mitochondria. Only a few microvilli are observed on the luminal surface of gas gland cells (cf. Pelster, 1997).

The metabolism of gas gland cells is highly specialized for the production of acidic metabolites. It is fueled by blood glucose, which, even under hyperoxic conditions, is largely converted into lactic acid (D'Aoust, 1970; Boström et al., 1972; Ewart and Driedzic, 1990; Pelster and Scheid, 1993; Pelster, 1995a). Although aerobic metabolism appears to be almost negligible in gas gland cells, some of the glucose is decarboxylated by the enzyme 6-phosphogluconate dehydrogenase in the pentose phosphate shunt, forming CO_2 without concomitant consumption of oxygen (Walsh and Milligan, 1993; Pelster et al., 1994).

Lactic acid as well as CO_2 are released into the bloodstream (Steen, 1963; Pelster and Scheid, 1993) and acidify the blood. Though CO_2 easily diffuses into the blood and into the red cells, the situation is more complex for lactic acid. Experiments on gas gland cells in primary culture suggest that the acid secretion from these cells may involve sodium-dependent pathways and in part be due to the activity of a V-ATPase (Pelster, 1995b). This acid release results in a remarkable acidification of the blood (Fig. 4). In the European eel, pH values as low as $6.6-6.8$ have been measured in blood after passage of the gas gland cells (Steen, 1963; Kobayashi et al., 1990).

In vitro measurements characterizing the oxygen-binding properties of Root effect hemoglobins have shown that blood pH values below 7.0 usually are sufficient to provoke a maximal Root effect. In the European eel, for example, anodic components, which exhibit a Root effect, make up about 60% of total hemoglobin and can be deoxygenated at pH values below 7.0 (Pelster and Weber, 1990). Titration of blood with CO_2 indeed revealed that about $40-50$% of eel blood is deoxygenated at an extracellular pH of 7.1 or below (Pelster et al., 1990). Given a hematocrit of about $20-30$%, typically observed in fish, deoxygenation of 40% of the respiratory pigment

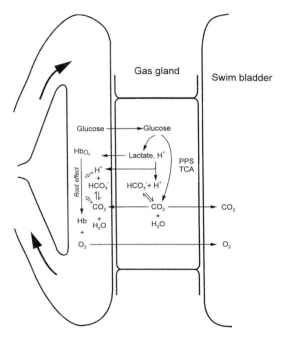

Fig. 4. Metabolic end products of glucose in gas gland cells are lactate (lactic acid) and CO_2, formed in the glycolytic pathway and in the pentose phosphate shunt (PPS) and tricarboxylic acid cycle (TCA), respectively. Both metabolites are released into the bloodstream, causing an acidification. This acidification induces the Root shift and thus the release of oxygen from the hemoglobin, resulting in an increase in blood P_{O_2}. According to Pelster and Scheid (1992), with permission.

results in a remarkable increase in oxygen partial pressure (Pelster and Weber, 1991).

The time courses of the initiation of the Root effect caused by the release of CO_2 and of lactic acid should be different. Diffusion of CO_2 and the presence of carbonic anhydrase activity in the extracellular space (Pelster, 1995b), as well as in the erythrocyte, will allow for a rapid increase in red cell proton concentration, and thus for a rapid activation of the Root effect. On the other hand, membranes are not easily penetrated by protons so that ion transfer is required to bring the protons out of the gas gland cells and into the red cell. In consequence, proton transfer into the red cell is much slower than diffusion of CO_2, possibly resulting in a transient disequilibrium in extracellular pH. In this case, liberation of oxygen from the hemoglobin would occur only after leaving the capillary system of the secretory bladder, when the blood is already on its way to the venous rete mirabile. Gas deposition into the swim bladder initially would be diminished, but oxygen back-diffusion in the rete mirabile would be enhanced so that the oxygen is not lost for gas deposition.

Acid production and release from the gas gland cells cause a significant acidification of the blood, but careful analysis of the acid–base status of the blood during passage of the swim bladder revealed that, in the European eel, blood pH decrased from 7.82 ± 0.06 to 7.33 ± 0.04 during arterial passage of the rete mirabile (Kobayashi et al., 1990); i.e., the blood was acidified even before reaching the gas gland cells. This acidification is caused by back-diffusion of acid—mainly CO_2—in the rete mirabile (Pelster et al., 1990; Kobayashi et al., 1990). The high rate of CO_2 production and release from gas gland cells assures that the P_{CO_2} in blood returning to the venous rete mirabile exceeds arterial P_{CO_2} and establishes partial pressure gradient for CO_2 from the venous to the arterial rete capillaries. The formation of CO_2 in the pentose phosphate shunt of the gas gland cells not only contributes to the acidification of blood during passage of the gas gland cells, but also sets the stage for back-diffusion of CO_2 and acidification of the blood in the countercurrent system.

In analyzing the Root effect, both steps have to be taken into account, as shown in Fig. 5. At pH 7.8, observed in the arterial blood supplying the swim bladder, almost complete saturation of the hemoglobin can be expected. Acidification of the blood down to pH 7.3, induced by back-diffusion of acid during passage of the rete, is sufficient to reduce the oxygen-carrying capacity of the hemoglobin by about 20%. The release of acid from gas gland cells into the blood causes a further decrease in pH to 7.1 and adds another 15–20% reduction (Fig. 5). Thus, the acidification of blood in two steps during passage of the swim bladder ensures an almost maximal reduction in the hemoglobin oxygen-carrying capacity.

The importance of acid back-diffusion in the rete mirabile for the generation of high P_{O_2} values was demonstrated by experiments of Kobayashi et al. (1990), in which, due to a high rate of oxygen deposition into the swim bladder, only a very small P_{O_2} gradient was measured between the venous blood returning to the rete mirabile and arterial blood leaving it. Nevertheless, P_{O_2} increased remarkably during arterial passage of the rete and dropped in the venous capillaries. These P_{O_2} changes, however, were not accompanied by changes in oxygen content, precluding oxygen back-diffusion as the underlying mechanism. Thus, the observed sevenfold increase in P_{O_2} during arterial passage of the rete was induced solely by acid back-diffusion in the rete, switching on the Root effect and partially deoxygenating the hemoglobin (Kobayashi et al., 1990). According to the basic concept of countercurrent concentration, the acidification of blood achieved by acid release from gas gland cells represents the single concentrating

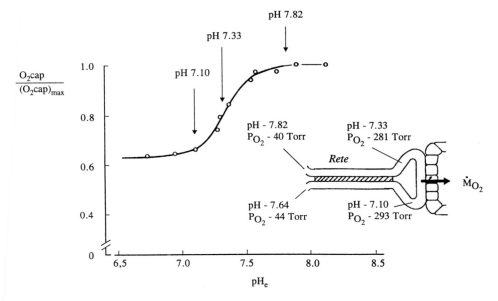

Fig. 5. The decrease in the ratio O_2 cap/$(O_2$ cap$)_{max}$ with decreasing pH in whole blood of the European eel (*Auguilla anguilla*) *in vitro,* and pH and P_{O_2} values in blood samples taken from swim bladder vessels of an *in situ* preparation of the eel. According to the pH values actually measured in swim bladder blood and the *in vitro* oxygen-binding characteristics of the blood, a severe decrease in the oxygen-carrying capacity can be predicted to occur in the blood during passage of the swim bladder. Modified after Pelster and Weber (1991).

effect (Kuhn *et al.,* 1963), the initial increase in gas partial pressure induced by a decrease in the effective gas transport capacity of hemoglobin (without change in the whole blood gas content). In a second step, this initial increase in gas partial pressure is then multiplied by back-diffusion in the countercurrent system, which results in increasing gas partial pressures and gas concentrations on the arterial side of the countercurrent system (Kuhn *et al.,* 1963; Pelster and Scheid, 1992). The magnitude of the final gas partial pressure that can be achieved largely depends on the magnitude of the single concentrating effect (Kobayashi *et al.,* 1989). Due to the action of the Root effect, the single concentrating effect for oxygen by far exceeds that of inert gases and CO_2 (Pelster *et al.,* 1990). As model calculations have shown, extremely high P_{O_2} values can be achieved by such a large single concentrating effect and subsequent countercurrent multiplication, certainly sufficient to explain the presence of fishes with a gas-filled swim bladder at a water depth of several thousand meters (Kuhn *et al.,* 1963; Kobayashi *et al.,* 1989, 1990).

V. OXYGEN-CONCENTRATING MECHANISMS IN THE FISH EYE

The fish eye is another organ that has attracted much attention with regard to the Root effect. A capillary network has been known to exist in the fish eye for almost two centuries (Albers, 1806). Jones (1838) called this structure the choroid rete mirabile, and the countercurrent arrangement of blood vessels in the choroid rete was confirmed subsequently (Müller, 1839; Barnett, 1951). An additional structure, which has long been discussed in context with the eye, is the pseudobranch. This gill-like hemibranch receives arterialized blood from the first gill arch. The efferent pseudo-branch artery gives rise to the arteria ophthalmica magna; i.e., the fish eye receives its blood supply from the pseudobranch. Analyzing 282 species, Müller (1839) realized that there was a connection between these two organs, that is, fish with a pseudobranch usually also had a choroid rete mirabile, an observation later confirmed by Wittenberg and Haedrich (1974). These authors found a choroid rete in a larger number of teleosts, but only in *Amia calva* among the nonteleosts analyzed. Although swim bladder rete mirabile and choroid rete mirabile appear to exist indepen-dently, the presence of the choroid rete mirabile was found to be linked to the presence of the pseudobranch.

The functional significance of this puzzling arrangement was not obvi-ous and is still open to speculation today. Even though P_{O_2} measurements in the vitreous humor of the eye of fish without a choroid rete usually did not exceed arterial P_{O_2}, in species with choroid rete P_{O_2} values of up to 1 atm have been reported and, in *Gadus morhua* and *Echeneis naucrates,* P_{O_2} values of almost 2 atm were found (Wittenberg and Wittenberg, 1962, 1974; Fairbanks *et al.,* 1969). This observation can be explained only if it is accepted that the Root effect is involved. Before discussing our scanty knowledge on the physiology of this organ, we need to consider the mor-phology of this system.

A. Circulation to the Fish Eye

The ophthalmic artery springs from the pseudobranch, pierces the sclera of the eyeball, and passes through the lumen of the wide ophthalmic venous sinus (Fig. 6). There it branches into two vessels, each supplying one limb of the horseshoe-shaped rete. The arterial capillaries of the choroid rete mirabile arise almost directly from each of these limbs. Occasionally giving off collateral branches, the capillaries radiate toward the periphery of the gland and coalesce to form a capillary network, the choriocapillaris, from

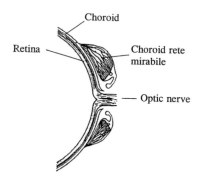

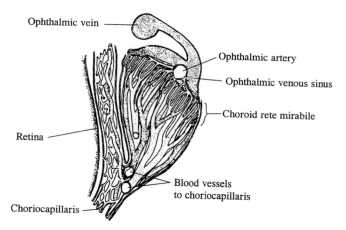

Fig. 6. Schematic drawing of the choroid rete mirabile and its location within the eye of the bluefish. According to Wittenberg and Wittenberg (1974), with permission.

which the retina is supplied. In contrast to other vertebrates, the retina of most teleosts is not vascularized and diffusion distances for nutrient supply to the retina therefore can be more than six times as high as in primate retinas (Wittenberg and Wittenberg, 1974; Nicol, 1989). The blood is returned from the choriocapillaris network by venous capillaries running parallel to the arterial capillaries in a sort of checkerboard arrangement with countercurrent flow, finally draining into the large venous sinus, from which the ophthalmic vein emerges (Fig. 6).

B. Oxygen Supply to the Fish Eye

Experimental evidence demonstrates the importance of hyperoxic conditions in the fish eye: elimination of the high oxygen tensions perturbs the electroretinogram characteristic of the teleost retina within minutes (Fonner *et al.*, 1973). Morphological similarities of the swim bladder and the choroid rete mirabile, together with the presence of hyperoxic P_{O_2} values in the vitreous humor of the fish eye, have led to the assumption that the choroid rete mirabile, just like the swim bladder rete mirabile, was essential for the generation of high P_{O_2} values in order to assure the oxygen supply to the avascular retina (Wittenberg and Wittenberg, 1974).

To initiate the Root effect and to use the choroid rete as a countercurrent exchanger, blood must be acidified in the eye after passing through the arterial choroid rete and before returning to the venous vessels of the rete. Studies on the metabolism of the teleost retina revealed that the tissue produces lactic acid even under atmospheric oxygen pressures, although to a lesser extent than under anoxia, indicating the existence of a Pasteur effect (Baeyens *et al.*, 1971). Furthermore, in addition to aerobic glucose oxidation and anaerobic lactate production, part of the glucose is metabolized in the pentose phosphate shunt (Hoffert and Fromm, 1970), which is also of importance in the swim bladder. Thus, analogous to the situation in the swim bladder, it would appear that the retina produces lactic acid and CO_2, which are released into the choriocapillaris vessels. This acidification decreases the effective oxygen-carrying capacity of the blood by initiating the Root effect. The resulting initial increase in P_{O_2} (single concentrating effect) may then be multiplied by oxygen back-diffusion in the choroid rete.

The role of CO_2 is not fully understood. Administration of carbonic anhydrase inhibitors rapidly abolishes the high oxygen partial pressure in the eye (Fairbanks *et al.*, 1969, 1974; Fonner *et al.*, 1973). The retinal cells, however, have been reported to be damaged rapidly at low pH values. Inhibition of carbonic anhydrase with an acidification down to pH 6.4 resulted in blindness (Maetz, 1956). The buildup of high P_{O_2} values to assure an oxygen supply to the retina requires acidification of the blood to initiate the Root effect, but pH must not become too acidic in order not to impair the proper functioning of the retina. This may preclude the possibility of acidifying the blood in the retina by releasing large quantities of CO_2, which because of its high diffusivity, would be concentrated in a countercurrent system, resulting in very high P_{CO_2} and low pH levels in the retina, in a manner similar to that demonstrated in the rete mirabile of the swim bladder (Kobayashi *et al.*, 1990).

On the other hand, high concentrations of carbon dioxide, *per se*, may not be the problem because tissues can adjust bicarbonate levels to regulate pH. It is sudden increases in carbon dioxide that reduce pH in tissues. Many teleost hemoglobins have a reduced buffering capacity, decreasing the amount of carbon dioxide required to cause a Root shift. This, coupled with a high capacity to regulate retinal intracellular pH, may protect the retina from acid damage. In addition, carbon dioxide levels in the retina could be high as long as they do not change very much, and such stable levels are possible in an organ with a constant oxygen uptake, carbon dioxide production, and blood flow. If carbon dioxide production matched carbon dioxide loss through the circulation, then levels in the eye could be high and stable due to the presence of the choroid rete. The blood flowing through the organ would see changes in P_{CO_2} but the CO_2 level, and therefore pH, in the retina would be relatively stable over time. If this is the case, then oxygen secretion in the eye could be similar to that observed in the swim bladder.

If the eye cannot be exposed to high carbon dioxide levels, however, the choroid rete in conjunction with the pseudobranch must develop high oxygen tensions without the simultaneous accumulation of high concentrations of CO_2 (Wittenberg and Haedrich, 1974; Fairbanks et al., 1974). According to the model of Fairbanks et al. (1974) carbonic anhydrase, located on the luminal surface of the endothelium of the venous rete capillaries, hydrates CO_2 to give H^+ and bicarbonate, thus diminishing the countercurrent concentration of CO_2. The HCO_3^- is then transferred to the arterial capillaries by anion (chloride) exchange, where it becomes available for buffering part of the retinal lactic acid, thus preventing inordinate pH decreases. Opposite to the model reported for the eel swim bladder rete mirabile, this would imply only a one-step acidification in the choriocapillaris. Furthermore, though the swim bladder rete mirabile appears to be a passive exchanger, where concentration gradients and diffusion of gases and ions dominate, this model would require anion exchange, which has not yet been described for the choroid rete. The idea that the pseudobranch preconditions the blood before it reaches the fish eye was forwarded many years ago (Leiner, 1938; Wittenberg and Haedrich, 1974), but only recent studies have shed some light on possible mechanisms. Using C_1-^{14}C- and C_6-^{14}C-labeled glucose, Hoffert and Fromm (1970) demonstrated that in the pseudobranch of the trout some of the glucose was decarboxylated in the pentose phosphate shunt, and measurement of enzyme activities indeed revealed significant activities of glucose-6-phosphate dehydrogenase, a key enzyme of the pentose phosphate shunt, in this tissue (Bridges et al., 1998). In saline- or blood-perfused preparations of the trout pseudobranch, CO_2 production largely exceeded the concomitant O_2 consumption, giving RQ values significantly higher than 1.0 (Berenbrink, 1994). In consequence, the blood is acidified during passage of the pseudobranch. The hypotheses emerging from these findings predict that in the pseudobranch blood is acidified by release of CO_2 to pH values just above the threshold for the onset of the Root effect (Berenbrink, 1994). Given the present findings on the physiology of the fish eye, an ideal situation would be that the retina would have to release only a small amount of acid to switch on the Root effect. The resulting increase in P_{O_2} would then be multiplied by countercurrent concentration in the choroid rete mirabile. Countercurrent concentration of acid, especially of CO_2 would be limited, because only a slight acidification was necessary.

In order to be able to acidify the blood to just above the threshold for the onset of the Root effect, blood gases must be measured. Chemosensitivity responding to P_{O_2}, P_{CO_2}, and pH has indeed been located in the pseudobranch (Laurent and Rouzeau, 1972; Randall and Jones, 1973; Laurent and Dunel-Erb, 1984). Berenbrink (1994) and Bridges *et al.* (1998) favor the idea that P_{O_2} should be measured and a slight increase in P_{O_2} would stop further acidification. As the onset of the Root effect is pH dependent, pH also could be an effective signal, but both strategies have their disadvantages. Given the variability in environmental water conditions, arterial P_{O_2} typically is quite variable in fish, interfering with a hypothetical P_{O_2} sensor. pH and P_{CO_2} also are variable, and adrenergic stimulation of red cells modifies the H^+ balance between plasma and red cells, shifting the threshold for the onset of the Root effect toward lower extracellular pH values (see Chapter 2; (Nikinmaa)).

The pseudobranch is located anterior to the choroid rete and, ideally, should lower blood pH without changes in blood gas partial pressures. This is because a countercurrent exchange system, which allows for an effective concentration of gases and metabolites in an organ, is an equally effective shunt when coming from the other side. For example, if the acidification in the pseudobranch surpasses the threshold for the Root effect, oxygen would be released and P_{O_2} would increase. Due to the resulting high P_{O_2} in arterial blood entering the choroid rete, O_2 would diffuse to the venous vessels of the rete and thus return to the central cardiac system before even reaching the retina (= diffusion shunt). The same applies to CO_2. Acidification of the blood by release of CO_2 clearly results in an increase in blood P_{CO_2}. In consequence, if the arterial P_{CO_2} exceeds venous P_{CO_2} in blood leaving the choroid rete, back-diffusion of CO_2 from the arterial to the venous capillaries of the choroid rete will realkalinize the blood before it reaches the retina.

Obviously, there is a difference in the need to establish very high P_{O_2} values in the swim bladder and the eye. In the swim bladder, high P_{O_2} values are necessary to establish a diffusion gradient toward the swim bladder, and the further the fish descends, the higher the P_{O_2} values that are needed. In the eye, high P_{O_2} values are necessary for satisfying the oxygen demand of the retina. However, as soon as the P_{O_2} in the choriocapillaris is high enough to meet the oxygen demand of the whole retina via diffusion, no further increase in P_{O_2} is necessary. A crucial parameter limiting the maximum gas partial pressure that can be achieved in a countercurrent system is the length of the capillaries in a countercurrent exchanger (Kobayashi *et al.*, 1989). A comparison of the choroid rete and the swim bladder rete reveals interesting differences. Whereas the capillaries of the choroid rete are relatively short and have similar lengths in all species analyzed (Wittenberg and Wittenberg, 1974), the length of the swim bladder rete capillaries usually is several times greater and exhibits a positive correlation with the depth of habitat, i.e., with the pressure inside the bladder (Marshall, 1960; Denton, 1961). Therefore, a large Root effect allowing for a large single concentrating effect, in combination with a short rete mirabile limiting the effectiveness of the countercurrent multiplication, may be the solution for the fish eye. It allows for the generation of hyperoxic pressures, but alleviates the problems associated with acid back-diffusion that might otherwise result in an overacidification of the retinal tissue, causing damage.

VI. ADRENERGIC EFFECTS

A. Catecholamines and Erythrocytic pH

Although hemoglobin oxygenation is influenced by erythrocytic pH, many studies, because of ease of measurement, report blood rather than erythrocytic pH. Most of the time, however, erythrocytic pH, although more acidic, varies with plasma pH. Thus, in most cases, plasma pH is a good indicator of erythrocytic pH. In many fish, however, increased levels of circulating catecholamines raise erythrocytic pH and decrease plasma pH, as discussed in detail in Chapter 2; (Nikinmaa) and in Nikinmaa (1990). This action of catecholamines effectively resets the relationship between erythrocytic and plasma pH, to maintain blood oxygen saturation during a period of general blood plasma acidosis.

The functional significance of having a Root effect hemoglobin is that local acid production can cause elevated oxygen levels in tissues such as the swim bladder and eye. If the acidosis is general, however, then the presence of a Root effect could be a liability. Acid conditions in the plasma, if transferred to the red blood cell, could initiate a Root shift that would prevent the full oxygen-carrying capacity of the hemoglobin from being utilized. Burst activity in fishes is maintained by anaerobic metabolism and results in a large and general blood acidosis. These acidic conditions in the plasma, if transferred to the red blood cell, would initiate a Root shift. Thus, burst activity might be expected to reduce the oxygen-carrying capacity of the blood in those fish with Root effect hemoglobins, which in turn would be expected to impair aerobic swimming ability. Burst swimming, however, had no effect on subsequent aerobic swimming capacity in coho salmon (Randall *et al.*, 1987). In fact, burst swimming can result in an elevation of blood oxygen-carrying capacity, even though there is a marked blood acidosis (Primmett *et al.*, 1986). This increase in blood oxygen-carrying capacity is due to an increase in circulating catecholamines that raise red blood cell pH and cause the release of red blood cells from the spleen (Randall and Perry, 1992).

Elevated blood catecholamines shut down blood flow to the rete in the eel (Pelster, 1994) and gas secretion to the swim bladder is curtailed, but in the face of a general acidosis, the oxygen-carrying capacity of the blood is maintained by an adrenergically mediated rise in erythrocytic pH. There is no information on the effects of adrenergically elevated red blood cell pH on oxygen levels in the eye, but interestingly, the adrenergically regulated increase in red blood cell pH is inhibited by high oxygen tensions so that oxygen delivery to the eye is probably maintained, even when circulating levels of catecholamines are high. In this light, any cessation of oxygen secretion into the swim bladder during a general acidosis must be viewed as conserving oxygen supplies for other purposes, presumably exercise, rather than an inability to secrete oxygen.

B. The Root and Haldane Effects

Hemoglobins with a large Root effect often possess a large Haldane effect (see Chapter 6); that is, oxygenation of the hemoglobin molecule results in a large production of protons. In trout, carp, and tench, this large Haldane effect is also associated with a low hemoglobin-buffering capacity (Jensen, 1989). Thus, oxygenation of erythrocytes will lead to a marked release of protons from hemoglobin and a fall in intracellular pH. This has the potential to limit oxygenation of the hemoglobin via the Root shift. Under physiological conditions, however, bicarbonate flux through the erythrocyte removes the protons and prevents a significant fall in intracellular pH. That is, to prevent the Root shift from limiting hemoglobin oxygen-carrying capacity, oxygen flux must be coupled to bicarbonate flux in the erythrocyte. This is the case in trout, where there is tight coupling of oxygen and carbon dioxide flux through the red blood cell (see Chapter 6), enhanced by the fact that there is almost no bicarbonate dehydration within the plasma as blood flows through the gills, due to the absence of any carbonic anhydrase activity on the walls of the gill blood vessels (see

Chapter 9). In fact, the large Haldane effect and low hemoglobin-buffering capacity ensure that oxygenation produces sufficient protons for bicarbonate dehydration within the erythrocyte. In addition, the reduced hemoglobin-buffering capacity maximizes the effectiveness of the Root shift, decreasing the acid production required for full expression of the Root effect. This may afford some protection to the retina against the damaging effects of low pH.

On the other hand, dogfish hemoglobin lacks a Root shift and has a much higher buffering capacity and a much smaller Haldane effect than hemoglobins from trout and carp (Jensen, 1989). Carbonic anhydrase is embedded in the dogfish branchial endothelium and, unlike the situation in teleost gills, catalyzes plasma bicarbonate dehydration in blood perfusing the gills (Swenson *et al.*, 1995, 1996). This branchial carbonic anhydrase in dogfish plays a significant role in carbon dioxide excretion, reducing flux through the red blood cell and uncoupling the linkage between carbon dioxide and oxygen transport (see Chapter 6). There is even carbonic anhydrase activity in dogfish plasma (Wood *et al.*, 1994), as well as on the branchial endothelial lining. Carbonic anhydrase is available to the plasma perfusing the respiratory circulation in both mammals and dogfish, and both have hemoglobins that lack a Root shift!

Thus, the presence of a Root shift, at least in trout blood, is associated with the absence of any carbonic anhydrase activity available to the plasma flowing through the gills. The distribution of carbonic anhydrase activity in the gills is associated with the mechanisms of ion exchange across the gills. Freshwater trout gills consist of pavement, mucus, and chloride cells. The pavement cells, which make up most of the gill epithelium, have an apical proton ATPase (Laurent *et al.*, 1994; Lin *et al.*, 1994; Lin and Randall, 1995) associated with an amiloride-sensitive sodium channel. The proton ATPase appears to energize sodium uptake, which is then removed from the cell by a basal Na^+/K^+-ATPase. The pavement cell has an apical distribution of carbonic anhydrase that decreases toward the baso-lateral membrane (Rahim *et al.*, 1988) and parallels the proton ATPase activity. Carbonic anhydrase activity is unavailable to the plasma in trout (Gilmour *et al.*, 1994). Acetazolamide (a carbonic anhydrase inhibitor) inhibits proton excretion across the gills of trout, but not proton ATPase *in vitro* (Lin and Randall, 1991), indicating that carbonic anhydrase serves to supply protons to the apical proton ATPase. Thus a small part of the carbon dioxide excreted from the blood across the gills is hydrated within the gill epithelium, supplying protons for the proton pump. If carbon dioxide is the source of protons, then an equivalent flux of bicarbonate must also exit the cell. There is a net acid excretion across the gills associated with the activity of the proton ATPase (Lin and Randall, 1991), indicating that the removal of bicarbonate probably occurs across the basal, rather than apical, membrane of the cell. There is, however, no direct evidence for anion transfer across the basolateral membrane of the teleost gill pavement cell.

If there is an anion exchange mechanism in the basal membrane of the pavement cell, however, acid can be transferred across the baso-lateral membrane via the Jacob–Stewart cycle (Lessard et al., 1995). The absence of carbonic anhydrase in the basal regions of the cell and in the membrane facing the plasma will slow the transfer of protons between the plasma and the gill epithelium because the rate-limiting step in the transfer of acid will be the uncatalysed carbon dioxide hydration/dehydration reaction in the plasma, which is very slow. If carbonic anhydrase were more generally distributed around the basal membrane, then pH changes in the plasma would be readily transferred to the gill epithelium and the activity of the proton pump, which is pH sensitive, would be a slave of plasma pH.

The absence of carbonic anhydrase on the gill endothelium protects the epithelial proton ATPase from the oscillations in plasma pH and at the same time promotes coupling of oxygen and carbon dioxide flux through the erythrocyte. This tight coupling of oxygen and carbon dioxide movements within the erythrocyte prevents hemoglobin oxygenation from limiting oxygen capacity via a Root shift, and any positive attributes of a Root shift could now be realized. That is, conditions for gas and ion exchange across the freshwater teleost gill favor the selection of Root shift hemoglobins, whereas in most other vertebrates conditions are against the retention of Root shift hemoglobins.

The presence of a Root shift is advantageous because it enhances oxygen transfer into the eye and swim bladder. This effect, however, need not be restricted to the eye and swim bladder. If carbon dioxide entry into the blood is more rapid then oxygen loss to the tissues, then carbon dioxide entering capillary blood could reduce erythrocytic pH and drive up P_{O_2} to values higher than that in arterial blood. These large oxygen gradients generated by the presence of a Root shift may explain the reduced capillary to mitochondrial density reported for some aerobic fish tissues compared with those in mammals (Brauner and Randall, 1996). The large P_{O_2} gradient ensures oxygen supply over large diffusion distances, as in the teleost eye. Reduced capillary exchange in the face of large oxygen gradients may permit normal rates of oxygen uptake, but may restrict ion and water transfer, particularly between muscle and blood following exhaustive exercise.

REFERENCES

Albers, J. A. A. (1806). Über das Auge des Kabeljau Gadus morhua und die Schwimmblase der Seeschwalbe, Trigla hirundo. Göttinger gelehrte Anzeiger **2,** 681–682.

Baeyens, D. A., Hoffert, J. R., and Fromm, P. O. (1971). Aerobic glycolysis and its role in maintenance of high O_2 tensions in the teleost retina. *Proc. Soc. Exp. Biol. Med.* **137,** 740–744.

Baines, G. W. (1975). Blood pH effects in eight fishes from the teleostean family scorpaenidae. *Comp. Biochem. Physiol. A* **51,** 833–843.

Barnett, C. H. (1951). The structure and function of the choroidal gland of teleostean fish. *J. Anat.* **85,** 113–119.

Berenbrink, M. (1994). Die Kontrolle des intrazellulären pH in den Erythrozyten von Knochenfischen. Dissertation, Math.-Nat. Fakultät Univ. Düsseldorf, Düsseldorf: Verlag Shaker.

Binotti, I., Giovenco, S., Giardina, D., Antonini, E., Brunori, M., and Wyman, J. (1971). Studies on the functional properties of fish hemoglobins. II. The oxygen equilibrium of the isolated hemoglobin components from trout blood. *Arch. Biochem. Biophys.* **142,** 274–280.

Bonaventura, C., Sullivan, B., and Bonaventura, J. (1976). Spot hemoglobin. Studies on the Root effect hemoglobin of a marine teleost. *J. Biol. Chem.* **251,** 1871–1876.

Boström, S. L., Fänge, R., and Johansson, R. G. (1972). Enzyme activity patterns in gas gland tissue of the swim bladder of the cod (*Gadus morrhua*). *Comp. Biochem. Physiol. B* **43,** 473–478.

Brauner, C. J., and Randall, D. J. (1996). The interaction between oxygen and carbon dioxide movements in fishes. *Comp. Biochem. Physiol. A* **113,** 83–90.

Bridges, C. R., Berenbrink, M., Müller, R., and Waser, W. (1998). The physiology and biochemistry of the pseudobranch—an unanswered question? *Comp. Biochem. Physiol.* 119A: 67–77.

Bridges, C. R., Hlastala, M. P., Riepl, G., and Scheid, P. (1983). Root effect induced by CO_2 and by fixed acid in the blood of the eel, *Anguilla anguilla. Respir. Physiol.* **51,** 275–286.

Bridges, C. R., Pelster, B., and Scheid, P. (1984). Root effect in vertebrate haemoglobins and invertebrate haemocyanin: Presence and function. *Pflügers Arch* **400,** R68.

Brittain, T. (1985). A kinetic and equilibrium study of ligand binding to a Root-effect haemoglobin. *Biochem. J.* **228,** 409–414.

Brittain, T. (1987). The Root effect. *Comp. Biochem. Physiol. B* **86,** 473–481.

Brix, O. (1982). The adaptive significance of the reversed Bohr and Root shift in blood from the marine gastropod, *Buccinum undatum. J. Exp. Zool.* **221,** 27–36.

Brunori, M., Giardina, B., Antonini, E., Benedetti, P. A., and Bianchini, G. (1974). Distribution of the haemoglobin components of trout blood among the erythrocytes: Observations by single-cell spectroscopy. *J. Mol. Biol.* **86,** 165–169.

Brunori, M. (1975). Molecular adaptation to physiological requirements: The hemoglobin system of trout *In* "Current Topics in Cellular Regulation" (B. L. Horecker and E. R. Stadtman, eds.), pp. 1–39. Academic Press, New York.

D'Aoust, B. G. (1970). The role of lactic acid in gas secretion in the teleost swim bladder. *Comp. Biochem. Physiol.* **32,** 637–668.

Dafré, A. L., and Wilhelm, D. (1989). Root effect hemoglobins in marine fish. *Comp. Biochem. Physiol. A* **92,** 467–471.

Denton, E. J. (1961). The buoyancy of fish and cephalopods. *Progr. Biophys. & Biophys. Chem.* **1,** 178–234.

Di Prisco, G. (1988). A study of hemoglobin in Antarctic fishes: Purification and characterization of hemoglobins from four species. *Comp. Biochem. Physiol. B* **90,** 631–637.

Enns, T., Douglas, E., and Scholander, P. F. (1967). Role of the swim bladder rete of fish in secretion of inert gas and oxygen. *Adv. Biol. Med. Phys.* **11,** 231–244.

Ewart, H. S., and Driedzic, W. R. (1990). Enzyme activity levels underestimate lactate production rates in cod (*Gadus morhua*) gas gland. *Can. J. Zool.* **68,** 193–197.

Fairbanks, M. B., Hoffert, J. R., and Fromm, P. O. (1969). The dependence of the oxygen-concentrating mechanism of the teleost eye (*Salmo gairdneri*) on the enzyme carbonic anhydrase. *J. Gen. Physiol.* **54,** 203–211.

Fairbanks, M. B., Hoffert, J. R., and Fromm, P. O. (1974). Short circuiting of the ocular oxygen-concentrating mechanism in the teleost *Salmo gairdneri* using carbonic anhydrase inhibitors. *J. Gen. Physiol.* **64**, 263–273.

Farmer, M., Fyhn, H. J., Fyhn, U. E. H., and Noble, R. W. (1979). Occurrence of Root effect hemoglobins in Amazonian fishes. *Comp. Biochem. Physiol. A* **62**, 115–124.

Fonner, D. B., Hoffert, J. R., and Fromm, P. O. (1973). The importance of the counter current oxygen multiplier mechanism in maintaining retinal function in the teleost. *Comp. Biochem. Physiol. A* **46**, 559–567.

Giardina, B., Brunori, M., Binotti, I., Giovenco, S., and Antonini, E. (1973). Studies on the properties of fish hemoglobins. Kinetics of reaction with oxygen and carbon monoxide of the isolated hemoglobin components from trout (*Salmo irideus*). *Eur. J. Biochem.* **39**, 571–579.

Giles, M. A., and Randall, D. J. (1980). Oxygenation characteristics of the polymorphic hemoglobins of coho salmon (*Oncorhynchus kitsuch*) at different developmental stages. *Comp. Biochem. Physiol. A* **65**, 265–271.

Gillen, R. G., and Riggs, A. (1973). Structure and function of the isolated hemoglobins of the American eel, *Anguilla rostrata. J. Biol. Chem.* **248**, 1961–1969.

Gilmour, K. M., Randall, D. J., and Perry, S. F. (1994). Acid–base disequilibrium in the arterial blood of rainbow trout. *Respir. Physiol.* **96**, 259–272.

Gronenborn, A. M., Clore, G. M., Brunori, M., Giardina, B., Falcioni, G., and Perutz, M. F. (1984). Stereochemistry of ATP and GTP bound to fish haemoglobins. A transferred nuclear overhauser enhancement, 31 P-nuclear magnetic resonance, oxygen equilibrium and molecular modelling study. *J. Mol. Biol.* **178**, 731–742.

Heidelberger, E., and Reeves, R. B. (1990). O_2 transfer kinetics in a whole blood unicellular thin layer. *J. Appl. Physiol.* **68**, 1854–1864.

Hoffert, R. J., and Fromm, P. O. (1970). Quantitative aspects of glucose catabolism by rainbow and lake trout ocular tissues including alterations resulting from various pathological conditions. *Exptl. Eye Res.* **10**, 263–272.

Ingermann, R. L., and Terwilliger, R. C. (1982). Presence and possible function of Root effect hemoglobins in fishes lacking functional swim bladders. *J. Exp. Zool.* **220**, 171–177.

Itada, N., Turitzin, S., and Steen, J. B. (1970). Root-shift in eel hemoglobin. *Respir. Physiol.* **8**, 276–279.

Jensen, F. B. (1989). Hydrogen ion equlibria in fish haemoglobins. *J. Exp. Biol.* **143**, 225–234.

Jones, T. W. (1838). On the so-called choroid gland or choroid muscle of the fish's eye. *London Med. Gaz.* **21**, 650–652.

Kobayashi, H., Pelster, B., and Scheid, P. (1989). Solute back-diffusion raises the gas concentrating efficiency in counter-current flow. *Respir. Physiol.* **78**, 59–71.

Kobayashi, H., Pelster, B., and Scheid, P. (1990). O_2 back-diffusion in the rete aids O_2 secretion in the swim bladder of the eel. *Respir. Physiol.* **79**, 231–242.

Kuhn, W., Ramel, A., Kuhn, H. J., and Marti, E. (1963). The filling mechanism of the swim bladder. Generation of high gas pressures through hairpin countercurrent multiplication. *Experientia* **19**, 497–511.

Laurent, P., and Dunel-Erb, S. (1984). The pseudobranch: Morphology and function. *In* "Fish Physiology" (W. S. Hoar and D. J. Randall, eds.), Vol. XB, pp. 285–323. Academic Press, Orlando.

Laurent, P., Goss, G. G., and Perry, S. F. (1994). Proton pumps in fish gill pavement cells? *Arch. Int. Physiol. Bioch. Bioph.* **102**, 77–79.

Laurent, P., and Rouzeau, J.-D. (1972). Afferent neural activity from pseudobranch of teleosts. Effects of P_{O_2}, pH, osmotic pressure and Na^+ ions. *Respir. Physiol.* **14**, 307–331.

Leiner, M. (1938). Die Augenkiemendrüse (Pseudobranchie) der Knochenfische. Experimentelle Untersuchungen über ihre physiolgische Bedeutung. *Z. Vergl. Physiol.* **26**, 416–466.

Lessard, J., Val, A. L., Aota, S., and Randall, D. J. (1995). Why is there no carbonic anhydrase activity available to fish plasma? *J. Exp. Biol.* **198**, 31–38.

Lin, H., Pfeiffer, D. C., Wayne Vogl, A., Pan, J., and Randall, D. J. (1994). Immunolocalization of proton-ATPase in the gill epithelia of rainbow trout. *J. Exp. Biol.* **195**, 169–183.

Lin, H., and Randall, D. J. (1991). Evidence for the presence of an electrogenic proton pump on the trout gill epithelium. *J. Exp. Biol.* **161**, 119–134.

Lin, H., and Randall, D. J. (1995). Proton pumps in fish gills. *In* "Fish Physiology: Ionoregulation: Cellular and Molecular" (C. M. Wood and T. Shuttleworth, eds.), pp. 229–255. Academic Press, New York.

Maetz, J. (1956). Le role biologique de l'anhydrase carbonique chez quelques téléostéens. *Suppléments au Bulletin biologique de France et de Belgique Les presses universitaires de France,* 1–129.

Marshall, N. B. (1960). Swim bladder structure of deep-sea fishes in relation to their systematics and biology. *Dis. Rep.* **31**, 1–122.

Miller, K. I., and Mangum, C. P. (1988). An investigation of the nature of Bohr, Root, and Haldane effects in *Octopus dofleini* hemocyanin. *J. Comp. Physiol. B* **158**, 522–547.

Müller, J. (1839). Vergleichende Anatomie der Myxinoiden III. Über das Gefässystem. *Abh. Dtsch. Akad. Wiss. Berlin* **1839**, 175–303.

Mylvaganam, S. E., Bonaventura, C., Bonaventura, J., and Getzoff, E. D. (1996). Structural basis for the Root effect in haemoglobin. *Nature Struct. Biol.* **3**, 275–283.

Nicol, J. A. C. (1989). "The Eyes of Fishes." Oxford Univ. Press, Cambridge.

Nikinmaa, M. (1990). Vertebrate red blood cells: Adaptations of function to respiratory requirements. *Zoophysiol.* **28**. Springer-Verlag, New York.

Noble, R. W., Kwiatkowski, L. D., De Young, A., Davis, B. J., Haedrich, R. L., Tam, L. T., and Riggs, A. F. (1986). Functional properties of hemoglobins from deep-sea fish: Correlations with depth distribution and presence of a swim bladder. *Biochim. Biophys. Acta* **870**, 552–563.

Noble, R. W., Parkhurst, L. J., and Gibson, Q. H. (1970). The effect of pH on the reactions of oxygen and carbon monoxide with the hemoglobin of the carp. *Cyprinus carpio. J. Biol. Chem.* **245**, 6628–6633.

Noble, R. W., Pennelly, R. R., and Riggs, A. (1975). Studies of the functional properties of the hemoglobin from the benthic fish, *Antimora rostrata. Comp. Biochem. Physiol. B* **52**, 75–81.

Parkhurst, L. J., Goss, D. J., and Perutz, M. F. (1983). Kinetic and equilibrium studies on the role of the β-147 histidine in the Root effect and cooperativity in carp hemoglobin. *Biochem.* **22**, 5401–5409.

Pelster, B. (1997). Buoyancy at depth. *In* "Fish Physiology" Vol. 16, (D. J. Randall and A. P. Farrell, eds.). Academic Press, San Diego. pp. 195–237.

Pelster, B. (1995a). Metabolism of the swim bladder tissue. *Biochem. Molec. Biol. Fishes* **4**, 101–118.

Pelster, B. (1995b). Mechanisms of acid release in isolated gas gland cells of the European eel *Anguilla anguilla. Am. J. Physiol.* **269**, R793–R799.

Pelster, B. (1994). Adrenergic control of swim bladder perfusion in the European eel *Anguilla anguilla. J. Exp. Biol.* **189**, 237–250.

Pelster, B., Hicks, J., and Driedzic, W. R. (1994). Contribution of the pentose phosphate shunt to the formation of CO_2 in swim bladder tissue of the eel. *J. Exp. Biol.* **197**, 119–128.

Pelster, B., Kobayashi, H., and Scheid, P. (1990). Reduction of gas solubility in the fish swim bladder. *In* "Oxygen Transport to Tissue XII" (J. Piiper, T. K. Goldstick, and M. Meyer, eds.), pp. 725–733. Plenum, New York, London.

Pelster, B., and Scheid, P. (1992). Countercurrent concentration and gas secretion in the fish swim bladder. *Physiol. Zool.* **65**, 1–16.

Pelster, B., and Scheid, P. (1993). Glucose metabolism of the swim bladder tissue of the European eel *Anguilla anguilla. J. Exp. Biol.* **185**, 169–178.

Pelster, B., Scheid, P., and Reeves, R. B. (1992). Kinetics of the Root effect and of O_2 exchange in whole blood of the eel. *Respir. Physiol.* **90**, 341–349.

Pelster, B., and Weber, R. E. (1990). Influence of organic phosphates on the Root effect in multiple fish haemoglobins. *J. Exp. Biol.* **149**, 425–437.

Pelster, B., and Weber, R. E. (1991). The physiology of the Root effect. *Adv. Comp. Environm. Physiol.* **8**, 51–77.

Perutz, M. F. (1996). Causes of the Root effect in fish haemoglobins. *Nature Struct. Biol.* **3**, 211–212.

Perutz, M. F., and Brunori, M. (1982). Stereochemistry of cooperative effects in fish and amphibian haemoglobins. *Nature* **299**, 421–426.

Powers, D. A., Fyhn, H. J., Fyhn, U. E. H., Martin, J. P., Garlick, R. L., and Wood, S. C. (1979). A comparative study of the oxygen equilibria of blood from 40 genera of Amazonian fishes. *Comp. Biochem. Physiol. A* **62**, 67–85.

Primmett, D. R. N., Randall, D. J., Mazeaud, M., and Boutilier, R. G. (1986). The role of catecholamines in erythrocyte pH regulation and oxygen transport in rainbow trout (*Salmo gairdneri*) during exercise. *J. Exp. Biol.* **122**, 139–148.

Rahim, S. M., Delaunoy, J.-P., and Laurent, P. (1988). Identification and immunocytochemical localization of two different carbonic anyhdrase isoenzymes in teleostean fish erythrocytes and gill epithelia. *Histochem.* **89**, 451–459.

Randall, D. J., and Jones, D. R. (1973). The effect of deafferentiation of the pseudobranch on the respiratory response to hypoxia and hyperoxia in the trout (*Salmo gairdneri*). *Respir. Physiol.* **17**, 291–301.

Randall, D. J., Mense, D., and Boutilier, R. G. (1987). The effects of burst swimming on aerobic swimming in chinook salmon (*Oncorhynchus tshawytscha*). *Mar. Behav. Physiol.* **13**, 77–88.

Randall, D. J., and Perry, S. F. (1992). Catecholamines. *In* "Fish Physiology. The Cardiovascular System" (W. S. Hoar and D. J. Randall, eds.), Vol. XII B, pp. 255–300. Academic Press, New York.

Riggs, A. (1979). Studies of the hemoglobins of Amazonian fishes: An overview. *Comp. Biochem. Physiol. A* **62**, 257–272.

Riggs, A. F. (1988). The Bohr effect. *Ann. Rev. Physiol.* **50**, 181–204.

Root, R. W. (1931). The respiratory function of the blood of marine fishes. *Biol. Bull.* **61**, 427–456.

Root, R. W., and Irving, L. (1943). The effect of carbon dioxide and lactic acid on the oxygen-combining power of whole and hemolyzed blood of the marine fish *Tautoga onitis* (*Linn*). *Biol. Bull.* **84**, 207–212.

Sauer, J., and Harrington, J. P. (1988). Hemoglobins of the sockeye salmon, *Oncorhynchus nerka. Comp. Biochem. Physiol. A* **91**, 109–114.

Scholander, P. F., and Van Dam, L. (1954). Secretion of gases against high pressures in the swim bladder of deep sea fishes. I. Oxygen dissociation in blood. *Biol. Bull.* **107**, 247–259.

Steen, J. B. (1963). The physiology of the swim bladder in the eel *Anguilla vulgaris*. III. The mechanism of gas secretion. *Acta Physiol. Scand.* **59**, 221–241.

Swenson, E. R., Lippincott, L., and Maren, T. H. (1995). Effect of gill membrane bound carbonic anhydrase inhibition on branchial bicarbonate excretion in the dogfish shark, *Squalus acanthias. Bull. Mount Desert Island Biol. Lab.* **34**, 94–95.

Swenson, E. R., Taschner, B. C., and Maren, T. H. (1996). Effect of membrane bound carbonic anhydrase (CA) inhibition on bicarbonate excretion in the shark, *Squalus acanthias. Bull. Mount Desert Island Biol. Lab.* **35,** 47.

Tan, A. L., and Noble, R. W. (1973). Conditions restricting allosteric transitions in carp hemoglobin. *J. Biol. Chem.* **248,** 2880–2888.

Vaccaro Torracca, A. M., Raschetti, R., Salvioli, R., Ricciardi, G., and Winterhalter, K. H. (1977). Modulation of the Root effect in goldfish by ATP and GTP. *Biochim. Biophys. Acta* **496,** 367–373.

Walsh, P. J., and Milligan, C. L. (1993). Roles of buffering capacity and pentose phosphate pathway activity in the gas gland of the gulf toadfish *Opsanus beta. J. Exp. Biol.* **176,** 311–316.

Weber, R. E. (1990). Functional significance and structural basis of multiple hemoglobins with special reference to ectothermic vertebrates. *In* "Animal Nutrition and Transport Processes. 2. Transport, Respiration and Excretion: Comparative and Environmental Aspects. Comp. Physiol. " (J. P. Truchot and B. Lahlou, eds.), pp. 58–75. Karger, Basel.

Weber, R. E., Lykkeboe, G., and Johansen, K. (1976a). Physiological properties of eel haemoglobin: Hypoxic acclimation, phosphate effects and multiplicity. *J. Exp. Biol.* **64,** 75–88.

Weber, R. E., Wood, S. C., and Lomholt, J. P. (1976b). Temperature acclimation and oxygen-binding properties of blood and multiple haemoglobins of rainbow trout. *J. Exp. Biol.* **65,** 333–345.

Wilhelm, D., and Reischl, E. (1981). Heterogeneity and functional properties of hemoglobins from South Brazilian freshwater fish. *Comp. Biochem. Physiol. B* **69,** 463–470.

Wittenberg, J. B, and Haedrich, R. L. (1974). The choroid rete mirabile of the fish eye. II. Distribution and relation to the pseudobranch and to the swim bladder rete mirabile. *Biol. Bull.* **146,** 137–156.

Wittenberg, J. B., and Wittenberg, B. A. (1962). Active secretion of oxygen into the eye of fish. *Nature* **194,** 106–107.

Wittenberg, J. B., and Wittenberg, B. A. (1974). The choroid rete mirabile of the fish eye. I. Oxygen secretion and structure. Comparison with the swim bladder rete mirabile. *Biol. Bull.* **146,** 116–136.

Wood, C. M., Perry, S. F., Walsh, P. J., and Thomas, S. (1994). HCO_3^- dehydration by the blood of an elasmobranch in the absence of a Haldane effect. *Respir. Physiol.* **98,** 319–337.

OXYGEN TRANSPORT IN FISH

MIKKO NIKINMAA AND ANNIKA SALAMA

I. Introduction
II. Hemoglobin Function: Basic Principles
III. Regulation of Hemoglobin Function by Changes in Erythrocytic Organic Phosphate Concentrations
IV. Effects of Cellular Hemoglobin Concentration and Red Cell Volume on Oxygen Transport
V. Regulation of Erythrocyte Volume
   A. Volume Regulation after Osmotic Disturbances
   B. Adrenergic Volume Changes
   C. Effects of Oxygenation-Sensitive Ion Transport Pathways
VI. Effects of Protons on Hemoglobin Function
VII. Control of Erythrocyte pH
   A. Agnathans
   B. Elasmobranchs
   C. Teleosts
VIII. Hemoglobin Oxidation
IX. Responses of Hemoglobin Function to Changes in the External and the Internal Environment of Fish
   A. Temperature
   B. Hypoxia
   C. Hypercapnia and Hypercapnic Hypoxia
   D. Hyperoxia
   E. Environmental Acidification
   F. Salinity Changes
   G. Exercise
   H. Anemia
   I. Environmental Pollutants
References

I. INTRODUCTION

The solubility of oxygen in water is only ca. 1/30, and the rate of oxygen diffusion only 1/10,000, of that in air. Because of these properties of water,

Fish Physiology, Volume 17:
FISH RESPIRATION

the oxidation of organic materials and the respiration of organisms in the water and in the sediments can result in a marked depletion of oxygen from freshwater bodies. In eutrophic waters with dense green vegetation, not only hypoxia is common, but active photosynthesis by green plants can also cause oxygen supersaturation. Life in aquatic environments thus requires efficient respiratory and metabolic adjustments to changes in oxygen availability. In addition, the oxygen demand by animals varies markedly. The oxygen consumption of fishes increases with increasing temperature, often more than doubling for every $10°$ increase in temperature (e.g., Brett and Glass, 1973). The oxygen consumption of fish also increases markedly with exercise: at optimum temperature, the maximal oxygen consumption can be more than ten times the standard oxygen consumption (Brett and Glass, 1973).

The flux of oxygen from the environment to the sites of consumption in the tissues consists of the following steps: (1) Breathing movements continuously bring new oxygen molecules in contact with the respiratory surface. (2) Oxygen diffuses down its partial pressure gradient from the ambient water into the capillaries of respiratory organs. (3) Oxygen binds to hemoglobin in the erythrocytes. The amount of oxygen bound per unit volume of blood depends on the number of erythrocytes, the concentration of hemoglobin within the erythrocyte, the prevailing oxygen partial pressures, and the oxygen-binding properties of the hemoglobin molecule. (4) Oxygen is transported in the bloodstream from the gills to the sites of consumption. (5) In tissue capillaries, the partial pressure of oxygen decreases and, consequently, oxygen dissociates from hemoglobin. (6) Oxygen diffuses from the capillaries to the oxygen-requiring sites, mainly mitochondria, within the cells. Since the mitochondrial oxygen tension is very close to zero, the rate of oxygen diffusion per unit area in a given tissue (with a unique diffusion coefficient for oxygen) is the function of the diffusion distance between the capillaries and the mitochondria and the oxygen tension of capillary blood.

Hemoglobin plays a decisive role in the oxygen transport cascade. In the blood spaces of gill epithelium, hemoglobin has to bind oxygen effectively, and in the tissues, the release of oxygen should take place at a high partial pressure to produce a large oxygen partial pressure gradient from capillaries to mitochondria. Furthermore, both oxygen loading in gills and oxygen unloading in tissues should respond in an appropriate manner to changes in oxygen availability and oxygen demand.

The erythrocyte provides the environment in which hemoglobin functions. Thus, hemoglobin function within the animal can be modified by changing the properties of the erythrocyte. Indeed, one major advantage of having an intracellular respiratory pigment (as in vertebrates) as opposed to an extracellular one (as in many invertebrates) is that the oxygen-binding properties of the pigment can be adjusted at the cellular level without the need to affect the properties of the whole blood volume. As a consequence, the oxygen-binding properties of blood can be adjusted very rapidly, as seen in exercised or hypoxic teleost fish (Nikinmaa *et al.,* 1984; Tetens and Christensen, 1987). The presence of an intracellular oxygen-carrying pigment also provides other advantages over the pigments that are free in solution. First, the number of erythrocytes (and oxygen-carrying capacity) can be changed by sequestration in or liberation from storage organs such as the spleen (Stevens, 1968; Nilsson and Grove, 1974; Yamamoto *et al.,* 1980). Second, the flow of erythrocytes (and oxygen) in different capillary beds and, in the case of teleost fish, in the secondary circulation, can be modulated by selective opening and closing of precapillary sphincters and/ or arterio–venous anastomoses (Olson, 1984; Steffensen *et al.,* 1986).

In view of the critical role of the intraerythrocytic environment in controlling hemoglobin function, this chapter discusses how the oxygen affinity of hemoglobin is controlled by the properties of the erythrocyte. Throughout the chapter, similarities and differences in the responses of different groups of fish, from hagfish to teleosts, are presented.

II. HEMOGLOBIN FUNCTION: BASIC PRINCIPLES

The oxygen-binding properties of hemoglobin can be described using the oxygen equilibrium curve (Fig. 1), which relates the prevailing oxygen tension to the oxygen saturation of hemoglobin. Oxygen transport to tissues can be affected both by changes in the arterial oxygen tension and by changes in the oxygen affinity of hemoglobin.

There are several examples of situations in which the *oxygen tension* of postbranchial blood is regulated by fish. For example, tench (*Tinca tinca*) and carp (*Cyprinus carpio*) breathe intermittently under normoxic, resting conditions (e.g., Hughes 1981), and the arterial oxygen tension of blood is low (20–40 torr; Jensen *et al.,* 1983, 1987; Jensen, 1987). In exercised animals, the oxygen tension increases markedly (up to 100 torr) owing to hyperventilation (Jensen *et al.,* 1983; Jensen, 1987). Thus, oxygen saturation of hemoglobin can be maintained even if the oxygen equilibrium curve is shifted to the right (Fig. 1). Another mechanism of elevating the oxygen tension of postbranchial blood utilizes the fact that arterial blood is in a closed system. Any acidification of blood that decreases the hemoglobin–oxygen affinity will liberate hemoglobin-bound oxygen in solution and increase the arterial oxygen tension. Pronounced acidification-induced increases in the

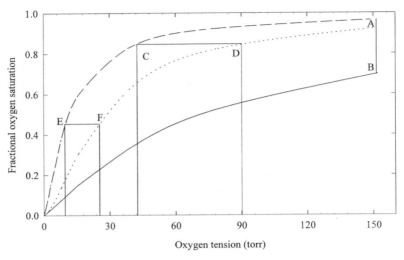

Fig. 1. Oxygen equilibrium curves for fish hemoglobin with the fractional oxygen saturation on the *y* axis and blood oxygen tension on the *x* axis. A decrease in pH or an increase
in cellular NTP concentration shifts the equilibrium curve to the right, thus reducing oxygen
affinity. In the case of teleost fish and lampreys, the oxygen saturation at low pH values is
reduced even at atmospheric oxygen tension (A–B). A shift of the oxygen equilibrium curve
to the right can be compensated for by fish that have a low resting blood oxygen tension: by
ventilatory adjustments the arterial oxygen tension can be increased and arterial saturation
maintained (C–D). As long as the arterial saturation can be maintained, a rightward shift of
the oxygen equilibrium curve is beneficial to the fish because the same amount of oxygen is
given up at a higher oxygen tension (E–F).

blood oxygen tension are observed in the blood entering the retina and
swim bladder of teleost fish (Fairbanks *et al.,* 1969; Pelster and Scheid, 1992).

The oxygen affinity of hemoglobin within the erythrocyte is a function
of the intrinsic oxygen affinity of hemoglobin, the sensitivity of hemoglobin–
oxygen affinity to heterotrophic ligands (molecules that bind to hemoglobin
at a site different from the oxygen-binding site), the concentration of hemoglobin within the cell, and the concentration of heterotrophic ligands within
the erythrocyte. The heterotrophic ligands that are most important in physiological regulation of hemoglobin–oxygen affinity in fishes are protons and
organic phosphates [mainly adenosine triphosphate (ATP) and guanosine
triphosphate (GTP); e.g., Jensen *et al.,* 1998]. The intracellular pH and
the concentrations of organic phosphates within the cell are regulated in
response to environmental changes, and therefore, their effects on hemoglobin function are discussed later in this chapter. In addition, several other
effectors such as lactate (Guesnon *et al.,* 1979), chloride (Fronticelli *et al.,*

1984), carbon dioxide (Jensen and Weber, 1982), and urea (Aschauer *et al.*, 1985) have been shown to affect the oxygen affinity of various vertebrate hemoglobins.

Both protons and organic phosphates bind preferentially to the deoxygenated, low-affinity conformation of hemoglobin and stabilize it (see Weber and Jensen, 1988; Jensen *et al.*, 1998). Consequently, an increase in organic phosphate concentration or a decrease in pH will decrease the hemoglobin–oxygen affinity (shift the oxygen equilibrium curve to the right). In many teleosts and in lampreys, the stabilization of the low-affinity conformation by protons is so strong that the maximal hemoglobin–oxygen saturation can be drastically reduced at atmospheric oxygen tension (Root effect; see Fig. 1 and, e.g., Jensen *et al.*, 1998). Since the reaction between hemoglobin and oxygen is exothermic under physiological conditions, the oxygen equilibrium curve is shifted to the right as temperature increases. However, there are pronounced differences between species in the temperature sensitivity of the reaction between hemoglobin and oxygen (e.g., Carey and Gibson, 1977; Weber *et al.*, 1976b; Jensen and Weber, 1982). The temperature dependence of the reaction between hemoglobin and oxygen is modulated by heterotrophic ligands. Both a decrease in pH and an increase in organic phosphate concentration diminish the effect of temperature on the oxygen affinity of hemoglobin because the liberation of these allosteric co-factors from hemoglobin, which takes place during oxygenation, is an endothermic reaction and thus reduces the overall heat of oxygenation (e.g., Jensen and Weber, 1982).

III. REGULATION OF HEMOGLOBIN FUNCTION BY CHANGES IN ERYTHROCYTIC ORGANIC PHOSPHATE CONCENTRATIONS

Organic phosphates (mainly ATP and GTP) affect the hemoglobin function of fishes (1) by direct binding preferentially to the deoxygenated form of hemoglobin and (2) by an indirect effect on intracellular pH (Wood and Johansen, 1973). Organic phosphates, however, do not bind to agnathan hemoglobins, which do not form stable tetramers (Bauer *et al.*, 1975; Nikinmaa and Weber, 1993). The binding of organic phosphates to hemoglobin is affected by complex formation with other intracellular components. ATP readily complexes with magnesium ions, and consequently, the effect of ATP on the oxygen equilibrium curve is reduced (Bunn *et al.*, 1971). In contrast, a considerable effect of GTP on the oxygen equilibrium curve of carp remains after GTP interacts with magnesium ions (Weber, 1978). The erythrocytic ATP and GTP concentrations vary markedly between species. Erythrocyte GTP concentration varies from nearly undetectable levels as in rainbow trout (e.g., Tetens, 1987) to values clearly exceeding those of ATP as in eel (e.g., Geoghegan and Poluhowich, 1974). ATP in fish erythrocytes is mainly produced by the aerobic metabolism of the erythrocytes (Ferguson and Boutilier, 1988), using glucose, monocarboxylic acids, or glutamine as substrates (Walsh *et al.*, 1990; Sephton *et al.*, 1991; Tiihonen and Nikinmaa, 1991a). Both monocarboxylic acids and glutamine appear to be effectively transported across the erythrocyte membrane via specific carriers in all fishes from hagfish to teleosts (Tiihonen, 1995), whereas glucose transport appears to be exceedingly slow in most teleost erythrocytes (Ingermann *et al.*, 1985; Tse and Young, 1990; Tiihonen and Nikinmaa, 1991b). Despite this, the high plasma glucose concentration generates an adequate flux of glucose that supports a significant utilization of glucose as a metabolic fuel.

Very little is known about the metabolic reasons for the high GTP concentration in fish erythrocytes. The activities of key synthetic or catabolic enzymes in the guanosine phosphate pathway are not known for fish. Based on information from rabbit erythrocytes (Hershko *et al.*, 1967; Fig. 2), the

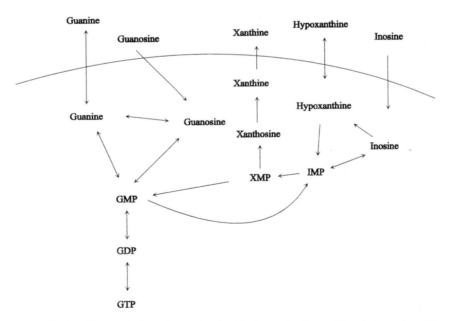

Fig. 2. Possible metabolic pathways involved in the production and breakdown of guanosine phosphates within the erythrocytes.

initial formation of guanine nucleotides may be limited by the availability of intracellular phosphoribose pyrophosphate concentration, the activity of guanine phosphoribosyl transferase (which adds the phosphoribosyl group to guanine), or the activities of inosinate dehydrogenase and guanylate synthetase (which catalyze the conversion of inosine monophosphate to guanosine monophosphate). A high concentration of intracellular guanosine nucleotides could also be caused by a low activity of the catabolic pathway, mainly the GMP reductase, which converts guanosine monophosphate to inosine monophoshate and initiates the further catabolism of purine nucleotides (Hershko et al., 1967; see Nikinmaa, 1990, for discussion). Once guanosine monophosphate is formed, it appears that in species with high GTP concentration, the guanosine monophosphate kinase/adenosine monophosphate kinase activity ratio is greater than that in species with low GTP concentration. At least in the eel, Anguilla rostrata, which has a high GTP concentration, the GMPK/AMPK ratio is much higher than that in hagfish and dogfish with low or negligible GTP concentration (Parks et al., 1973). Guanosine diphosphate is subsequently converted to GTP in a reaction catalyzed by nucleoside diphosphokinase, which has a much higher activity than GMPK (Parks et al., 1973), suggesting that the step from monophosphate to diphosphate is rate limiting in the synthesis of GTP.

Despite the importance of organic phosphates in regulating hemoglobin function, the mechanisms by which the ATP and GTP levels are adjusted to respond to respiratory requirements are not known. Initially, it was suggested (Greaney and Powers, 1978) that the hypoxia-induced decrease in red cell ATP concentration would be a direct consequence of decreased oxidative phosphorylation because of inadequate oxygen supply. However, this is unlikely since the erythrocyte ATP levels in vivo decrease at relatively high oxygen tensions (e.g., 35–40 torr in rainbow trout, Oncorhynchus mykiss; Soivio et al., 1980) at which more than 50% of the total hemoglobin-bound oxygen stores are still available for erythrocyte metabolism. In other cell types without such oxygen stores, e.g., liver cells, oxidative phosphorylation is not affected until oxygen tension is reduced below 5 torr (DeGroot and Noll, 1987). Furthermore, the red cell ATP concentration in vitro decreases only in nearly complete anoxia (Greaney and Powers, 1978; Tetens and Lykkeboe, 1981).

Tetens and Lykkeboe (1981) suggested that a humoral factor would induce the decrease in the erythrocyte ATP concentration in hypoxia. Catecholamines were prime candidates for such hormones since their concentrations increase in hypoxia (Tetens and Christensen, 1987) and since adrenaline decreases erythrocytic ATP concentration (Nikinmaa, 1983). The decrease is probably due to the activation of sodium/proton exchange and a subsequent increase in the activity of the sodium pump, which increases ATP consumption (Tufts and Boutilier, 1991). However, a significant role for catecholamines in the regulation of erythrocytic NTP concentration is unlikely for two reasons. First, the catecholamine-induced reduction of cellular ATP concentration is entirely due to the conversion of ATP to ADP; i.e. the total cellular pool of adenylates remains constant, but in hypoxia the total pool of nucleotide phosphates is reduced (Tetens, 1987; Fig. 3). Second, catecholamines do not affect the GTP concentration in carp (Salama and Nikinmaa, 1988), although GTP concentration is preferentially decreased in hypoxia *in vivo* (Weber and Lykkeboe, 1978; Lykkeboe and Weber, 1978) and *in vitro* (Jensen and Weber, 1985). The decrease of NTP concentration can also be due to an elevated level of cortisol in the blood: an injection of cortisol caused a reduction of erythrocyte ATP concentration in red snapper, *Pagrus auratus* (Bollard *et al.,* 1993). However, it is not known whether the GTP concentration is affected, and the relationship between cortisol and ATP levels under physiological conditions (hypoxia, etc.) has not been, to our knowledge, examined. Thus, the nature of the humoral factor controlling the cellular organic phosphate levels is still unknown.

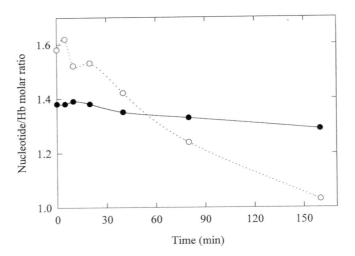

Fig. 3. The nucleotide phosphate/hemoglobin molar ratio as a function of time in rainbow trout erythrocytes subjected to hypoxia *in vivo* (oxygen tension 30 torr; empty circles) or stimulated with 10^{-5} M noradrenaline *in vitro* (filled circles). Data from Tetens (1987).

IV. EFFECTS OF CELLULAR HEMOGLOBIN CONCENTRATION AND RED CELL VOLUME ON OXYGEN TRANSPORT

The erythrocytic concentration of agnathan hemoglobins, which do not form stable tetramers, affects their affinity for oxygen. In the lamprey, *Lampetra fluviatilis,* intracellular hemoglobin concentrations higher than normal decrease the hemoglobin–oxygen affinity (Airaksinen and Nikinmaa, 1995). An increase in hemoglobin concentration within intact erythrocytes shifts the monomer–oligomer equilibrium toward oligomers. Since the oligomeric forms of hemoglobin have a much lower oxygen affinity than the monomers, the overall oxygen affinity will be decreased (for detailed mechanisms of oxygen binding in agnathan erythrocytes, see Perutz, 1990; Nikinmaa *et al.,* 1995). Such direct effects of hemoglobin concentration on oxygen affinity appear to be uncommon in other fishes, although dissociation–association reactions have been described for elasmobranch hemoglobins (Fyhn and Sullivan, 1975). The observed concentration effects—an increase in hemoglobin–oxygen affinity with a decrease in hemoglobin concentration within the cell (Soivio and Nikinmaa, 1981) and a decrease in hemoglobin–oxygen affinity with an increase in hemoglobin concentration (Jensen, 1990)—may be caused either via the concomitant changes in the intracellular pH or via alterations in the interaction between organic phosphates and hemoglobin. A reduction in the concentration of both hemoglobin and ATP/GTP will reduce the likelihood of complex formation between the two, and thus the oxygen affinity will increase (Lykkeboe and Weber, 1978).

The red cell volume may also affect oxygen transport by influencing the deformability of erythrocytes, and thus their behavior in circulation (e.g., entrance and movement in capillaries). The deformability of erythrocytes is influenced by the surface-to-volume ratio and the internal viscosity (which is a function of internal hemoglobin concentration; see Nikinmaa, 1990). If the red cell volume is reduced, there is an increase in the surface-to-volume ratio of the cells that tends to increase deformability. However, the simultaneously occurring increase in internal viscosity tends to decrease deformability. The net effect depends on which of the two factors predominates. This depends on the size of the pore through which the cell must traverse. Reinhart and Chien (1985) have shown that when human erythrocytes enter pores with a diameter close to the critical size for red cell passage, a reduced volume is beneficial (indicating a major role for surface-to-volume ratio), but when the cells traverse larger pores, swollen cells have smaller resistance than normal-sized cells (indicating that internal viscosity dominates in determining the deformability of the cells). Thus, the observation (Hughes and Kikuchi, 1984) that swollen, hypoxic erythrocytes of rainbow trout traversed through 8-μm Nuclepore filters faster than the normal-sized erythrocytes of control fish suggests that the behavior of fish erythrocytes in this instance is mainly determined by changes in internal viscosity.

V. REGULATION OF ERYTHROCYTE VOLUME

The intracellular impermeable polyions (mainly hemoglobin and organic phosphates that are negatively charged at the physiological pH range of 6.5 to 8) and their counterions generate an osmotic pressure difference across the erythrocyte membrane. Unopposed, this would lead to continuous influx of water (and permeable solutes) into the cell, until the cell would burst. The effect of impermeable polyions is counterbalanced by the sodium pump, which actively extrudes sodium ions from the erythrocyte at the same rate as they enter the cells. Thus, sodium can be treated as a "functionally impermeable" solute and provides the osmotic force outside the cell that is required for maintenance of steady-state volume (Tosteson and Hoffman, 1960).

The volume of erythrocytes is influenced by the charge of organic phosphates and hemoglobin. If the negative charge of hemoglobin and organic phosphates decreases, either permeable cations must leave the cell or permeable anions enter the cell to maintain electroneutrality. Similarly, if the negative charge of hemoglobin increases, permeable cations must enter or permeable anions leave the cell. In teleost and elasmobranch fish, the permeability of erythrocyte membrane to small anions, chloride and bicarbonate, is much greater than the permeability to potassium or sodium (see, e.g., Nikinmaa, 1992a). Thus, changes in the charge of hemoglobin and organic phosphates are mainly compensated for by a net influx or efflux of chloride. In lampreys, on the other hand, it appears that potassium permeability may predominate (Virkki and Nikinmaa, 1995). Furthermore, in lamprey erythrocytes, association–dissociation reactions of hemoglobin occur under physiological conditions (Nikinmaa et al., 1995). The volume of lamprey erythrocytes, therefore, responds to changes in the charge of hemoglobin or organic phosphates in a fashion different from that of other fish erythrocytes.

Physiologically, the major changes that affect the charge of impermeable polyions are variations of pH and oxygen tension. A decrease in extracellular pH will cause a decrease in erythrocyte pH, which will reduce the negative charge of hemoglobin (mainly the charge of histidine imidazole) and organic phosphates. Consequently, chloride will enter the cell, and the cell swells (Hladky and Rink, 1977). A reduction of oxygen tension increases erythrocyte volume since hemoglobin takes up protons upon deoxygenation, its negative charge decreases, and chloride enters the cells, (Hladky and Rink, 1977). In contrast to other fishes, the volume of lamprey erythrocytes is hardly affected by a decrease in extracellular pH or oxygen tension (e.g., Tufts and Boutilier, 1989; Nikinmaa and Mattsoff, 1992).

A. Volume Regulation after Osmotic Disturbances

A reduction in the osmolality of the medium will immediately result in cell swelling and an increase in osmolality will result in cell shrinking because the rapid transport of water tends to abolish differences in the osmotic pressure between the cell and plasma. These volume changes are sensed by the cells because of either the stretch of the cell membrane (Sackin, 1989) or the dilution/concentration of intracellular macromolecules (Colclasure and Parker, 1991, 1992). In most fish erythrocytes, transport pathways are then activated and tend to restore the original cell volume. A notable exception are the erythrocytes of myxinoids, which do not seem to respond to osmotically induced volume changes (Nikinmaa et al., 1993). The recovery of erythrocyte volume after the cells are initially swollen in hypoosmotic medium is called regulatory volume decrease (RVD), and the recovery of volume after the cells are shrunk in hyperosmotic medium is called regulatory volume increase (RVI). The mechanisms of RVD and RVI are depicted in Fig. 4. The activation pathways for RVI and RVD are reciprocal and involve a complex cascade of phosphorylation–dephosphorylation reactions (Cossins and Gibson, 1997).

RVD is achieved either by the loss of organic osmolytes, mainly taurine, or by the loss of potassium and chloride from the erythrocytes, followed by osmotically obliged water. In the case of lampreys, RVD is solely due to the loss of potassium and chloride (Nikinmaa et al., 1993) via electrically coupled conductive pathways (Virkki and Nikinmaa, 1995). In elasmobranch and teleost fish, the loss of taurine is an important component of RVD (Boyd et al., 1977; Fugelli and Zachariassen, 1976; Fugelli and Rohrs, 1980). Erythrocytes accumulate taurine via a sodium-dependent pathway (Fugelli and Thoroed, 1986; Fincham et al., 1987). When the cell volume is increased, a channel-like transport pathway is opened, and there is a pronounced efflux of taurine down its concentration gradient (Fugelli and Thoroed, 1986; Fincham et al., 1987; Kirk et al., 1992; Goldstein and Musch, 1994). The loss of potassium during RVD in teleost fish appears to take place

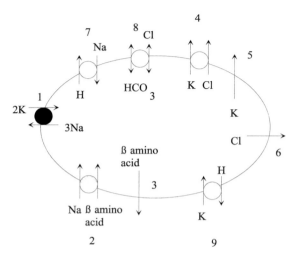

Fig. 4. Transport pathways involved in the control of erythrocyte volume in fish. (1) The sodium pump is involved in the maintenance of steady-state volume and generates the sodium and potassium gradients utilized in volume regulation. (2). A sodium-dependent transport system generates the high intraerythrocytic taurine concentration. In regulatory volume decrease (RVD), taurine is transported via a channel-like pathway (3) and potassium and chloride either via a K–Cl co-transport pathway (4) or via separate channels (5, 6). Regulatory volume increase (RVI) occurs mainly by activation of sodium/proton exchange (7) and consecutive net transport of chloride into the cell via the anion exchanger (8). A potassium/proton exchange pathway (9) has also been described for fish erythrocytes. However, its role in volume regulation is not clear. In every case, the net movement of osmotically active particles (taurine, potassium, chloride and sodium) is followed by osmotically obliged water, and hence cell volume is changed.

via two separate pathways. One of these pathways is chloride-independent (Garcia-Romeu *et al.*, 1991; Cossins *et al.*, 1994; Bursell and Kirk, 1996), and could be the channel involved in taurine transport (Bursell and Kirk, 1996). The other pathway for chloride efflux is a swelling-activated potassium/chloride co-transporter (Lauf, 1982; Garcia-Romeu *et al.*, 1991; Guizouarn *et al.*, 1993; Cossins *et al.*, 1994).

RVI involves the osmotic activation of sodium/proton exchange. The sodium/proton exchange is usually coupled to chloride/bicarbonate exchange resulting in the uptake of sodium and chloride (and osmotically obliged water). Net transport of chloride into the cell via the anion exchanger is not due to the increase in the turnover rate of the exchanger (which far exceeds that of the osmotically activated sodium/proton exchanger under all conditions). Rather, it is due to the following sequence of events. First, there is a net flux of protons from the cell via the sodium/proton exchange. The proton efflux shifts the reaction catalyzed by carbonic anhydrase toward bicarbonate and protons. Thus, there is a transient increase in bicarbonate concentration within the cells. An increase in intracellular bicarbonate concentration reduces the concentration gradient for bicarbonate, and consequently, the anion exhanger is temporarily out of equilibrium. The equilibrium is regained by a net transport of chloride into the cell and bicarbonate out of the cell. This net transport continues as long as the proton extrusion via the sodium/proton exchange persists.

The degree to which volume recovery after osmotic shrinkage occurs differs markedly between species. In lamprey erythrocytes, although sodium/proton exchange is activated by osmotic shrinking, there is no net transport of sodium into the cell because of the lack of driving force for the exchanger (Virkki and Nikinmaa, 1994). Among teleost fish, at least the erythrocytes of the winter flounder (*Pseudopleuronectes platessa;* Cala, 1977), eel (*Anguilla anguilla;* Gallardo Romero *et al.,* 1996), and carp (*Cyprinus carpio;* Orlov and Skryabin, 1993) effectively regulate red cell volume after osmotic shrinkage. In the salmonids, brown trout (*Salmo trutta*), and rainbow trout (*Oncorhynchus mykiss*), volume activation of the sodium/ proton exchange is weaker (Orlov *et al.,* 1994; Gallardo Romero *et al.,* 1996), and volume recovery is not complete.

B. Adrenergic Volume Changes

Adrenergic stimulation of rainbow trout erythrocytes leads to cell swelling (Nikinmaa, 1982). Cell swelling is caused by the activation of sodium/ proton exchange (Nikinmaa and Huestis, 1984; Cossins and Richardson, 1985). The adrenergically activated sodium/proton exchange functions in parallel with anion exchange (Nikinmaa and Huestis, 1984; Borgese *et al.,* 1986), resulting in net sodium and chloride influx. Catecholamine stimulation of sodium/proton exchange has since been observed in many teleost fish (Salama and Nikinmaa, 1989; Cossins and Kilbey, 1991) with notable exceptions such as the eel (e.g., Hyde and Perry, 1990). Apart from teleost fish, it appears that there is a slight activation of the sodium/proton exchanger by catecholamines in lamprey (Gusev *et al.,* 1992; Virkki and Nikinmaa, 1994). However, the activation is too weak to cause significant changes in cell volume under physiological conditions (Tufts, 1991; Virkki and Nikinmaa, 1994). Up to the present, data for elasmobranch fish do not indicate that there are catecholamine-stimulated transport pathways affecting erythrocyte volume (see Tufts and Randall, 1989).

C. Effects of Oxygenation-Sensitive Ion Transport Pathways

There is a pronounced efflux of potassium, together with chloride, from fish erythrocytes at high oxygen tensions even in isoosmotic medium (Jensen, 1990). The efflux of potassium and chloride causes a decrease in erythrocyte volume far below that expected on the basis of the oxygenation-dependent change in the charge of hemoglobin (Borgese *et al.*, 1991; Fig. 5). The transport pathway is potassium/chloride co-transport, as indicated by the nearly absolute requirement of oxygenation-induced potassium efflux on chloride (Borgese *et al.*, 1991; Nielsen *et al.*, 1992). As discussed earlier, the potassium/chloride co-transporter is also involved in RVD. Because this pathway is oxygenation-sensitive, RVD is much more effective at high than at low oxygen tension (Nielsen *et al.*, 1992; Jensen, 1995). In addition to oxygenation, treatment of cells with CO or nitrite (which oxidizes heme groups) activates the transporter (Jensen, 1990; Nielsen and Lykkeboe, 1992). These findings indicate that a heme group is involved in the activation step. The actual activation sequence is not known, although interaction between hemoglobin and band 3 protein could be involved (Jensen, 1990; Nielsen *et al.*, 1992; Jensen, 1995).

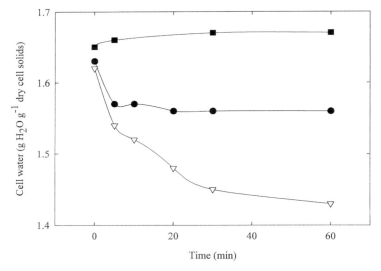

Fig. 5. Changes in erythrocyte water content of rainbow trout brought about by oxygenation as a function of time (data from Borgese *et al.*, 1991). Squares: erythrocytic water content in deoxygenated erythrocytes. Triangles: erythrocytic water content in cells oxygenated at time 0. Circles: erythrocyte water content in cells maintained in a medium in which nitrate has been substituted for chloride precluding the activation of potassium/chloride co-transport pathway and oxygenated at time 0. The difference in water content between squares and circles indicates the effect of the oxygenation-induced change in the charge of hemoglobin on cellular water content. The further decrease shown by the curve with triangles shows the effect of the oxygenation-activated potassium/chloride co-transport pathway on the cellular water content.

Since the activation of the potassium/chloride co-transporter by oxygenation occurs under isotonic conditions, it is probable that the potassium/chloride co-transport has other physiological functions in addition to osmotic volume regulation. Effects on oxygen transport are possible: a reduction of volume at high blood oxygen tensions, caused, e.g., by diurnal hyperoxia, would decrease the hemoglobin–oxygen affinity by increasing complex formation between hemoglobin and organic phosphates, thus facilitating the initial phases of oxygen delivery to tissues. Alternatively, a reduction of volume at high oxygen tensions could facilitate the entrance of erythrocytes to capillaries if the surface-to-volume ratio of the erythrocyte is the major determinant of the overall deformability of the erythrocyte.

VI. EFFECTS OF PROTONS ON HEMOGLOBIN FUNCTION

Protons affect the hemoglobin–oxygen affinity of practically all fishes. One important point to remember is that hemoglobin function must be related to intraerythrocytic pH. This point is demonstrated by data on lampreys—early reports (Bird et al., 1976; Nikinmaa and Weber, 1984) gave low Bohr factors (-0.1 to -0.3). However, these values were measured as a function of extracellular pH. Later studies (Nikinmaa, 1986) showed that the erythrocyte pH of lamprey decreased only slightly with a decrease in extracellular pH (Fig. 6). Thus, when intracellular pH was used in determining the pH dependence of hemoglobin–oxygen affinity of lamprey, it was found that the hemoglobins were highly sensitive to protons, with Bohr factors above -0.6 (Ferguson et al., 1992; Nikinmaa, 1993). In lampreys, the effect of protons on hemoglobin–oxygen affinity is due to the stabilization of low-affinity oligomers that dissociate to high-affinity monomers upon oxygenation (see Perutz, 1990; Nikinmaa et al., 1995). In contrast to lampreys, the effect of protons on oxygen affinity of hemoglobins within the erythrocytes of myxinoids is marginal (as are the effects caused by changes in the hemoglobin concentration; see Hardisty, 1979), suggesting that, in these animals, the aggregation state of hemoglobin is hardly affected by changes in the red cell pH or volume within the physiological range.

In elasmobranch and teleost fish, protons stabilize the low-affinity conformation of the tetrameric hemoglobins. Thus, a decrease in pH moves the oxygen equilibrium curve to the right. The effect of protons (Bohr effect) varies markedly between species, and even within species, as exemplified by the fact that in rainbow trout and eel there are hemoglobin components both with very large Bohr factors and with reversed Bohr factors (oxygen affinity increases with decreasing pH) within the pH range

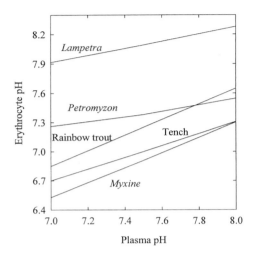

Fig. 6. Relationship between erythrocyte pH and plasma pH in a myxinoid (*Myxine glutinosa*), two lampreys (*Petromyzon marinus* and *Lampetra fluviatilis*), and two teleosts (tench, *Tinca tinca*, and rainbow trout, *Oncorhynchus mykiss*). Figure modified from Nikinmaa (1997).

7–8 in the absence of organic phosphates (Gillen and Riggs, 1973; Weber *et al.*, 1976a,b). As a broad generalization, however, it can be stated that the effect of protons on hemoglobin function is greater in teleost than in elasmobranch fish (cf. Nikinmaa, 1997).

The effect of protons on hemoglobin–oxygen affinity provides the animals with a means to rapidly respond to the conflicting demands for an oxygen equilibrium curve at the respiratory surface and within the tissues—at the gills a high oxygen affinity is required for effective oxygen loading, whereas within the tissues a low oxygen affinity ensures a high unloading partial pressure for oxygen. Proton excretion across the gills (either via ion exchanges or as carbon dioxide excretion) will result in an increase in erythrocyte pH at a constant oxygen saturation, thus increasing the hemoglobin–oxygen affinity. In the tissues, production of metabolic acids and carbon dioxide will decrease the erythrocyte pH at a constant oxygen saturation. This shifts the oxygen equilibrium curve to the right, which will increase the amount of oxygen dissociating from hemoglobin at a given oxygen tension. The greater the Bohr effect, and the decrease of pH in the capillary blood at a constant oxygen saturation, the more oxygen will be delivered. Thus, the physiological oxygen equilibrium curve will be much steeper than the oxygen equilibrium curve in the absence of proton loads from the tissues and proton excretion across the gills.

It is the blood pH at constant oxygen saturation that is important in terms of the physiological oxygen equilibrium curve. Changes in the oxygen saturation of hemoglobin also have a marked effect on blood pH, especially in teleost fish and lampreys, the hemoglobins of which often have large Haldane effects (i.e., oxygenation-dependent proton uptake and release). The deoxygenation-dependent proton uptake can be so large that, despite the carbon dioxide/metabolic acid load from the tissues, venous blood has a higher pH than arterial blood (see Milligan and Wood, 1986; Nikinmaa et al., 1990; Tufts et al., 1992). However, the proton uptake upon deoxygenation cannot affect hemoglobin–oxygen affinity since it is a part of the hemoglobin mechanism and the protons taken up will be released as soon as hemoglobin is again oxygenated.

VII. CONTROL OF ERYTHROCYTE pH

In the simplest case, the erythrocyte pH is determined by (1) the concentration and charge of the impermeable polyions, (2) the electrically silent one-to-one exchange of chloride for bicarbonate, and (3) the diffusion of carbon dioxide across the erythrocyte membrane and its hydration–dehydration reactions in the erythrocyte and blood plasma (see, e.g., Nikinmaa, 1990). Any extracellular proton load (decrease in pH) is transferred into the intracellular compartment via the Jacobs–Stewart cycle. Extracellular protons react with bicarbonate, forming carbon dioxide. Carbon dioxide diffuses into the erythrocyte and is hydrated to bicarbonate and protons. The protons are buffered by hemoglobin, the charge of which decreases. The bicarbonate ions formed leave the cell in exchange for chloride. Because the charge of hemoglobin decreases, the distribution ratio ($[A^-]_i/[A^-]_e$) for the permeable anions, chloride and bicarbonate, must increase in order to maintain electroneutrality. Since the proton distribution ratio is the inverse of the anion distribution ratio, the pH gradient across the erythrocyte membrane decreases.

The effect of extracellular acid loads on both the plasma and erythrocyte pH is determined largely by the buffering capacity of hemoglobin. The major determinant of the buffering capacity of hemoglobins at physiological pH values is the number of histidine residues per hemoglobin chain. The number of histidine residues per hemoglobin chain varies greatly in fishes, being small (2–6) in agnathans and teleosts and large (around 10) in elasmobranchs (see Jensen, 1989; Nikinmaa, 1990). Correspondingly, the buffering capacity of hemoglobin molecules, as measured by direct titration, is much lower in lampreys and teleosts than in elasmobranchs (Jensen, 1989; F. B. Jensen, unpublished data).

Oxygenation of hemoglobin also affects the erythrocyte pH. Hemoglobin binds protons upon deoxygenation, its charge decreases, and the anion ratio and intracellular pH increase. Conversely, protons are released upon oxygenation, whereby the anion ratio and intracellular pH decrease. The Haldane effect is very large in lamprey and many teleost fish—the intraerythrocytic pH at a constant extracellular pH increases up to 0.3–0.4 unit in tench and in the lampreys *Petromyzon marinus* and *Lampetra fluviatilis* when hemoglobin is deoxygenated (Jensen, 1986, 1989; Nikinmaa and Mattsoff, 1992; Ferguson *et al.*, 1992). In contrast, the Haldane effect of elasmobranch hemoglobins is very small—when determined by direct titration, the maximal proton uptake of carp hemoglobin upon deoxygenation is 0.95 proton per hemoglobin chain, but that of dogfish hemoglobin only 0.19 proton per chain (Jensen, 1989).

A. Agnathans

The control of pH in hagfish erythrocytes is influenced by the fact that the traditional anion exchange pathway is absent (Ellory *et al.*, 1987). Thus, the buffering of extracellular fixed acid loads by hemoglobin is slowed down. However, although the cells lack DIDS-sensitive chloride and bicarbonate transport, some equilibration of bicarbonate between plasma and erythrocytes occurs (Tufts and Boutilier, 1990a). One possible route for bicarbonate is the sodium-dependent carboxylic acid transporter of hagfish erythrocytes (Tiihonen, 1995).

Because the equilibration of anions across the erythrocyte membrane of hagfish is slow, even relatively slow secondarily active proton transport could influence erythrocyte pH. However, there is no indication that either pH or volume disturbances would activate ion transport pathways in the red blood cell membrane (Nikinmaa *et al.*, 1993). Thus, the intraerythrocytic pH of hagfish at the physiological pH range is much lower than the extracellular pH (Tufts and Boutilier, 1990a; Fig. 6).

Lamprey erythrocytes also lack a functional anion exchange, although a protein immunologically related to the anion exchanger has been described (Kay *et al.*, 1995; Cameron *et al.*, 1996). The erythrocyte membrane of lampreys thus has a very low permeability to bicarbonate and acid equivalents (Nikinmaa and Railo, 1987; Tufts and Boutilier, 1989). Lamprey erythrocytes are able to maintain a high intracellular pH (Nikinmaa, 1986; Fig. 6) by secondarily active sodium/proton exchange. The sodium/proton exchange is markedly activated by acidification (Virkki and Nikinmaa, 1994).

Since the intrinsic hemoglobin–oxygen affinity of lampreys is low and the Bohr effect large, the high erythrocyte pH is required to achieve effective oxygen loading in gills (Nikinmaa *et al.*, 1995). If the intracellular pH values were similar to those of rainbow trout, hemoglobin would reach only 60% oxygen saturation in the gills of the lamprey, *Lampetra fluviatilis*, in normoxia. With the high intracellular pH, the oxygen affinity of hemoglobin within lamprey erythrocytes is in the range expected for active teleost fish such as the salmonids (Nikinmaa *et al.*, 1995). The high intracellular pH is also required to ensure that the dissociation–association reactions of hemoglobin occur in intact erythrocytes since these reactions are the basis of any cooperative phenomena and of Bohr and Haldane effects of lamprey hemoglobins (see Perutz, 1990; Nikinmaa *et al.*, 1995).

B. Elasmobranchs

Elasmobranch erythrocytes conform to the basic pattern of erythrocytic pH control. The erythrocytes are characterized by a very rapid anion exchange (Obaid *et al.*, 1979) and high buffering capacity of hemoglobin (Jensen, 1989, 1991). Both of these factors facilitate the buffering of extracellular acid loads, but reduce the possibility of regulating hemoglobin function by secondarily active ion transport pathways such as sodium/proton exchange (see Nikinmaa, 1997). If changes in the intracellular pH play a role in the physiological adjustments of hemoglobin function, they are probably caused by changes in the organic phosphate concentrations within the erythrocyte and consecutive changes in the anion and proton distribution ratio since no published data are available indicating a role for secondarily active transport in the control of intracellular pH.

C. Teleosts

Similar to elasmobranch erythrocytes, the pH of unstimulated teleost erythrocytes conforms to the basic pattern of pH control. The major difference between teleost and elasmobranch fish is that the buffering capacity of teleost hemoglobins is generally much smaller, and the Haldane effects much larger, than those of elasmobranch fish (see, e.g., Jensen, 1991). As a consequence, extracellular acid loads should cause larger changes in the erythrocyte and plasma pH of teleosts than of elasmobranchs. Furthermore, the effect of oxygenation on the intracellular pH should be larger in teleost than in elasmobranch fish.

In many instances, the catecholamine-activated sodium/proton exchange influences the erythrocyte pH of teleost fish. The effects of catecholamine-sensitive sodium/proton exchange on the intraerythrocytic pH are known in great detail (for reviews, see, e.g., Motais *et al.*, 1992; Nikinmaa, 1992a; Thomas and Perry, 1992; Fig. 7). A reduction in the oxygen availability (normally a reduction in the arterial oxygen content; see Perry *et al.*,

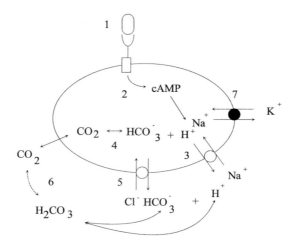

Fig. 7. Adrenergic effects on teleost erythrocytes. (1) Binding of catecholamines to β-adrenergic receptors on the erythrocyte membrane activates adenylate cyclase, and (2) the cAMP concentration increases. This leads to the activation of sodium/proton exchange (3) with the consequence that intraerythrocytic pH increases. As a result (4), more bicarbonate is formed from carbon dioxide, causing a disequilibrium for the anion exchanger. Thus, (5) chloride enters the cell in exchange for bicarbonate. The increase in intracellular pH and the decrease in extracellular pH, observed after adrenergic stimulation, occur because the extracellular buffering of protons is slow owing to the slow rate of dehydration of extracellular carbonic acid to carbon dioxide (6). The increase in intracellular sodium concentration increases the activity of the sodium pump (7), leading to a reduction in the cellular ATP concentration.

1989) /oxygen demand ratio at the level of chromaffin tissue causes liberation of catecholamines to the bloodstream (Perry *et al.,* 1991). Catecholamines bind to the β-adrenergic receptors on the red blood cell membrane. Binding of catecholamines to the receptors causes an accumulation of cyclic AMP (Mahé *et al.,* 1985) and activation of the sodium/proton exchange. Early studies (mainly on rainbow trout and carp) suggested that noradrenaline would always be a more potent activator of the system than adrenaline (e.g., Tetens *et al.,* 1988; Salama and Nikinmaa, 1990; Nikinmaa, 1992a). However, recent data (Berenbrink and Bridges, 1994) suggest that in some species such as the cod, *Gadus morhua,* adrenaline can be more potent than noradrenaline.

The sodium/proton exchange carries out proton extrusion and displaces protons from electrochemical equilibrium during the initial minutes of activation. This is possible even though the initial efflux of protons and influx of sodium via the sodium/proton exchange is only ca. 1/200 of the chloride and bicarbonate exchange fluxes via anion exchange (see Nikinmaa and Boutilier, 1995), as long as the turnover rate of the sodium/proton exchange approaches the extracellular uncatalyzed rate of dehydration of bicarbonate and protons to carbon dioxide, which is the speed by which protons are buffered extracellularly (Motais et al., 1989; Nikinmaa et al., 1990b; Nikinmaa, 1992a; Nikinmaa and Boutilier, 1995). Thus, a large number of protons can be extruded via the sodium/proton exchanger, and can accumulate in the incubation medium, before the carbonic acid formed from the accumulated protons and bicarbonate is dehydrated to carbon dioxide to a significant extent. After this, the proton and bicarbonate movements are coupled, and no further changes in the erythrocyte pH occur until the sodium/proton exchange is inactivated. However, during this time, there is a continuous net influx of sodium and chloride into the cell (for further details on the mechanism of adrenergic pH changes, see Nikinmaa, 1992a; Nikinmaa and Boutilier, 1995). Notably, the low buffering capacity of teleost hemoglobins contributes to the observed adrenergic pH changes: in elasmobranch fish, much more pronounced proton fluxes would be required for a similar change in intracellular pH because of the large buffering capacity.

There are pronounced species differences in the activity of the sodium/proton exchanger at any given temperature, and in the pH dependence of the activity. In salmonids, the adrenergic net sodium influx is maximal at around extracellular pH 7.3, whereas in carp and in the percid, Stizostedion lucioperca, the activity continues to increase at least down to pH 7 (Salama and Nikinmaa, 1989; Cossins and Kilbey, 1991). At physiological extracellular pH values under normoxic conditions, the activity of the sodium/proton exchange is much smaller in the cyprinids and percids than in the salmonids, and practically nonexistent in the eel (Hyde and Perry, 1990). Whenever the activity of the exchanger is small, there is almost exactly one-to-one coupling of the net sodium and chloride influxes after adrenergic stimulation, but when the activity of the sodium/proton exchange is large, the ratio between net sodium influx and net chloride influx clearly exceeds one, and can be more than two (Fig. 8). Thus the ratio of net sodium influx/net chloride influx can be used to obtain an idea of the activity of adrenergic sodium/proton exchanger.

In the case of the adrenergic sodium/proton exchange, the loose coupling of this transport pathway to the anion exchanger diminishes the intra- and extracellular pH changes. However, in the case of the coupling of the potassium/chloride co-transporter and the anion exchanger, pH changes will be generated. The potassium/chloride co-transporter reduces the intracellular and increases the extracellular chloride concentration. This causes a disequilibrium for the anion exchanger, and chloride reenters the cell in exchange for bicarbonate. A consequence of the net bicarbonate efflux is

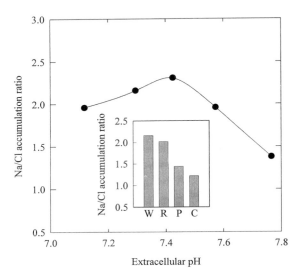

Fig. 8. The molar accumulation ratio of sodium and chloride during a 30-min incubation with 10^{-5} M isoproterenol as a function of pH in whitefish. An increase in the ratio indicates an increase in the activity of the sodium/proton exchanger and, consequently, larger intra- and extracellular pH changes. The inset shows the molar accumulation ratio of sodium and chloride during a 30-min incubation with 10^{-5} M isoproterenol at extracellular pH of 7.1 in whitefish (W), rainbow trout (R), pikeperch (P), and carp (C), indicating that in normoxic carp only slight pH effects are produced because of the modest activation of the sodium/proton exchange. Data from Salama and Nikinmaa (1989).

a reduction of intracellular pH. To date, changes in the intracellular pH as a consequence of the potassium and chloride efflux have not been measured. However, alkalinization of the extracellular compartment has been observed in nitrite-treated and hypotonically treated carp erythrocytes (Jensen, 1990, 1995), and in both cases, net potassium and chloride effluxes are seen. In view of this, one possible reason for the oxygenation-sensitive potassium/chloride efflux could be to reduce the intraerythrocytic pH at high oxygen tensions. This could, by reducing the oxygen affinity of hemoglobin, facilitate oxygen unloading in the arterial end of the capillary bed whenever fish are exposed to diurnally hyperoxic conditions as in many eutrophic lakes.

VIII. HEMOGLOBIN OXIDATION

Hemoglobin subunits that have been oxidized (the ferrous ion of heme is oxidized to ferric ion) are not capable of binding oxygen. Thus, methemoglobin formation either by autoxidation or during exposure to nitrite, reactive oxygen species, or thiols causes a reduction in the oxygen-carrying capacity of blood. In addition, partial oxidation of hemoglobin in fishes appears to decrease the oxygen affinity of the remaining functional hemoglobin molecules (Jensen et al., 1987; Jensen, 1990). A major reason for this decrease appears to be the concomitant reduction of red cell volume, and, possibly, decrease in intraerythrocytic pH (Jensen, 1990; Jensen et al., 1993). At the cellular level, methemoglobin reduction back to functional hemoglobin involves two systems (see Nikinmaa, 1990): the NADPH methemoglobin reductase system and NADH methemoglobin reductase system (Fig. 9). At present, the regulation of these reductive pathways is not known.

IX. RESPONSES OF HEMOGLOBIN FUNCTION TO CHANGES IN THE EXTERNAL AND THE INTERNAL ENVIRONMENT OF FISH

A. Temperature

An increase in temperature decreases the hemoglobin–oxygen affinity. This affinity decrease is beneficial as long as oxygen loading can be secured in the gills because a reduction in the hemoglobin–oxygen affinity increases the amount of oxygen given up at any capillary oxygen tension (or increases the oxygen partial pressure in the capillaries for a given amount of oxygen delivered). It appears that the oxygen affinity of hemoglobin is further decreased with an increasing temperature by an increase in the erythrocytic organic phosphate concentration, as long as arterial oxygen saturation can be secured (Nikinmaa et al., 1980; Laursen et al., 1985). However, at high temperatures, where the temperature-induced decrease in hemoglobin–oxygen affinity would significantly reduce oxygen loading in gills, the cellular NTP concentration is reduced (Nikinmaa et al., 1980; Albers et al., 1983; Laursen et al., 1985). Furthermore, the red cell magnesium concentration is increased (Houston and Koss, 1984), and since magnesium complexes with NTP, the binding of NTP to hemoglobin is reduced. Both responses increase hemoglobin–oxygen affinity. Indeed, Grigg (1969) reported that the hemoglobin–oxygen affinity of fish acclimated to high temperatures was higher than that of fish acclimated to low temperatures, when measurements were made at the same temperature.

An increase in temperature is usually associated with a decrease in blood (and erythrocyte) pH (Heisler, 1984). As long as the decrease in pH is such that the protein charge of hemoglobin is not affected, it does not

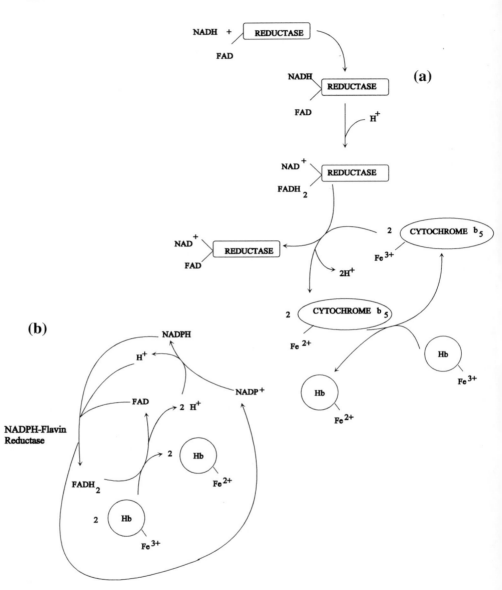

Fig. 9. Reduction of methemoglobin within the erythrocytes. (a) Cytochrome b_5 reductase pathway. (b) NADPH-flavin reductase pathway. Modified from Nikinmaa (1990).

cause an additional decrease in hemoglobin–oxygen affinity on top of the intrinsic temperature sensitivity of hemoglobin (see Nikinmaa, 1990). Only if the erythrocyte pH decreases more with a temperature increase than the pK value of histidine imidazole will the Bohr effect contribute to the observed decrease in hemoglobin–oxygen affinity.

In addition to the effects of temperature on hemoglobin function as such, seasonal phenomena may modify gas transport. It appears that erythrocyte methemoglobin concentrations are higher in temperate fish in winter than in summer (Graham and Fletcher, 1986), and that the β-adrenergic responsiveness of rainbow trout erythrocytes is reduced in winter (Nikinmaa and Jensen, 1986; Cossins and Kilbey, 1989). These responses reduce the effectiveness of oxygen transport by blood in cold environments and reduce the scope of activity in winter when food availability may be limited. In spring, it is probable that production of new erythrocytes increases, cellular methemoglobin levels fall, and the β-adrenergic activity of the sodium/proton exchange increases with increasing temperature. Thus, blood oxygen transport can respond appropriately to the large increases in activity brought about by the increase in temperature.

B. Hypoxia

The occurrence of environmental hypoxia is common especially in freshwater environments. Hypoxic conditions may be intermittent, occurring in eutrophied waters during the night, in which case the water may be hyperoxic during the day. In most cases, hypoxic conditions occur together with hypercapnia since oxygen depletion is normally due to the oxidation of organic matter and to the respiration of organisms with carbon dioxide as an end product. In addition, internal hypoxia that results from an increased diffusion distance between water and blood across the gills, as occurs during exposure to many pollutants, may develop in fishes (see Hughes, 1981). This section deals with responses to pure hypoxia; other forms of hypoxia are treated in a later sections.

When fish are subjected to a hypoxic environment, the hemoglobin oxygen saturation decreases, and the erythrocyte pH increases owing to the Haldane effect. As discussed earlier, this increase of erythrocyte pH will not affect the oxygen-binding properties of hemoglobin, since it is part of the hemoglobin mechanism. Fish also respond to hypoxia with immediate hyperventilation (e.g., Holeton and Randall, 1967). Thus, hypoxia will result in an increase in plasma pH as long as the repiratory alkalosis is greater than the metabolic acidosis that results from the inadequate oxygen supply to the tissues and consequent lactacidosis. The increase in plasma pH will, to some degree, be transmitted to erythrocytes, and the hemoglobin–oxygen affinity will increase.

In many teleost fish, the general alkalinization of plasma in acute hypoxia is interrupted 1–2 min after the onset of hypoxia by a marked acidification (Thomas and Hughes, 1982)—up to 0.5 pH unit. This acidification represents the activation of adrenergic sodium/proton exchange and is associated with a rapid increase in erythrocyte pH (Tetens and Christensen, 1987). The adrenergic increase in erythrocytic pH can be differentiated from the increase caused by deoxygenation of hemoglobin: in the adrenergic increase, there is an amiloride- and propranolol-inhibitable increase in cellular sodium concentration and an increase in chloride concentration that is partially inhibited by the aforementioned inhibitors; in the deoxygenation-induced increase only the chloride concentration increases, and the increase is not blocked by amiloride or propranolol (Nikinmaa et al., 1987).

The adrenergic sodium/proton exchange is uniquely suited to modulate erythrocyte pH under hypoxic conditions since most of the components of the receptor–effector cascade (Fig. 7) are hypoxia-sensitive. Catecholamines are released to the bloodstream in response to hypoxia (Tetens and Christensen, 1987). The number of β-adrenergic receptors on the erythrocyte membrane increases under hypoxic conditions (Marttila and Nikinmaa, 1988; Reid and Perry, 1991). The concentration of cAMP formed under hypoxic conditions appears to be higher than the formation under normoxic conditions in some species (Salama and Nikinmaa, 1990; Salama, 1993). The activity of sodium/proton exchange as such is increased in hypoxia (Motais et al., 1987; Salama and Nikinmaa, 1988), and the number of exchangers on the cell surface may also increase (Reid et al., 1993; Guizouarn et al., 1995). Notably, in normoxic tench and carp erythrocytes, catecholamines do not cause an increase in intracellular pH either in vivo or in vitro at normal plasma pH values (Jensen, 1987; Nikinmaa et al., 1987; Salama and Nikinmaa, 1988). However, when the arterial oxygen tension of carp is reduced, a pronounced increase in erythrocyte pH that is associated with a propranolol-inhibitable increase in intracellular sodium concentration and water content is seen (Nikinmaa et al., 1987). The increase in erythrocyte pH is the major reason for the rapid increase in hemoglobin–oxygen affinity of teleost fish in acute hypoxia (Nikinmaa, 1983; Tetens and Lykkeboe, 1985), although the dilution of hemoglobin and organic phosphates may also play a role (Soivio and Nikinmaa, 1981). In chronic hypoxia, other mechanisms must take over since a prolonged elevation of blood catecholamine levels diminishes the red cell adrenergic response (Thomas et al., 1991).

Elasmobranchs and agnathans do not respond to adrenergic stimulation via a significant activation of sodium/proton exchanger. However, in lamprey it is possible that hypoxia affects erythrocyte pH via activating sodium/proton exchange by some other mechanism—at least hypoxic conditions lead to an elevation of cellular sodium concentration, water content, and intracellular pH (Nikinmaa and Weber, 1984).

In chronic hypoxia, the most important mechanism for regulating hemoglobin function of fish (apart from agnathans) is the reduction of erythrocyte NTP concentrations (Wood and Johansen, 1972; Weber and Lykkeboe, 1978; Soivio et al., 1980). A reduction in cellular ATP and GTP concentrations increases the blood oxygen affinity both by reducing allosteric interaction of phosphates with hemoglobin and by increasing intraerythrocytic pH. The erythrocytic pH increases only if the total pool of nucleotide phosphates decreases since the negative charge of the hydrolysis products of NTPs is similar to that of NTPs themselves (see Nikinmaa, 1990). A reduction in the total pool of nucleotide phosphates does, indeed, take place under hypoxic conditions (Tetens, 1987; Fig. 3). A considerable decrease in NTP concentrations occurs within a few hours from the onset of hypoxia (Tetens and Lykkeboe, 1985; Jensen and Weber, 1985). Whenever both ATP and GTP are present in fish erythrocytes, the GTP concentrations decrease more rapidly and to a greater extent than the ATP concentrations (Weber and Lykkeboe, 1978). It appears that high red cell concentrations of GTP have been selected for in species, such as carp, tench, and eel, that encounter large fluctuations in oxygen tension in their natural habitats (Weber and Jensen, 1988).

C. Hypercapnia and Hypercapnic Hypoxia

An elevation of environmental carbon dioxide tension causes a respiratory acidosis (Heisler et al., 1976; Toews et al., 1983). Respiratory acidosis causes liberation of catecholamines to the bloodstream of rainbow trout under both hypoxic and normoxic conditions, but not under hyperoxic conditions (Perry and Kinkead, 1989; Perry et al., 1989). However, if hyperoxic hypercapnia is induced to fish made anemic before the exposure, a liberation of catecholamines to the bloodstream also occurs, indicating that a reduction in arterial oxygen content is the primary reason for catecholamine release (Perry et al., 1989). The liberation of catecholamines to the bloodstream causes an activation of the adrenergic sodium/proton exchanger, as indicated by the propranolol-inhibitable elevation of erythrocyte pH in normoxic hypercapnic fish compared to hyperoxic hypercapnic fish in which plasma catecholamine levels did not increase (Perry and Kinkead, 1989). As a consequence, the oxygen affinity of hemoglobin can be maintained. The situation in carp is somewhat different since the animals can significantly increase the arterial oxygen tension in response to hypercapnic conditions (Takeda, 1991). In hypercapnic carp, the change in oxygen tension is the major mechanism for maintaining constant oxygen saturation of blood (Takeda, 1991).

Hypercapnic hypoxia reduces arterial oxygen saturation, and despite the hyperventilation, a respiratory acidosis is induced owing to the elevated carbon dioxide tension of water (Jensen and Weber, 1982). Thus, opposite to "pure" hypoxia there is a tendency for a reduction of erythrocyte pH and for a consecutive decrease of hemoglobin–oxygen affinity (e.g., Jensen *et al.,* 1993). Apart from this difference, the responses to hypoxia in hypercapnic conditions are the same as in normocapnic hypoxia, i.e., initial activation of the sodium/proton exchange, which elevates intraerythrocytic pH (Thomas and Perry, 1994), and a slower reduction of erythrocytic NTP concentration, which, in the case of the tench *Tinca tinca,* is caused by the reduced GTP content (Jensen and Weber, 1985). Thus, there is a leftward shift of the oxygen equilibrium curve in hypercapnic, hypoxic animals compared to normoxic, normocapnic animals.

D. Hyperoxia

When fish experience environmental hyperoxia, their ventilatory drive is reduced since the regulation of ventilation is mainly governed by the oxygen demand of fish (Dejours, 1973). As a consequence of the reduced ventilatory volume, there is an accumulation of carbon dioxide in the blood and a decrease in plasma pH (Randall and Jones, 1973; Wood and Lemoigne, 1991). However, blood oxygenation can easily be maintained despite the acidification since blood oxygen tension increases (Fig. 10). The fact that oxygen saturation in the arterial blood does not increase despite the pronounced increase in arterial oxygen tension in rainbow trout suggests that the oxygen affinity of hemoglobin within the erythrocytes decreases in hyperoxia. It is possible that the oxygenation-dependent potassium/chloride co-transport and consecutive efflux of bicarbonate from the erythrocytes in exchange for chloride contribute to this, as there is an increase in plasma potassium concentration (Wheatly *et al.,* 1984), and the erythrocyte pH decreases during initial stages of hyperoxia (Wood and Lemoigne, 1991).

E. Environmental Acidification

In the absence of other disturbances, a decrease of water pH to values below 5 causes only small changes in the blood pH or oxygen tension of teleost fish, although there are reports of gill damage in acid-exposed animals (Daye and Garside, 1976). Malte (1986) observed a 0.15-unit decrease

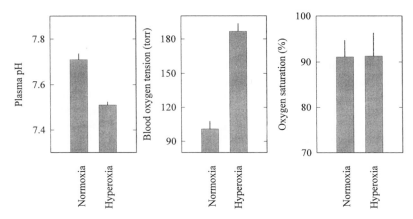

Fig. 10. Plasma pH, blood oxygen tension and blood oxygen saturation of six rainbow trout held in normoxic water (oxygen tension 140–155 torr) or subjected to 3-h hyperoxia (oxygen tension 240 torr) at 15°C. Hyperoxia induced a significant ($P < 0.05$; t test was used for comparisons) decrease in plasma pH and an increase in blood oxygen tension, but did not affect arterial oxygen saturation. M. Nikinmaa, unpublished data.

in the pH of rainbow trout during a 102-h exposure to pH 5; Wood *et al.* (1988a) observed a 0.1-unit decrease in the plasma pH of brook trout (*Salvelinus fontinalis*) during the first 40 h of a 10-day exposure to pH 4.8; Nikinmaa *et al.* (1990a), working on rainbow trout, and Van Dijk *et al.* (1993a), working on carp, did not see any effect of acidification on the pH of arterial blood during short-term exposures to pH 5 and 4, respectively. In contrast, similar treatment of the lamprey, *Lampetra fluviatilis,* caused a marked reduction in plasma pH, by 0.6 unit when the water pH was decreased from 7.8 to 5 for 24 h, and by 0.8 unit when the pH was decreased to 4 (Mattsoff and Nikinmaa, 1988). This difference between lamprey and teleost fish is largely because, in contrast to teleost fish in which extracellular acid loads can be buffered by hemoglobin, lamprey hemoglobin does not participate in rapid buffering of extracellular acid loads owing to the low permeability of erythrocyte membrane to acid equivalents (for a review, see Nikinmaa *et al.,* 1995).

If additional stresses occur simultaneously with acidification, the respiratory physiology of teleost fish is markedly affected. For example, the aluminum concentration of water commonly increases with acidification. In this case, the diffusion distance between water and blood increases markedly both because of gill damage and because of precipitation of aluminum hydroxide on the gill surface (see McDonald and Wood, 1993). Consequently, there is a marked reduction in arterial oxygen tension, oxygen saturation, and pH (Malte, 1986; Jensen and Weber, 1987). Similar gill damage occurs if the iron concentration of water is high, as is the case in many water bodies in Finland (Peuranen *et al.*, 1994).

Fish respond to this internal hypoxia as to an externally induced hypercapnic hypoxia: there is a liberation of catecholamines to the bloodstream (Witters *et al.*, 1991) and the erythrocyte volume increases (Milligan and Wood, 1982). The erythrocyte pH increases or is maintained in the face of marked extracellular acidification. However, it is likely that the erythrocyte pH at a constant oxygen saturation would be decreased, since, owing to the decrease in hemoglobin oxygen saturation, the change in erythrocyte pH includes the contribution of the marked deoxygenation-induced proton uptake by hemoglobin. The NTP concentration of the erythrocytes decreases, which in the case of tench is mainly due to a selective decrease on GTP concentration (Jensen and Weber, 1987). Thus, the oxygen affinity of hemoglobin at a given pH is increased. However, the increase cannot fully offset the effects of a marked decrease in oxygen tension and a reduction of erythrocyte pH at a given oxygen saturation that occur in acid-exposed fish since the blood oxygen saturation of acid-exposed fish remains much lower than that of nonexposed fish (Malte, 1986; Jensen and Weber, 1987; Malte and Weber, 1988). To some degree, preexposure to the metal results in increased tolerance (see MacDonald and Wood, 1993): there was no reduction in arterial oxygen tension in fish preacclimated to aluminum-containing (150 μg L^{-1}) water at pH 5.2, when they were further exposed to pH 4.8 and 333 μg L^{-1} aluminum (Wood *et al.*, 1988b).

The respiratory function of blood is also markedly affected if acid-exposed fish are exercised or exposed to hypoxia. Nikinmaa *et al.* (1990a) exposed rainbow trout to either pH 5 for 24 h or to hypoxia (environmental oxygen tension 55–60 torr) or their combination. Neither treatment alone caused any changes in the respiratory function of the erythrocytes. However, hypoxia in acid water caused a pronounced release of catecholamines to blood, a marked reduction in blood NTP concentration, and an increase in erythrocytic sodium concentration. Furthermore, there was a pronounced loss of sodium and chloride from plasma. Similarly, Van Dijk *et al.* (1993a,b) exposed carp to a gradual decrease of pH from 7 to 4 in 4 h. In resting animals, this did not influence plasma pH or sodium and chloride concentrations. However, when animals were exercised, there was a decrease in plasma pH and sodium and chloride concentrations. These results indicate that hyperventilation, when occurring in acid-exposed fish, plays an important role in proton accumulation into the animal and ion loss to the environment and that, in addition to aluminum, many other physiological stresses influence "acid toxicity."

F. Salinity Changes

A transfer of fish from freshwater to seawater (rainbow trout) or to brackish water (2 or 2.5% salinity; whitefish) causes both a transient reduction of arterial oxygen content and a respiratory acidosis, probably because of a reduced diffusion conductance in gills (Maxime *et al.*, 1991; Larsen and Jensen, 1993; Madsen *et al.*, 1996). In the whitefish (*Coregonus lavaretus*), the acidosis persists for at least 48 h, but shifts from being predominantly respiratory to predominantly metabolic (Madsen *et al.*, 1996). It is, as yet, uncertain whether the respiratory acidosis observed in seawater elicits the normal rapid response to hypoxia, i.e., the compensatory increase in erythrocyte pH via catecholamine stimulation of the sodium/proton exchange. In the whitefish, some protection of the erythrocyte pH may have occurred since the slope of the relationship between erythrocyte pH and plasma pH ($\Delta pH_i/\Delta pH_e$) was only 0.26 (Madsen *et al.*, 1996), i.e., much shallower than in teleost fish in general (0.6–0.9). However, no data are available on catecholamine concentrations nor on the effects of β-adrenergic blockade on the responses to seawater-induced changes in the oxygen transport parameters.

G. Exercise

The responses of hemoglobin function to exercise depend on the severity of exercise. As long as the increased oxygen demand of muscle work can be fulfilled by an increase in the ventilatory volume and in the cardiac output, a rightward shift of the oxygen equilibrium curve in the capillary bed is beneficial since more oxygen can be unloaded from hemoglobin at a given oxygen tension. A facilitation of oxygen delivery requires that the pH of capillary blood at a given oxygen saturation of hemoglobin decreases. Such a decrease is caused by the flux of carbon dioxide and metabolic acid from the working muscles into the capillaries.

During exhaustive exercise, a mixed respiratory and metabolic acidosis develops (e.g., Jensen *et al.*, 1983). In the absence of compensatory responses in the oxygen transport system, this would seriously affect the oxygen loading in gills. Teleost fish have two types of responses enhancing oxygen transport. In species with high resting blood oxygen tensions, such as the salmonids and the striped bass, *Morone saxatilis,* the plasma catecholamine concentration is increased (Ristori and Laurent, 1985; Butler *et al.*, 1986), activating the adrenergic sodium/proton exchange. This prevents or reduces the decrease in the erythrocyte pH that would otherwise take place (Nikinmaa *et al.*, 1984; Primmett *et al.*, 1986; Milligan and Wood, 1987). Plasma catecholamine concentration increases much more when there is a "psychological" component (tail grabbing or chasing with a stick) in the stress than in burst swimming alone (cf. Ristori and Laurent, 1985; Butler et al., 1986; Milligan et al., 1989). Repeated stresses (e.g., chasing until exhaustion daily for a week) cause an increase in the responsiveness of erythrocytes to catecholamines (Perry et al., 1996). It appears that this effect is due to an increase in the sensitivity of the sodium/proton exchanger to cAMP, since the accumulation of cAMP was not affected, but the dose–response curve for the cAMP-sensitive sodium accumulation was shifted to lower cAMP concentrations (Perry et al., 1996).

In species with a low resting oxygen tension of blood, such as tench and carp, the major blood response to normoxic exercise is a marked elevation of arterial oxygen tension, from 20–40 up to 100 torr (e.g., Jensen et al., 1983). Thus, although the oxygen equilibrium curve is shifted to the right owing to the exercise-induced acidification, the increase in arterial oxygen tension limits the reduction in hemoglobin–oxygen saturation. In these species, catecholamines are not capable of activating the sodium/proton exchange at normal blood pH values under normoxic conditions (Jensen, 1987; Salama and Nikinmaa, 1988).

In lampreys, the permeability of the red blood cell membrane to acid equivalents is very low (e.g., Nikinmaa and Railo, 1987). Thus, the metabolic proton production that takes place during exhaustive exercise will only acidify the plasma compartment. Indeed, exhaustive exercise causes a pronounced reduction in both arterial and venous plasma pH in *Petromyzon marinus* by 0.36 and 0.46 pH unit, respectively, but does not affect erythrocyte pH (Tufts et al., 1992). However, exhaustive exercise also causes a pronounced carbon dioxide load, as evident by the increase of carbon dioxide tension by 0.39 kPa in venous and by 0.17 kPa in arterial blood (Tufts et al., 1992), which should be transmitted to the erythrocyte in lamprey and cause a reduction in erythrocyte pH. This is the case if the erythrocyte pH is extrapolated to a constant oxygen saturation. The data of Tufts et al. (1992) show that the arterial oxygen saturation decreased from ca. 95 to ca. 75%, and the venous saturation from ca. 75 to ca. 18% following exhaustive exercise. In the absence of a carbon dioxide load, these decreases in oxygen saturation would have caused approximately 0.1- and 0.25-unit increases in the pH of arterial and venous erythrocytes, respectively, owing to the pronounced Haldane effect of lamprey hemoglobin (calculations based on the data of Ferguson et al., 1992). Thus, the carbon dioxide load causes a 0.1- to 0.25-pH-unit decrease in the erythrocyte pH, relative to that in resting lampreys. This decrease causes a reduction in arterial oxygen saturation (Tufts et al., 1992). Thus, it appears that the sodium/proton exchange of lamprey erythrocytes, which is activated by acidification (Virkki and Nikinmaa, 1994), is not able to correct the decrease in erythrocyte pH, at least during short-term exhaustive exercise.

H. Anemia

When fish become anemic, they hyperventilate, the carbon dioxide tension of dorsal aortic blood decreases, and pH increases (Fig. 11). These changes tend to increase the oxygen affinity of blood. However, in anemic fish such a response would be maladaptive since a reduction of blood oxygen capacity as such will, in addition to reducing the amount of oxygen carried by unit volume of blood, reduce the partial pressure of oxygen at which a given amount of oxygen is given up in the tissues. Thus, if a further increase in oxygen affinity took place, the capillary oxygen tension would be further reduced, as would the diffusion of oxygen from capillary blood to tissues. In view of this, a decrease of oxygen affinity would be beneficial, and this is, indeed, what is observed in anemic fish. Anemia is associated with an increase in erythrocytic NTP concentration (Lane *et al.*, 1981; Vorger and Ristori, 1985; Jensen *et al.*, 1990), and consequently, a significant increase in the P_{50} value of blood (Vorger and Ristori, 1985).

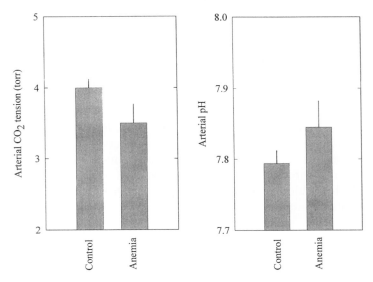

Fig. 11. The arterial carbon dioxide tension and pH of normal (Hct value 28%; $N = 14$) and anemic (Hct value 15%; $N = 8$) rainbow trout at 10°C. Anemia induced a significant ($P < 0.05$; t test was used for comparisons) decrease in arterial carbon dioxide tension and an increase in pH. A. Soivio, K. Nyholm, K. Westman, and M. Nikinmaa, unpublished data.

I. Environmental Pollutants

Various toxicants affect blood oxygen transport by causing gill damage and consecutive internal hypoxia (e.g., Mallatt, 1985; McDonald and Wood, 1993). As illustrated by the aluminum and acidification-induced internal hypoxia, the responses are similar to the ones caused by environmental hypoxia (see also Hughes, 1981). In other cases, effects on blood oxygen transport occur independently from effects on other tissues. The effects of pollutants on erythrocyte function have been discussed by Nikinmaa (1992b). *Nitrite* and other oxidizing agents convert functional hemoglobin to methemoglobin, which does not bind oxygen (e.g., Jensen *et al.*, 1987; Jensen, 1990). This decreases the oxygen capacity of blood markedly. Furthermore, there is a significant reduction in the oxygen affinity of the remaining functional heme groups (Jensen *et al.*, 1987; Jensen, 1990). Both effects reduce oxygen transport to the tissues. Nitrite also inhibits the adrenergically activated sodium/proton exchange of erythrocyte membranes (Nikinmaa and Jensen, 1992), thus reducing the possibility of utilizing the β-adrenergic increase in erythrocyte pH in acute hypoxia or exercise. *Tributyltin chloride* similarly inhibits sodium/proton exchange across the erythrocyte membrane (Virkki and Nikinmaa, 1993). Alkyltin compounds also function as chloride/hydroxyl ion exchangers, causing a marked reduction of the erythrocyte pH in lamprey (Tufts and Boutilier, 1990b; Nikinmaa *et al.*, 1995) and consecutive reduction in hemoglobin–oxygen affinity. *Resin acids*, important constituents of paper and pulp mill effluents, decrease erythrocyte pH in lampreys (A. Bogdanova and M. Nikinmaa, unpublished data) since they behave as protonophores. Resin acids also reduce the energy production of fish erythrocytes (Bushnell *et al.*, 1985), which leads to a reduction in cellular ATP levels, and in the long term to altered red cell shape and red cell breakdown (see Nikinmaa, 1992b). These examples indicate that many toxicants affect red cell function and, consequently, oxygen transport. These changes, when occurring together with natural variations in environmental temperature and oxygen tension, may be a decisive factor behind fish deaths and reduction in fish growth and reproduction in contaminated environments.

REFERENCES

Airaksinen, S., and Nikinmaa, M. (1995). Effect of haemoglobin concentration on the oxygen affinity of intact lamprey erythrocytes. *J. Exp. Biol.* **198,** 2393–2396.

Albers, C., Goetz, K.-H., and Hughes, G. M. (1983). Effect of acclimation temperature on intraerythrocytic acid–base balance and nucleoside triphosphates in the carp. *Cyprinus Carpio. Respir. Physiol.* **54**, 145–159.

Aschauer, H., Weber, R. E., and Braunitzer, G. (1985). The primary structure of the hemoglobin of the dogfish shark (*Squalus acanthias*): Antagonist effects of ATP and urea on oxygen affinity on an elasmobranch hemoglobin. *Biol. Chem. Hoppe-Seyler.* **366**, 589–599.

Bauer, C., Engels, U., and Paleus, S. (1975). Oxygen binding to haemoglobins of the primitive vertebrate *Myxine glutinosa* L. *Nature* **256**, 66–68.

Berenbrink, M., and Bridges, C. R. (1994). Catecholamine-activated sodium/proton exchange in the red blood cells of the marine teleost *Gadus morhua. J. Exp. Biol.* **192**, 253–267.

Bird, D. J., Lutz, P. L, and Potter, I. C. (1976). Oxygen dissociation curves of the blood of larval and adult lampreys (*Lampetra fluviatilis*). *J. Exp. Biol.* **65**, 449–458.

Bollard, B. A., Pankhurst, N. W., and Wells, R. M. G. (1993). Effects of artificially elevated plasma cortisol levels on blood parameters in the teleost fish *Pagrus auratus* (*Sparidae*). *Comp. Biochem. Physiol. A* **106**, 157–162.

Borgese, F., Garcia-Romeu, F., and Motais, R. (1986). Catecholamine-induced transport systems in trout erythrocyte: Na^+/H^+ countertransport or NaCl cotransport? *J. Gen. Physiol.* **87**, 551–566.

Borgese, F., Motais, R., and Garcia-Romeu, F. (1991). Regulation of chlorine-dependent potassium transport by oxy–deoxyhemoglobin transitions in trout red cells. *Biochim. Biophys. Acta* **1066**, 252–256.

Boyd, T. A., Cha, C. J., Forster, R. P., and Goldstein, L. (1977). Free amino acids in tissues of the skate *Raja erinacea* and the stingray *Dasyatis sabina*: Effects of environmental dilution. *J. Exp. Zool.* **199**, 435–442.

Brett, J. R., and Glass, N. R. (1973). Metabolic rates and critical swimming speeds of sockeye salmon (*Oncorhynchus nerca*) in relation to size and temperature. *J. Fish. Res. Board Can.* **30**, 379–387.

Bunn, H. F. Ransil, B. J., and Chao, A. (1971). The interaction between erythrocyte organic phosphates, magnesium ion and hemoglobin. *J. Biol. Chem.* **246**, 5273–5279.

Bursell, J. D. H., and Kirk, K. (1996). Swelling-activated K^+ transport via two functionally distinct pathways in eel erythrocytes. *Am. J. Physiol.* **270**, R61–R70.

Bushnell, P. G., Nikinmaa, M., and Oikari, A. (1985). Metabolic effects of dehydroabietic acid on rainbow trout erythrocytes. *Comp. Biochem. Physiol. C* **81**, 391–394.

Butler, P. J., Metcalfe, J. D., and Ginley, S. A. (1986). Plasma catecholamines in the lesser spotted dogfish and in rainbow trout at rest and during different levels of exercise. *J. Exp. Biol.* **123**, 409–421.

Cala, P. M. (1977). Volume regulation by flounder red blood cells in anisotonic media. *J. Gen. Physiol.* **69**, 537–552.

Cameron, B. A., Perry, S. F. II, Wu, C., Ko, K., and Tufts, B. L. (1996). Bicarbonate permeability and immunological evidence for an anion exchange-like protein in the red blood cells of the sea lamprey, *Petromyzon marinus. J. Comp. Physiol. B* **166**, 197–204.

Carey, F. G., and Gibson, Q. H. (1977). Reverse temperature dependence of tuna hemoglobin oxygenation. *Biochem. Biophys. Res. Commun.* **78**, 1376–1382.

Colclasure, G. C., and Parker, J. C. (1991). Cytosolic protein concentration is the primary volume signal in dog red cells. *J. Gen. Physiol.* **98**, 881–892.

Colclasure, G. C., and Parker, J. C. (1992). Cytosolic protein concentration is the primary volume signal for swelling-induced [K-Cl] cotransport in dog red cells. *J. Gen. Physiol.* **100**, 1–10.

Cossins, A. R., and Gibson, J. S. (1997). Volume-sensitive transport systems and volume homeostasis in vertebrate red blood cells. *J. Exp. Biol.* **200**, 343–352.

Cossins, A. R., and Kilbey, R. V. (1989). The seasonal modulation of Na$^+$/H$^+$ exchanger activity in trout erythrocytes. *J. Exp. Biol.* **144**, 463–478.

Cossins, A. R., and Kilbey, R. V. (1991). Adrenergic responses and the Root effect in erythrocytes of freshwater fish. *J. Fish. Biol.* **38**, 421–429.

Cossins, A. R., and Richardson, P. A. (1985). Adrenalin-induced Na$^+$/H$^+$ exchange in trout erythrocytes and its effects upon oxygen-carrying capacity. *J Exp. Biol.* **118**, 229–246.

Cossins, A. R., Weaver, Y. R., Lykkeboe, G., and Nielsen, O. B. (1994). Role of protein phosphorylation in control of K flux pathways of trout red blood cells. *Am. J. Physiol.* **267**, C1641–C1650.

Daye, P. G., and Garside, E. T. (1976). Histopathologic changes in surficial tissue of brook trout, *Salvelinus fontinalis* (Mitchill), exposed to acute and chronic levels of pH. *Can. J. Zool.* **54**, 2140–2146.

DeGroot, H., and Noll, T. (1987). Oxygen gradients: The problem of hypoxia. *Biochem. Soc. Trans.* **15**, 363–365.

Dejours, P. (1973). Problems of control of breathing in fishes. *In* "Comparative Physiology: Locomotion, Respiration, Transport and Blood" (L. Bolis, K. Schmidt-Nielsen, and S. H. P. Maddrell, eds.), pp. 117–133. North-Holland/American Elsevier, Amsterdam.

Ellory, J. C., Wolowyk, M. W., and Young, J. D. (1987). Hagfish (*Eptatretus stouti*) erythrocytes show minimal chloride transport activity. *J. Exp. Biol.* **129**, 377–383.

Fairbanks, M. B., Hoffert, J. R., and Fromm, P. O. (1969). The dependence of the oxygen-concentrating mechanism of the teleost eye (*Salmo gairdneri*) on the enzyme carbonic anhydrase. *J. Gen. Physiol.* **54**, 203–211.

Ferguson, R. A., Sehdev, N., Bagatto, B., and Tufts, B. L. (1992). *In vitro* interactions between oxygen and carbon dioxide transport in the blood of the sea lamprey (*Petromyzon marinus*). *J. Exp. Biol.* **173**, 25–41.

Ferguson, R. A., and Boutilier, R. G. (1988). Metabolic energy production during adrenergic pH regulation in red cells of the Atlantic salmon, *Salmo salar. Respir. Physiol.* **74**, 65–76.

Fincham, D. A., Wolowyk, M. W., and Young, J. D. (1987). Volume-sensitive taurine transport in fish erythrocytes. *J. Membr. Biol.* **96**, 45–56.

Fronticelli, C., Bucci, E., and Orth, C. (1984). Solvent regulation of oxygen affinity in haemoglobin. Sensitivity of bovine haemoglobin to chloride ions. *J. Biol. Chem.* **259**, 10841–10844.

Fugelli, K., and Rohrs, H. (1980). The effect of Na$^+$ and osmolality on the influx and steady state distribution of taurine and γ-aminobutyric acid in flounder (*Platichthys flesus*) erythrocytes. *Comp. Biochem. Physiol. A* **67**, 545–551.

Fugelli, K., and Thoroed, S. M. (1986). Taurine transport associated with cell volume regulation in flounder erythrocytes under anisosmotic conditions. *J. Physiol. (London)* **374**, 245–261.

Fugelli, K., and Zachariassen, K. E. (1976). The distribution of taurine, γ-aminobutyric acid and inorganic ions between plasma and erythrocytes in flounder (*Platichthys flesus*) at different plasma osmolalities. *Comp. Biochem. Physiol. A* **55**, 173–177.

Fyhn, U. E. H., and Sullivan, B. (1975). Elasmobranch hemoglobins: dimerization and polymerization in various species. *Comp. Biochem. Physiol. B* **50**, 119–129.

Gallardo Romero, M., Guizouarn, H., Pellissier, B., Garcia-Romeu, F., and Motais, R. (1996). The erythrocyte Na$^+$/H$^+$ exchangers of eel (*Anguilla anguilla*) and rainbow trout (*Oncorhynchus mykiss*): A comparative study. *J. Exp. Biol.* **199**, 415–426.

Garcia-Romeu, F., Cossins, A. R., and Motais, R. (1991). Cell volume regulation by trout erythrocytes: Characteristics of the transport systems activated by hypotonic swelling. *J. Physiol. (London)* **440**, 547–567.

Geoghegan, W. D., and Poluhowich, J. J. (1974). The major erythrocytic organic phosphates of the American eel, *Anguilla rostrata. Comp. Biochem. Physiol. B* **49**, 281–290.

Gillen, R. G., and Riggs, A. (1973). Structure and function of the isolated hemoglobins of the American eel. *J. Biol. Chem.* **246,** 1961–1969.

Goldstein, L., and Musch, M. W. (1994). Volume-activated amino acid transport and cell signaling in skate erythrocytes. *J. Exp. Zool.* **268,** 133–138.

Graham, M. S., and Fletcher, G. L. (1986). High concentrations of methemoglobin in five species of temperate marine teleosts. *J. Exp. Zool.* **239,** 139–142.

Greaney, G. S., and Powers, D. A. (1978). Allosteric modifiers of fish hemoglobins: In vitro and in vivo studies of the effect of ambient oxygen and pH on erythrocyte ATP concentrations. *J. Exp. Zool.* **203,** 339–350.

Grigg, G. C. (1969). Temperature-induced changes in the oxygen equilibrium curve of the blood of the brown bullhead, *Ictalurus nebulosus. Comp. Biochem. Physiol.* **28,** 1203–1223.

Guesnon, P., Poyart, C., Bursaux, E., and Bohn, B. (1979). The binding of lactate and chloride ions to human adult hemoglobin. *Respir. Physiol.* **38,** 115–129.

Guizouarn, H., Harvey, B. J., Borgese, F., Gabillat, N., Garcia-Romeu, F., and Motais, R. (1993). Volume-activated Cl^--independent and Cl^--dependent K^+ pathways in trout red blood cells. *J. Physiol. (London)* **462,** 609–626.

Guizouarn, H., Borgese, F., Pellissier, B., Garcia-Romeu, F., and Motais, R. (1995). Regulation of Na^+/H^+ exchange activity by recruitment of new Na^+/H^+ antiporters: Effect of calyculin A. *Am. J. Physiol.* **268,** C434–C441.

Gusev, G. P., Sherstobitov, A. O., and Bogdanova, A.Y. (1992). Sodium transport in red blood cells of lamprey *Lampetra fluviatilis. Comp. Biochem. Physiol. A* **103,** 763–766.

Hardisty, M. W. (1979). "Biology of the Cyclostomes." pp. 428. Chapman and Hall, London.

Heisler, N. (1984). Acid–base regulation in fishes. In *"Fish Physiology"* (W. S. Hoar and D. J. Randall, eds.), Vol. XA, pp. 315–401. Academic Press, New York.

Heisler, N., Weitz, H., and Weitz A. M. (1976). Hypercapnia and resultant bicarbonate transfer processes in an elasmobranch fish (*Scyliorhinus stellaris*). *Bull. Eur. Physiopathol. Respir.* **12,** 77–86.

Hershko, A., Razin, A., Shoshani, T., and Mager, J. (1967). Turnover of purine nucleotides in rabbit erythrocytes. II. Studies in vitro. *Biochim. Biophys. Acta* **149,** 59–73.

Hladky, S. B., and Rink, T. J. (1977). pH equilibrium across the red cell membrane. In "Membrane Transport in Red Cells." (J. C. Ellory and V. L. Lew, eds.), pp. 115–135. Academic Press, London.

Holeton, G. F., and Randall, D. J. (1967). The effect of hypoxia upon the partial pressure of gases in the blood and water afferent and efferent to the gills of rainbow trout. *J. Exp. Biol.* **46,** 317–327.

Houston, A. H., and Koss, T. F. (1984). Erythrocytic haemoglobin, magnesium and nucleoside triphosphate levels in rainbow trout exposed to progressive heat stress. *J. Therm. Biol.* **9,** 159–164.

Hughes, G. M. (1981). Effects of low oxygen and pollution on the respiratory systems of fish. In "Stress and Fish" (A. D. Pickering, ed.), pp. 121–146. Academic Press, London.

Hughes, G. M., and Kikuchi, Y. (1984). Effect of *in vivo* and *in vitro* changes in pO_2 on the deformability of red blood cells of rainbow trout (*Salmo gairdneri* R.). *J. Exp. Biol.* **111,** 253–257.

Hyde, D. A., and Perry, S. F. (1990). Absence of adrenergic red cell pH and oxygen content regulation in American eel (*Anguilla rostrata*) during hypercapnic acidosis *in vivo* and *in vitro. J. Comp. Physiol. B* **159,** 687–693.

Ingermann, R. L., Bissonnette, J. M., and Hall, R. E. (1985). Sugar uptake by red blood cells. In "Circulation, Respiration and Metabolism" (R. Gilles, ed.), pp. 290–300. Springer-Verlag, Berlin.

Jensen, F. B. (1986). Pronounced influence of Hb–O_2 saturation on red cell pH in tench blood *in vivo* and *in vitro. J. Exp. Zool.* **238**, 119–124.

Jensen, F. B. (1987). Influences of exercise-stress and adrenaline upon intra- and extracellular acid–base status, electrolyte composition and respiratory properties of blood in tench (*Tinca tinca*) at different seasons. *J. Comp. Physiol. B* **157**, 51–60.

Jensen, F. B. (1989). Hydrogen ion equilibria in fish haemoglobins. *J. Exp. Biol.* **143**, 225–234.

Jensen, F. B. (1990). Nitrite and red cell function in carp: Control factors for nitrite entry, membrane potassium ion permeation, oxygen affinity and methaemoglobin formation. *J. Exp. Biol.* **152**, 149–166.

Jensen, F. B. (1991). Multiple strategies in oxygen and carbon dioxide transport by erythrocytes. *In* "Physiological Strategies for Gas Exchange and Metabolism" (A. J. Woakes, M. K. Grieshaber, and C. R. Bridges, eds.), pp. 55–78. Cambridge Univ. Press, Cambridge.

Jensen, F. B. (1995). Regulatory volume decrease in carp red blood cells: Mechanisms and oxygenation-dependency of volume-activated potassium and amino acid transport. *J. Exp. Biol.* **198**, 155–165.

Jensen, F. B., and Weber, R. E. (1982). Respiratory properties of tench blood and hemoglobin: Adaptation to hypoxic-hypercapnic water. *Molec. Physiol.* **2**, 235–250.

Jensen, F. B., and Weber, R. E. (1985). Kinetics of the acclimational responses of tench to combined hypoxia and hypercapnia. I. Respiratory responses. *J. Comp. Physiol. B* **156**, 197–203.

Jensen, F. B., and Weber, R. E. (1987). Internal hypoxia-hypercapnia in tench exposed to aluminium in acid water: Effects on blood gas transport, acid–base status and electrolyte composition in arterial blood. *J. Exp. Biol.* **127**, 427–442.

Jensen, F. B., Nikinmaa, M., and Weber, R. E. (1983). Effects of exercise stress on acid–base balance and respiratory function in blood of the teleost *Tinca tinca. Respir. Physiol.* **51**, 291–301.

Jensen, F. B., Andersen, N. A., and Heisler, N. (1987). Effects of nitrite exposure on blood respiratory properties, acid–base and electrolyte regulation in the carp (*Cyprinus carpio*). *J. Comp. Physiol. B* **157**, 533–541.

Jensen, F. B., Andersen, N. A., and Heisler, N. (1990). Interrelationships between red cell nucleoside triphosphate content, and blood pH, O_2-tension and haemoglobin concentration in the carp, *Cyprinus carpio. Fish Physiol. Biochem.* **8**, 459–464.

Jensen, F. B., Nikinmaa, M., and Weber, R. E. (1993). Environmental perturbations of oxygen transport in teleost fishes: Causes, consequences and compensations. *In* "Fish Ecophysiology" (J. C. Rankin and F. B. Jensen, eds.), pp. 161–179. Chapman and Hall, London.

Jensen, F. B., Fago, A., and Weber, R. E. (1998). Haemoglobin structure and function. *In* "Fish Physiology, Vol. XVII" (S. F. Perry and B. Tufts, eds.). Academic Press, San Diego.

Kay, M. M., Cover, C., Schluter, S. F., Bernstein, R. M., and Marchalonis, J. J. (1995). Band 3, the anion transporter, is conserved during evolution: Implications for aging and vertebrate evolution. *Cell Mol. Biol.* **41**, 833–842.

Kirk, K., Ellory, J. C., and Young, J. D. (1992). Transport of organic substrates via a volume-activated channel. *J. Biol. Chem.* **267**, 23475–23478.

Lane, H. C., Rolfe, A. E., and Nelson, J. R. (1981). Changes in the nucleotide triphosphate/hemoglobin and nucleotide triphosphate/red cell ratios of rainbow trout, *Salmo gairdneri* Richardson, subjected to prolonged starvation and bleeding. *J. Fish Biol.* **18**, 661–668.

Larsen, B. K., and Jensen, F. B. (1993). Arterial PO_2, acid–base status, and red cell nucleoside triphosphates in rainbow trout transferred from fresh water to 20-percent sea water. *J. Fish Biol.* **42**, 611–614.

Lauf, P. K. (1982). Evidence for chloride dependent potassium and water transport induced by hyposmotic stress in erythrocytes of the marine teleost, *Opsanus tau. J. Comp. Physiol. B* **146**, 9–16.

Laursen, J. S., Andersen, N. A., and Lykkeboe, G. (1985). Temperature acclimation and oxygen binding properties of the European eel, *Anguilla anguilla*. *Comp. Biochem. Physiol. A* **81**, 79–86.

Lykkeboe, G., and Weber, R. E. (1978). Changes in the respiratory properties of the blood in the carp, *Cyprinus carpio*, induced by diurnal variation in ambient oxygen tension. *J. Comp. Physiol.* **128**, 117–125.

Madsen, S. S., Larsen, B. K., and Jensen, F. B. (1996). Effects of freshwater to seawater transfer on osmoregulation, acid–base balance and respiration in river migrating whitefish (*Coregonus lavaretus*). *J. Comp. Physiol. B* **166**, 101–109.

Mahé, Y., Garcia-Romeu, F., and Motais, R. (1985). Inhibition by amiloride of both adenylate cyclase activity and the Na^+/H^+ antiporter in fish erythrocytes. *Eur. J. Pharmacol.* **116**, 199–206.

Mallatt, J. (1985). Fish gill structural changes induced by toxicants and other irritants: A statistical review. *Can. J. Fish. Aquat. Sci.* **42**, 630–648.

Malte, H. (1986). Effects of aluminium in hard, acid water on metabolic rate, blood gas tensions and ionic status in the rainbow trout. *J. Fish. Biol.* **29**, 187–198.

Malte, H., and Weber, R. E. (1988). Respiratory stress in rainbow trout dying from aluminium exposure in soft, acid water, with or without added sodium chloride. *Fish. Physiol. Biochem.* **5**, 249–256.

Marttila, O. N. T., and Nikinmaa, M. (1988). Binding of β-adrenergic antagonists ^{3}H-DHA and ^{3}H-CGP 12177 to intact rainbow trout (*Salmo gairdneri*) and carp (*Cyprinus carpio*) red blood cells. *Gen. Comp. Endocrinol.* **70**, 429–435.

Mattsoff, L., and Nikinmaa, M. (1988). Effects of external acidification on the blood acid–base status and ion concentrations of lamprey. *J. Exp. Biol.* **136**, 351–361.

Maxime, V., Pennec, J. P., and Peyraud, C. (1991). Effects of direct transfer from freshwater to seawater on respiratory and circulatory variables and acid–base status in rainbow trout. *J. Comp. Physiol. B* **161**, 557–568.

McDonald, D. G., and Wood, C. M. (1993). Branchial mechanisms of acclimation to metals in freshwater fish. *In* "Fish Ecophysiology" (J. C. Rankin and F. B. Jensen, eds.), pp. 297–321. Chapman and Hall, London.

Milligan, C. L., and Wood, C. M. (1982). Disturbances in haematology, fluid volume distribution and circulatory function associated with low environmental pH in the rainbow trout, *Salmo gairdneri*. *J. Exp. Biol.* **99**, 397–415.

Milligan, C. L., and Wood, C. M. (1986). Intracellular and extracellular acid–base status and H^+ exchange with the environment after exhaustive exercise in the rainbow trout. *J. Exp. Biol.* **123**, 93–121.

Milligan, C. L., and Wood, C. M. (1987). Regulation of blood oxygen transport and red cell pHi after exhaustive activity in rainbow trout (*Salmo gairdneri*) and starry flounder (*Platichthys stellatus*). *J. Exp. Biol.* **133**, 263–282.

Milligan, C. L., Graham, M. S., and Farrell, A. P. (1989). The response of trout red cells to adrenaline during seasonal acclimation and changes in temperature. *J. Fish Biol.* **35**, 229–236.

Motais, R., Garcia-Romeu, F., and Borgese, F. (1987). The control of Na^+/H^+ exchange by molecular oxygen in trout erythrocytes: A possible role of hemoglobin as a transducer. *J. Gen. Physiol.* **90**, 197–207.

Motais, R., Fievet, B., Garcia-Romeu, F., and Thomas, S. (1989). Na^+-H^+ exchange and pH regulation in red blood cells: Role of uncatalyzed H_2CO_3 dehydration. *Am. J. Physiol.* **256**, C728–C735.

Motais, R., Borgese, F., Fievet, B., and Garcia-Romeu, F. (1992). Regulation of Na^+/H^+ exchange and pH in erythrocytes of fish. *Comp. Biochem. Physiol. A* **102**, 597–602.

Nielsen, O. B., and Lykkeboe, G. (1992). *In vitro* effects of pH and hemoglobin–oxygen saturation on plasma and erythrocyte potassium levels in blood from trout. *J. Appl. Physiol.* **72**, 1291–1296.

Nielsen, O. B., Lykkeboe, G., and Cossins, A. R. (1992). Oxygenation-activated K-fluxes in trout red blood cells. *Am. J. Physiol.* **263**, C1057–C1064.

Nikinmaa, M. (1982). Effects of adrenaline on red cell volume and concentration gradient of protons across the red cell membrane in the rainbow trout, *Salmo gairdneri*. *Mol. Physiol.* **2**, 287–297.

Nikinmaa, M. (1983). Adrenergic regulation of haemoglobin oxygen affinity in rainbow trout red cells. *J. Comp. Physiol. B* **152**, 67–72.

Nikinmaa, M. (1986). Red cell pH of lamprey (*Lampetra fluviatilis*) is actively regulated. *J. Comp. Physiol. B* **156**, 747–750.

Nikinmaa, M. (1990). "Vertebrate Red Blood Cells." Springer-Verlag, Berlin.

Nikinmaa, M. (1992a). Membrane transport and the control of haemoglobin–oxygen affinity in nucleated erythrocytes. *Physiol. Rev.* **72**, 301–321.

Nikinmaa, M. (1992b). How does environmental pollution affect red cell function in fish? *Aquat. Toxicol.* **22**, 227–238.

Nikinmaa, M. (1993). Haemoglobin function in intact *Lampetra fluviatilis* erythrocytes. *Respir. Physiol.* **91**, 283–293.

Nikinmaa, M. (1997). Oxygen and carbon dioxide transport in vertebrate erythrocytes: An evolutionary change in the role of membrane transport. *J. Exp. Biol.* **200**, 369–380.

Nikinmaa, M., and Boutilier, R. G. (1995). Adrenergic control of red cell pH, organic phosphate concentrations and haemoglobin function in teleost fish. *In* "Advances in Comparative and Environmental Physiology," Vol. 21, "Mechanisms of Systemic Regulation: Respiration and Circulation" (N. Heisler, ed.), pp. 107–133. Springer-Verlag, Berlin.

Nikinmaa, M., and Huestis, W. H. (1984). Adrenergic swelling in nucleated erythrocytes: Cellular mechanisms in a bird, domestic goose, and two teleosts, striped bass and rainbow trout. *J. Exp. Biol.* **113**, 215–224.

Nikinmaa, M., and Jensen, F. B. (1986). Blood oxygen transport and acid–base status of stressed trout (*Salmo gairdneri*): Pre- and postbranchial values in winter fish. *Comp. Biochem. Physiol. A* **84**, 391–396.

Nikinmaa, M., and Jensen, F. B. (1992). Inhibition of the adrenergic sodium/proton exchange activity in rainbow trout red cells by nitrite-induced methaemoglobinaemia. *J. Comp. Physiol. B* **162**, 424–429.

Nikinmaa, M., and Mattsoff, L. (1992). Effects of oxygen saturation on the CO_2 transport properties of *Lampetra* red cells. *Respir. Physiol.* **87**, 219–230.

Nikinmaa, M., and Railo, E. (1987). Anion movements across lamprey (*Lampetra fluviatilis*) red cell membrane. *Biochim. Biophys. Acta* **899**, 134–136.

Nikinmaa, M., and Weber, R. E. (1984). Hypoxic acclimation in the lamprey, *Lampetra fluviatilis*: Organismic and erythrocytic responses. *J. Exp. Biol.* **109**, 109–119.

Nikinmaa, M., and Weber, R. E. (1993). Gas transport in lamprey erythrocytes. *In* "The Vertebrate Gas Transfer Cascade: Adaptations to Environment and Mode of Life" (J. E. P. W. Bicudo, ed.), pp. 179–187. CRC Press, Boca Raton, FL.

Nikinmaa, M., Tuurala, H., and Soivio, A. (1980). Thermoacclimatory changes in blood oxygen binding properties and gill secondary lamellar structure of *Salmo gairdneri*. *J. Comp. Physiol. B* **140**, 255–260.

Nikinmaa, M., Cech, J. J., Jr., and McEnroe, M. (1984). Blood oxygen transport in stressed striped bass (*Morone saxatilis*): Role of beta-adrenergic responses. *J. Comp. Physiol. B* **154**, 365–369.

Nikinmaa, M., Cech, J. J., Jr., Ryhänen, E.-L., and Salama, A. (1987). Red cell function of carp (*Cyprinus carpio*) in acute hypoxia. *Exp. Biol.* **47,** 53–58.

Nikinmaa, M., Salama, A., and Tuurala, H. (1990a). Respiratory effects of environmental acidification in perch (*Perca fluviatilis*) and rainbow trout (*Salmo gairdneri*). In "Acidification in Finland" (P. Kauppi, P. Anttila, and K. Kenttämies, eds.), pp. 929–940. Springer-Verlag, Berlin.

Nikinmaa, M., Tiihonen, K., and Paajaste, M. (1990b). Adrenergic control of red cell pH in salmonid fish: Roles of the sodium/proton exchange, Jacobs–Stewart cycle and membrane potential. *J. Exp. Biol.* **154,** 257–271.

Nikinmaa, M., Tufts, B. L., and Boutilier, R. G. (1993). Volume and pH regulation in Agnathan erythrocytes—Comparisons between the hagfish, *Myxine glutinosa*, and the lampreys, *Petromyzon marinus* and *Lampetra fluviatilis*. *J. Comp. Physiol. B* **163,** 608–613.

Nikinmaa, M., Airaksinen, S., and Virkki, L. V. (1995). Haemoglobin function in intact lamprey erythrocytes: Interactions with membrane function in the regulation of gas transport and acid–base balance. *J. Exp. Biol.* **198,** 2423–2430.

Nilsson, S., and Grove, D. J. (1974). Adrenergic and cholinergic innervation of the spleen of the cod, *Gadus morhua*. *Eur. J. Pharmacol.* **28,** 135–143.

Obaid, A. L., McElroy Critz, A., and Crandall, E. D. (1979). Kinetics of bicarbonate/chloride exchange in dogfish erythrocytes. *Am. J. Physiol.* **237,** R132–R138.

Olson, K. R. (1984). Distribution of flow and plasma skimming in isolated perfused gills of three teleosts. *J. Exp. Biol.* **109,** 97–108.

Orlov, S. N., and Skryabin, G. A. (1993). Catecholamine-dependent and volume-dependent ion fluxes in carp (*Cyprinus-carpio*) red blood cells. *J. Comp. Physiol. B* **163,** 413–420.

Orlov, S. N, Cragoe, E. J., Jr., and Hänninen, O. (1994). Volume- and catecholamine-dependent regulation of Na/H antiporter and unidirectional potassium fluxes in *Salmo trutta* red blood cells. *J. Comp. Physiol. B* **164,** 135–140.

Parks, R. E., Jr., Brown, P. R., Cheng, Y.-C., Agarwal, K. C., Kong, C. M., Agarwal, R. P., and Parks, C. C. (1973). Purine metabolism in primitive erythrocytes. *Comp. Biochem. Physiol. B* **45,** 355–364.

Pelster, B., and Scheid, P. (1992). Countercurrent concentration and gas secretion in the fish swim bladder. *Physiol. Zool.* **65,** 1–16.

Perry, S. F., and Kinkead, R. (1989). The role of catecholamines in regulating arterial oxygen content during acute hypercapnic acidosis in rainbow trout (*Salmo gairdneri*). *Respir. Physiol.* **77,** 365–378.

Perry, S. F., Kinkead, R., Gallaugher, P., and Randall, D. J. (1989). Evidence that hypoxemia promotes catecholamine release during hypercapnic acidosis in rainbow trout (*Salmo gairdneri*). *Respir. Physiol.* **77,** 351–364.

Perry, S. F., Fritsche, R., Kinkead, R., and Nilsson, S. (1991). Control of catecholamine release *in vivo* and *in situ* in the Atlantic cod (*Gadus morhua*) during hypoxia. *J. Exp. Biol.* **155,** 549–566.

Perry, S. F., Reid, S. G., and Salama, A. (1996). The effects of repeated physical stress on the β-adrenergic response of the rainbow trout red blood cell. *J. Exp. Biol.* **199,** 549–562.

Perutz, M. (1990). "Mechanisms of Cooperativity and Allosteric Regulation in Proteins." Cambridge Univ. Press; Cambridge.

Peuranen, S., Vuorinen, P. J., Vuorinen, M., and Hollender, A. (1994). The effects of iron, humic acids and low pH on the gills and physiology of brown trout (*Salmo trutta*). *Ann. Zool. Fennici* **31,** 389–396.

Primmett, D. R. N., Randall, D. J., Mazeaud, M., and Boutilier, R. G. (1986). The role of catecholamines in erythrocyte pH regulation and oxygen transport in rainbow trout (*Salmo gairdneri*) during exercise. *J. Exp. Biol.* **122,** 139–148.

Randall, D. J., and Jones, D. R. (1973). The effect of deafferentation of the pseudobranch on the respiratory response to hypoxia and hyperoxia in the trout (*Salmo gairdneri*). *Respir. Physiol.* **17,** 291–301.

Reid, S. D., and Perry, S. F. (1991). The effects and physiological consequences of raised levels of cortisol on rainbow trout (*Oncorhynchus mykiss*) erythrocyte β-adrenoreceptors. *J. Exp. Biol.* **158,** 217–240.

Reid, S. D., Lebras, Y., and Perry, S. F. (1993). The *in vitro* effect of hypoxia on the trout erythrocyte β-adrenergic signal transduction system. *J. Exp. Biol.* **176,** 103–116.

Reinhart, W. H., and Chien, S. (1985). Roles of cell geometry and cellular viscosity in red cell passage through narrow pores. *Am. J. Physiol.* **248,** C473–C479.

Ristori, M. T., and Laurent, P. (1985). Plasma catecholamines and glucose during moderate exercise in the trout: Comparison with bursts of violent activity. *Exp. Biol.* **44,** 247–253.

Sackin, H. (1989). A stretch-activated K⁺ channel sensitive to cell volume. *Proc. Natl. Acad. Sci. USA* **86,** 1731–1735.

Salama, A. (1993). The role of cAMP in regulating the β-adrenergic response of rainbow trout (*Oncorhynchus mykiss*) red blood cells. *Fish Physiol. Biochem.* **10,** 485–490.

Salama, A., and Nikinmaa, M. (1988). The adrenergic responses of carp (*Cyprinus carpio*) red cells: Effects of Po_2 and pH. *J. Exp. Biol.* **136,** 405–416.

Salama, A., and Nikinmaa, M. (1989). Species differences in the adrenergic responses of fish red cells: Studies on whitefish, pikeperch, trout and carp. *Fish. Physiol. Biochem.* **6,** 167–173.

Salama, A., and Nikinmaa, M. (1990). Effect of oxygen tension on catecholamine-induced formation of cAMP and on swelling of carp red blood cells. *Am. J. Physiol.* **259,** C723–C726.

Sephton, D. H., Macphee, W. L., and Driedzic, W. R. (1991). Metabolic enzyme activities, oxygen consumption and glucose utilization in sea raven (*Hemitripterus americanus*) erythrocytes. *J. Exp. Biol.* **159,** 407–418.

Soivio, A., and Nikinmaa, M. (1981). The swelling of erythrocytes in relation to the oxygen affinity of the blood of the rainbow trout, *Salmo gairdneri* Richardson. *In* "Stress and Fish" (A. D. Pickering, ed.), pp. 103–119. Academic Press, London.

Soivio, A., Nikinmaa, M., and Westman, K. (1980). The blood oxygen binding properties of hypoxic *Salmo gairdneri*. *J. Comp. Physiol. B* **136,** 83–87.

Steffensen, J. F., Lomholt, J. P., and Vogel, W. O. P. (1986). *In vivo* observations on a specialized microvasculature, the primary and secondary vessels in fishes. *Acta Zool. (Stockholm)* **67,** 193–200.

Stevens, E. D. (1968). The effect of exercise on the distribution of blood to various organs in rainbow trout. *Comp. Biochem. Physiol.* **25,** 615–625.

Takeda, T. (1991). Regulation of blood oxygenation during short-term hypercapnia in the carp, *Cyprinus carpio*. *Comp. Biochem. Physiol. A* **98,** 517–522.

Tetens, V. (1987). Regulation of blood O_2 affinity during acute hypoxic exposure of rainbow trout, *Salmo gairdneri*: Organismal and cellular processes. Ph.D. thesis. Aarhus Univ., Denmark.

Tetens, V., and Christensen, N. J. (1987). Beta-adrenergic control of blood oxygen affinity in acutely hypoxia exposed rainbow trout. *J. Comp. Physiol. B* **157,** 667–675.

Tetens, V., and Lykkeboe, G. (1981). Blood respiratory properties of rainbow trout *Salmo gairdneri*: Responses to hypoxia acclimation and anoxic incubation of blood *in vitro*. *J. Comp. Physiol. B* **145,** 117–125.

Tetens, V., and Lykkeboe, G. (1985). Acute exposure of rainbow trout to mild and deep hypoxia: O_2 affinity and O_2 capacitance of arterial blood. *Respir. Physiol.* **61,** 221–235.

Tetens, V., Lykkeboe, G., and Christensen, N. J. (1988). Potency of adrenaline and noradrenaline for β-adrenergic proton extrusion from red cells of rainbow trout, *Salmo gairdneri*. *J. Exp. Biol.* **134**, 267–280.

Thomas, S., and Hughes, G. M. (1982). Effects of hypoxia on blood gas and acid–base parameters of sea bass. *J. Appl. Physiol.* **53**, 1336–1341.

Thomas, S., and Perry, S. F. (1992). Control and consequences of adrenergic activation of red blood cell Na$^+$/H$^+$ exchange on blood oxygen and carbon dioxide transport in fish. *J. Exp. Zool.* **263**, 160–175.

Thomas, S., and Perry, S. F. (1994). Influence of initial respiratory status on the short- and long-term activity of the trout red blood cell β-adrenergic Na$^+$/H$^+$ exchanger. *J. Comp. Physiol. B* **164**, 383–389.

Thomas, S., Kinkead, R., Walsh, P. J., Wood, C. M., and Perry, S. F. (1991). Desensitization of adrenaline-induced red blood cell H$^+$ extrusion *in vitro* after chronic exposure of rainbow trout to moderate environmental hypoxia. *J. Exp. Biol.* **156**, 233–248.

Tiihonen, K. (1995). Substrate transport and utilization in fish erythrocytes. Ph.D. thesis. Univ. of Helsinki, Finland.

Tiihonen, K., and Nikinmaa, M. (1991a). Substrate utilization by carp (*Cyprinus carpio*) erythrocytes. *J. Exp. Biol.* **161**, 509–514.

Tiihonen, K., and Nikinmaa, M. (1991b). D-Glucose permeability in river lamprey (*Lampetra fluviatilis*) and carp (*Cyprinus carpio*) erythrocytes. *Comp. Biochem. Physiol. A* **100**, 581–584.

Toews D. P., Holeton G. F., and Heisler, N. (1983). Regulation of the acid–base status during environmental hypercapnia in the marine teleost fish *Conger conger*. *J. Exp. Biol.* **107**, 9–20.

Tosteson, D. C., and Hoffman, J. F. (1960). Regulation of cell volume by active cation transport in high and low potassium sheep red cells. *J. Gen. Physiol.* **44**, 169–194.

Tse, C., and Young, J. D. (1990). Glucose transport in fish erythrocytes: Variable cytochalasin-B-sensitive hexose transport activity in the common eel (*Anguilla japonica*) and transport deficiency in the paddyfield eel (*Monopterus albus*) and rainbow trout (*Salmo gairdneri*). *J. Exp. Biol.* **148**, 367–383.

Tufts, B. L. (1991). *In vitro* evidence for sodium-dependent pH regulation in sea lamprey (*Petromyzon marinus*) red blood cells. *Can. J. Zool.* **70**, 411–416.

Tufts, B. L., and Boutilier, R. G. (1989). The absence of rapid chloride/bicarbonate exchange in lamprey erythrocytes: Implications for CO$_2$ transport and ion distributions between plasma and erythrocytes in the blood of *Petromyzon marinus*. *J. Exp. Biol.* **144**, 565–576.

Tufts, B. L., and Boutilier, R. G. (1990a). CO$_2$ transport properties of the blood of a primitive vertebrate *Myxine glutinosa*. *Exp. Biol.* **48**, 341–347.

Tufts, B. L., and Boutilier, R. G. (1990b). CO$_2$ transport in agnathan blood: Evidence of erythrocyte Cl$^-$/HCO$_3^-$ exchange limitations. *Respir. Physiol.* **80**, 335–348.

Tufts, B. L., and Boutilier, R. G. (1991). Interactions between ion exchange and metabolism in erythrocytes of the rainbow trout *Oncorhynchus mykiss*. *J. Exp. Biol.* **156**, 139–151.

Tufts, B. L. and Randall, D. J. (1989). The functional significance of adrenergic pH regulation in fish erythrocytes. *Can. J. Zool.* **67**, 235–238.

Tufts, B. L., Bagatto, B., and Cameron, B. (1992). *In vivo* analysis of gas transport in arterial and venous blood of the sea lamprey *Petromyzon marinus*. *J. Exp. Biol.* **169**, 105–119.

Van Dijk, P. L. M., Van Den Thillart, G. E. E. J. M., Balm, P., and Wendelaar Bonga, S. E. (1993a). The influence of gradual water acidification on the acid/base status and plasma hormone levels in carp. *J. Fish Biol.* **42**, 661–671.

Van Dijk, P. L. M., Van Den Thillart, G. E. E. J. M., and Wendelaar Bonga, S. E. (1993b). Is there a synergistic effect between steady-state exercise and water acidification in carp? *J. Fish Biol.* **42**, 673–681.

Virkki, L., and Nikinmaa, M. (1993). Tributyltin inhibition of adrenergically activated sodium-proton exchange in erythrocytes of rainbow trout (*Oncorhynchus mykiss*). *Aquat. Toxicol.* **25**, 139–146.

Virkki, L. V., and Nikinmaa, M. (1994). Activation and physiological role of Na$^+$/H$^+$ exchange in lamprey (*Lampetra fluviatilis*) erythrocytes. *J. Exp. Biol.* **191**, 89–105.

Virkki, L.V., and Nikinmaa, M. (1995). Regulatory volume decrease in lamprey erythrocytes: Mechanisms of K$^+$ and Cl$^-$ loss. *Am. J. Physiol.* **268**, R590–R597.

Vorger, P., and Ristori, M. T. (1985). Effects of experimental anemia on the ATP content and the oxygen affinity of the blood in the rainbow trout (*Salmo gairdneri*). *Comp. Biochem. Physiol.* **82**, 221–224.

Walsh, P. J., Wood, C. M., Thomas, S., and Perry, S. F. (1990). Characterization of red blood cell metabolism in rainbow trout. *J. Exp. Biol.* **154**, 475–489.

Weber, R. E. (1978). Functional interaction between fish hemoglobin, erythrocyte nucleoside triphosphates and magnesium. *Acta Physiol. Scand.* **102**, 20A–21A.

Weber, R. E., and Jensen, F. B. (1988). Functional adaptations in hemoglobins from ectothermic vertebrates. *Annu. Rev. Physiol.* **50**, 161–179.

Weber, R. E., and Lykkeboe, G. (1978). Respiratory adaptations in carp blood. Influences of hypoxia, red cell organic phosphates, divalent cations and CO$_2$ on hemoglobin–oxygen affinity. *J. Comp. Physiol. B* **128**, 127–137.

Weber, R. E., Lykkeboe, G., and Johansen, K. (1976a). Physiological properties of eel haemoglobin: Hypoxic acclimation, phosphate effects and multiplicity. *J. Exp. Biol.* **64**, 75–88.

Weber, R. E., Wood, S. C., and Lomholt, J. P. (1976b). Temperature acclimation and oxygen-binding properties of blood and multiple haemoglobins of rainbow trout. *J. Exp. Biol.* **65**, 333–345.

Wheatly, M. G., Höbe, H., and Wood, C. M. (1984). The mechanisms of acid–base and ionoregulation in the freshwater rainbow trout during environmental hyperoxia and subsequent normoxia. II. The role of the kidney. *Respir. Physiol.* **55**, 155–173.

Witters, H. E., Van Puymbroeck, S., and Vanderborght, O. L. J. (1991). Adrenergic response to physiological disturbances in rainbow trout, *Oncorhynchus mykiss,* exposed to aluminum at acid pH. *Can. J. Fish. Aquat. Sci.* **48**, 414–420.

Wood, C. M., and Lemoigne, J. (1991). Intracellular acid–base responses to environmental hyperoxia and normoxic recovery in rainbow trout. *Respir. Physiol.* **86**, 91–114.

Wood, C. M., Playle, R. C., Simons, B. P., Goss, G. G., and McDonald, D. G. (1988a). Blood gases, acid–base status, ions, and hematology in adult brook trout (*Salvelinus fontinalis*) under acid/aluminum exposure. *Can. J. Fish. Aquat. Sci.* **45**, 1575–1586.

Wood, C. M., Simons, B. P., Mount, D. R., Bergman, H. L. (1988b). Physiological evidence of acclimation to acid/aluminum stress in adult brook trout (*Salvelinus fontinalis*). 2. Blood parameters by cannulation. *Can. J. Fish. Aquat. Sci.* **45**, 1597–1605.

Wood, S. C., and Johansen, K. (1972). Adaptation to hypoxia by increased HbO$_2$ affinity and decreased red cell ATP concentration. *Nature* **237**, 278–279.

Wood, S. C., and Johansen, K. (1973). Organic phosphate metabolism in nucleated red cells: Influence of hypoxia on eel HbO$_2$ affinity. *Neth. J. Sea. Res.* **7**, 328–338.

Yamamoto, K., Itazawa, Y., and Kobayashi, H. (1980). Supply of erythrocytes into the circulating blood from the spleen of exercised fish. *Comp. Biochem. Physiol. A* **65**, 5–13.

6

HEMATOCRIT AND BLOOD
OXYGEN-CARRYING CAPACITY

P. GALLAUGHER AND A. P. FARRELL

I. Introduction
II. The Influence of Sampling Methodology on Hematocrit
    A. Red Blood Cell Swelling
    B. Plasma Skimming
    C. Changes in Plasma Volume
    D. Splenic Release of Red Blood Cells
    E. Serial Blood Sampling
III. Interspecific Diversity in Hematocrit Values
IV. Intraspecific Regulation of Hematocrit
    A. Erythropoiesis
    B. Long-Term Influences on Hct
    C. Short-Term Influences on Hct
V. Critique of the Optimal Hematocrit Theory
VI. Conclusions
    References

I. INTRODUCTION

The amount of O_2 taken up from the environment and transported via the cardiovascular system in vertebrates to meet tissue O_2 requirements (V_{O_2}) is a function of both blood flow and blood O_2-carrying capacity. The O_2-carrying capacity of the blood is primarily determined by the concentration of the O_2-binding protein hemoglobin (Hb), which is packaged into red blood cells. Thus, the percentage of red blood cells in blood (hematocrit, Hct) is not only a measure of the capacity of the blood to carry O_2, but a major determinant of arterial O_2 content. Remarkably, reported Hct values in fish species range from 0 to >50%, the largest range of any vertebrate group. In addition, reported intraspecific values vary considerably. The purpose of this chapter is to describe this variability and, where possible, identify factors known to influence Hct.

185

Fish Physiology, Volume 17:
FISH RESPIRATION

On a very broad scale, there are data to suggest what factors set the boundary conditions for Hct. For example, anemic states can limit metabolic scope in salmonids as measured by either a reduction in swimming performance (Jones, 1971) and/or maximum O_2 consumption (V_{O_2max}) (Thorarensen et al., 1993; Gallaugher et al., 1995). In this way, the lower limit for Hct is likely related to a reduction in metabolic scope. At the other extreme, polycythemia increases fish blood viscosity (Gallaugher et al., 1995) because red blood cells are a major determinant of blood viscosity (Graham and Fletcher, 1983). Polycythemia therefore will be associated with increased resistance to blood flow and, unless there is a compensatory vasodilatation, the heart must generate sufficient blood pressure to overcome this vascular resistance or blood flow will decrease (see Farrell and Jones, 1992). For the most part, these boundary conditions of anemia and polycythemia may not be commonplace, except under pathological, toxicological, and deficient nutritional states. Therefore, within a given species and under a particular environmental setting, Hct operates within a relatively narrow range. This general observation, together with some experimental work, has lead to the long-standing optimal Hct hypothesis: Vertebrate Hct is set to maximize the rate of O_2 transport in arterial blood (T_{O_2}), without compromising blood flow and cardiac work through elevated blood viscosity (Richardson and Guyton, 1961). However, recent experimental evidence involving polycythemia sheds doubt on the validity of this hypothesis for fish. A discussion of the inadequacy of the optimal hematocrit theory as a framework for explaining routine Hct values in fish is found in Section V at the end of this chapter.

On an evolutionary scale, one might expect that interspecific differences in Hct (and cardiac performance) are matched with V_{O_2}. For example, there are indications that the most active fishes with the greatest V_{O_2} have the highest Hct values. However, as shown in Section III, broad correlations do not show clear patterns, and only limited relationships exist between Hct and metabolic scope.

At the scale of the individual, although dynamic changes in Hct could theoretically match those in V_{O_2} (e.g., during aerobic swimming or exposure to hypoxia), they are often (but not always) of lesser magnitude than the dynamic changes in blood flow and O_2 extraction at the tissues. Nonetheless, physiological state (e.g., stress, life cycle, prolonged swimming) and environmental factors (e.g., temperature, hypoxia, photoperiod, season, nutrition, toxicology) influence Hct, and these situations are considered in Section IV under intraspecific variability in Hct.

Hct might be considered one of the simplest cardiovascular variables that can be measured in fishes. Yet, the variability in Hct resulting from different sampling methodologies is appreciable. To understand this measurement variability requires an appreciation of short-term, transient changes in Hct. Factors involved in these changes in Hct include changes in red blood cell size resulting from fluid shifts between plasma and the red blood cell, release of stores of red blood cells, and changes in plasma volume resulting from fluid shifts between plasma and the interstitial space. These factors are considered in Section II.

II. THE INFLUENCE OF SAMPLING METHODOLOGY ON HEMATOCRIT

There are two common methods for sampling blood in fishes. Acute sampling involves the puncture of either a blood vessel or the heart or the severance of the tail to drain blood from cut vessels. Acute sampling is common in older studies, field work, and whenever fish are too small for cannulation. Fish may be unanethestized, anesthetized, and/or killed with a blow to the head prior to acute sampling. These procedures produce a variable amount of stress depending on the degree of handling and air exposure involved. Repetitive sampling is obviously impossible with this method, and so statistical analysis of data cannot use a repeated-measures approach and is therefore open to the vagaries of individual variation in Hct. In adult fish, it is often possible to take several samples of blood. However, sequential sampling employing acute sampling techniques would tend to compound the effects of the associated stress factors on the measured Hct values.

The second method of blood sampling involves the chronic implantation of a cannula into a blood vessel. This technique allows for less stressful sampling of blood. In addition, repetitive blood sampling from a single fish allows for powerful statistical analysis using repeated measures. Cannulation, however, can lead to other sorts of complications such as anemia if there is excessive blood loss during surgery or leakage around the cannulation site post-surgery. Our experience has been that certain cannulation sites are more subject to bleeding than others; e.g., seepage of blood around the cannulation site in the ventral aorta is greater than that for the dorsal aorta. Repetitive blood sampling, if appreciable, can also significantly reduce Hct. Saline replacement, reinfusion of blood used for gas measurements, and, more recently, whole blood replacement from a donor fish have been used to minimize this problem. An adaptation of the cannulation method involves a loop between arterial and venous circulations, an extracorporeal circulation.

Both sampling methodologies have the potential for a large margin of error unless the quality control of the researcher is exemplary. Obviously, such errors could obscure more subtle physiological changes due to experimental treatments. Moreover, the errors are in opposite directions; whereas acute sampling invariably overestimates Hct, chronic sampling is more likely to underestimate Hct.

To illustrate this difference, a selected database on Hct for rainbow trout (Table 1) is arranged according to the two sampling techniques.

Table 1

Hct, [Hb], and MCHC for Selected Studies with Rainbow Trout (*Oncorhynchus mykiss*)[a]

Sampling method	Variable	Het (vol%)	[Hb](g/100 ml)	MCHC (g/L)	Ref.
Acute					
Hypoxia effects					
Kill, caudal puncture	Normoxia	25.1			Wells and Weber, 1990
	Hypoxia (chronic)	19.2			
	Hypoxia (acute)	32.8			
	Induced exercise	41.8			
Caudal puncture	Normoxia	28.2	7.95		Tetens and Lykkeboe, 1981
	Hypoxia (80 mm Hg)	28.5	8.02		
	Hypoxia (50 mm Hg)	35.3	7.15		
Caudal puncture	Normoxia	26.3			Bushnell *et al.*, 1984
	Chronic hypoxia (2w)	25.2			
Hypoxia, photoperiod, temperature effects:					
Kill, cardiac puncture	Normoxia (5°C, 16L:8D)	33	6.9	211	Tun and Houston, 1986
	Hypoxia (5°C, 16L:8D)	32.4	7.2	229	
	Normoxia (5°C, 8L:8D)	28	6.3	227	
	Hypoxia (5°C, 8L:8D)	41.2	8.7	213	
	Normoxia (20°C, 16L:8D)	32.4	6.7	209	
	Hypoxia (20°C, 16L:8D)	36.5	7.4	203	
	Normoxia (20°C, 8L:16D)	34.1	7.7	234	
	Hypoxia (20°C, 8L:16D)	45.9	9.8	214	
Kill, cardiac puncture	Acclimated 10–25°C	27.1–31.5	7.2–6.9		Houston and Koss, 1984
Kill, cardiac puncture	Winter	29	6.9		Houston *et al.*, 1996a
	Spring	24.5	5.5		
	Summer	24.5	4.7		
Section tail	11.9°C	36.2			Jones, 1971
	22.6°C	34.4			

188

(continues)

					Reference
Kill, cardiac puncture	Summer fish, 2°C	32.8	8.1	240	Houston and Smeda, 1979
	Summer fish, 10°C	33.7	8.2	211	
	Summer fish, 18°C	33.2	8.1	222	
	Winter fish (2°C)	31.5	8.1	215	
	Winter fish (10°C)	31.6	8.2	240	
	Winter fish (18°C)	30.8	7.6	207	
Swimming effects					
Caudal peduncle	Sham-operated, U_{crit}	42	9.92	236	Pearson and Stevens, 1991b
	Splenectomized, U_{crit}	33.1	7.96	229	
Exercise training effects					
Decerebrate	Untrained		9.1		Hochachka, 1961
	Exercise trained		11.4		
	Wild stock		11.9		
Heart puncture	Untrained	44.2	8.4	191	Davie et al., 1986
	Exercise trained	39.3	8.5	218	
Habitat and starvation effects					
Caudal puncture	Female, lake		9.9		Graham and Farrell, 1992
	Female, anadromous		10.9		
	Male, lake		10.1		
	Male, anadromous		10.1		
	Male, hatchery		7.7		
	Lake		8.8		
	Anadromous		7.9		
Cardiac puncture	Control	32.3	10.29	327	Lane et al., 1981
	Starvation (30d)	24	7.61	316	
Toxicant exposure effects					
MS222	Pair-fed control	38.3	7.91		Lanno and Dixon, 1996
Cervical dislocation	35 SCN mg/L	24.8	4.49		
	Pair-fed control	36.9	8.34		
	77 SCN mg/L	24.2	4.62		
	Pair-fed control	37.4	7.8		
	115 SCN mg/L	17,8	3.73		

Table 1 (*Continued*)

Sampling method	Variable	Hct (vol%)	[Hb](g/100 ml)	MCHC (g/L)	Ref.
Sever caudal peduncle	Control	31.7			McLeay and Gordon, 1977
	Pulpmill effluent	28.5			
Sampling technique effects					
Caudal puncture, MS222		41.3	7.2		Korcock *et al.*, 1988
Caudal puncture, stunned		39.5	9.2		
Heart puncture		22	5.41	246	Pearson and Stevens, 1991a
Sever caudal peduncle		19.6	3.83	202	
MS222 injection		20.3	4.15	205	
Inject MS222 and net		38.3	9.75	251	
Chronic Cannulation					
DA cannula		27.9	10.8		Tetens and Lykkeboe, 1985
DA cannula		20.6	6.1		Perry *et al.*, 1989
DA cannula		20.8	2.3		Vorger and Ristori, 1985
DA cannula		24.7	6.1		Val *et al.*, 1994
DA cannula		25.5			Wood *et al.*, 1979
DA & subintestinal		22.3			
DA cannula		26.8			Cameron and Davis, 1970
DA & van dam masks		22.8	7.2		
Swimming effects					
DA cannula	Aerobic swimming	19			Primmet *et al.*, 1986
	Burst swimming	27			
DA cannula	Resting	13.8	4.3	252	Wilson and Egginton, 1994
	Max aerobic swim	16	4.5	267	
Extracorporeal	Resting	24.5	10.5		Thomas *et al.*, 1987
	Near U_{crit}	28.3	11.4		
DA cannula	Resting	16.7			Kikuchi *et al.*, 1985
	U_{crit}	17.1			
DA cannula	Resting	22.2	8.7	383	Nielsen and Lykkeboe, 1992
	Swimming, 1.6bl/s	27.8	10.3	367	

Method	Condition				Reference
DA cannula	Resting	22.6			Kiceniuk and Jones, 1977
	Swimming at U_{crit}	25.7			
VA cannula	Resting	24.2			
	Swimming at U_{crit}	27.4			
DA cannula	Resting (freshwater, 18°C)	27.7	9.4	338	Gallaugher et al., 1992
	Swimming at 50% U_{crit}	26.2	8.9	338	
DA cannula	Resting (seawater, 10°C)	27.6	8.2	297	Thorarensen et al., 1996
	Swimming at U_{crit}	30.2	8.1	272	
DA cannula	Resting (freshwater, 5°C)	29.8	9.8	333	Gallaugher et al., 1995
	Swimming at U_{crit}	33.1	10.2	309	
DA cannula (+ Doppler)	Resting (seawater, 13°C)	27.2	9.1	336	Gallaugher et al., 1995
	Swimming at U_{crit}	30.1	10.4	347	
DA cannula	Resting (seawater, 10°C)	24.6	8.5	348	Gallaugher, 1994
	Swimming at U_{crit}	25.6	8.7	340	
VA cannula	Resting	24.2	8.5	349	
	Swimming at U_{crit}	26.6	8.4	317	
Temperature effects					
DA cannula	Resting (4°C)	30.7			Taylor et al., 1993
	Max sust exercise (4°C)	28.5			
	Resting (11°C)	27.3			
	Max sust exercise (11°C)	26.3			
	Resting (18°C)	16.5			
	Max sust exercise (18°C)	15.2			
DA cannula	5°C	30.4			Wood et al., 1979
	12°C	23.9			
	20°C	23			
DA cannula	7°C	28.5	8.2	287	Nikinmaa et al., 1981
	11°C	23.4	6.4	273	
	16°C	21.1	6	280	
DA cannula	6°C	27.6			Barron et al., 1987
	12°C	27.6			
	18°C	30.4			

(continues)

Table 1 (*Continued*)

Sampling method	Variable	Hct (vol%)	[Hb](g/100 ml)	MCHC (g/L)	Ref.
Hypoxia effects					
DA cannula	Normoxia		7.2		Perry and Reid, 1994
	Hypoxia, 11 kPa, 5°C		6.8		
	Hypoxia, 11 kPa, 15°C		6.1		
	Hypoxia, 4 kPa, 5°C		6		
	Hypoxia, 4 kPa, 15°C		6.5		
DA cannula	Normoxia	16.9			Hughes and Kikuchi, 1984
	Hypoxia, 20 mm Hg	28.9			
DA cannula	Normoxia	27.8			Aota *et al.*, 1990
	Hypoxia, 6 kPa	28.1			
	Hypoxia + saline	31.7			
	Hypoxia + propranolol	25.8			
DA cannula	155 Torr	22.5	7.7	340	Boutilier *et al.*, 1988
	120 Torr	22.7	8.1	350	
	90 Torr	23.5	8.2	350	
	50 Torr	30.2	9.6	320	
	30 Torr	31.2	9.2	290	
Extracorporeal	Normoxia	18.5	6.17	342	Perry and Gilmour, 1996
	Hypoxia	18.2	5.91	339	
	Hypoxia + cats bolus	19.9	5.82	309	
Hypercapnia effects					
Extracorporeal	Normocapnia	18.2	6.51	359	Perry and Gilmour, 1996
	Hypercapnia	19.1	6.66	355	
	Hypercapnia + cats bolus	20.1	6.82	327	
DA cannula	Normocapnia	22.8	8.7	379	Nielsen and Lykkeboe, 1992
	Hypercapnia, 200 m	36.2	12.5	345	
	Hypercapnia, 180 m	30.8	10.3	332	
DA cannula	Normocapnia	27.2	8.2		Ishimatsu *et al.*, 1992
	Hypercapnia, 2%, 1 h	32	8.4		
Induced (exhaustive) exercise effects					
DA and VA cannula	Resting	26.9			Currie and Tufts, 1993
(Values not different)	Induced exercise	31.6			

Method	Condition				Reference
DA cannula	Resting	30.1	8.1	300	Milligan and Wood, 1987
	Induced exercise	45.1	10.1	240	
DA cannula	Resting	21.1	6.5	320	Milligan and Wood, 1986
	Induced exercise	34.1	7.5	250	

Toxicant and acid exposure effects

Method	Condition				Reference
DA cannula	Control	21.9			Milligan and Wood, 1982
	Acid exposure	29.7			
DA cannula	Freshwater control	22.4	8.9	403	Wilson and Taylor, 1993a
	Copper exposure, 16 h	42.1	13.4	314	
	Copper exposure, 19 h	48.2	13.4	369	
DA cannula	100% seawater control	24.1	9.6	403	Wilson and Taylor, 1993b
	Copper exposure	19.1	7.8	403	
	33% seawater control	24.1	9.6	403	
	Copper exposure	19.1	7.8	437	
DA cannula	Control	23.1	2.4	186	Stormer et al., 1996
	Nitrite exposure	13.1	1.8	181	
DA cannula	Control	23.5			Jensen, 1990
	Nitrite exposure	15.5			

Catecholamine effects

Method	Condition				Reference
DA cannula	Splenectomy	22.1	5.1		Perry and Kinkead, 1989
	Splenectomy + cats	22.1	4.1		
	Sham-operated	18.1	5.5		
	Sham-operated + cats	24.1	6.2		

Compare acute and chronic cannulation sampling

Method	Condition				Reference
DA cannula	Control	21.9			Milligan and Wood, 1982
	Acid exposure	29.7			
Acute	Control	21.3	5.9		
	Acid exposure	47.3	8.9		
DA cannula		16.9		294	Sorensen and Weber, 1995
Acute		28.3		167	
DA cannula		19.4	2.1	123	Wells and Weber, 1991
Acute		25.1	2.6	97	

[a] Grouped according to sampling method (acute or cannulation) and study design.

Comparison of these data sets shows a trend toward higher Hct values with acute sampling. For example, it is clear from Fig. 1 that, for a given [Hb], acutely sampled rainbow trout have a higher Hct than cannulated fish. In addition, sampling methods have been specifically compared in a number of studies in a variety of species (e.g., white sucker, Wilkes *et al.,* 1981; Northern pike, Oikari and Soivio, 1975; Antarctic fishes, Wells *et al.,* 1990; starry flounder, Wood *et al.,* 1979; channel catfish, McKim *et al.,* 1994; rainbow trout, Wells and Weber, 1991). Again, the general trend is higher Hct values with acute sampling procedures than with chronic cannulation.

The effects of the "stress" of capture and handling, as well as air exposure during acute sampling, seem to be the primary reason for this elevated Hct. Several studies have examined the possibility that the use of anesthetic may "blunt" the stress response associated with acute sampling, but the results are contradictory and indicate that anesthesia itself may elevate Hct (e.g., see Oikari and Soivio, 1975; Korkock *et al,* 1988; Wells and Weber, 1991).

Since stress is essentially unavoidable with acute sampling techniques, Hct values derived using this methodology should be treated with some caution. Also, mean cell hemoglobin concentration (MCHC) will be under-

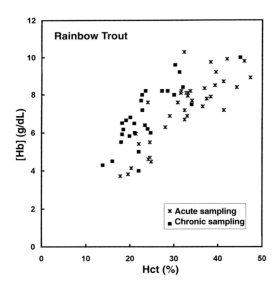

Fig. 1. The relationship between measured values of Hct and [Hb] in rainbow trout using blood sampled either acutely (via blood vessel puncture or bleeding) or by indwelling cannulae. The regression for acute sampling is [Hb] = 0.190 Hct + 1.15 (R^2 = 0.69). The regression for chronic sampling is [Hb] = 0.186 Hct + 2.42 (R^2 = 0.62). Data obtained from Table 1.

estimated in acutely sampled fish, probably as a result of the red blood cell swelling described later.

When fish are stressed, various factors may influence Hct in a transient way. Figure 2a summarizes the ways in which Hct can be altered theoretically. The most important mechanisms relate to (1) fluid movements between the various tissue compartments and (2) the release of red blood cells from a splenic store. The following subsections describe our present understanding of the main mechanisms that bring about transient changes in Hct.

A. Red Blood Cell Swelling

Red blood cell swelling results from fluid moving from the plasma into the erythrocyte. Chapter 2 details the mechanisms associated with adrenergically and CO_2-mediated red blood cell swelling. When fish are stressed and plasma catecholamines are elevated, the expectation is that Hct will be elevated. Several examples of this phenomenon are found in Tables 1 and 2. Red blood cell swelling is not universal among fish species, however (see Chapter 2).

An estimate of the degree to which red blood cell swelling contributes to an increase in Hct requires a concurrent measurement of hemoglobin concentration ([Hb]) and a calculation of MCHC. For example, an increase in Hct with no change in [Hb] (i.e., a decrease in MCHC) is taken as evidence of red blood cell swelling. Using this procedure, Primmett et al. (1986) demonstrated that one-half of a 40% increase in Hct following burst swimming in rainbow trout was most likely due to adrenaline-mediated red blood cell swelling. Similarly, Gallaugher et al. (1992) used MCHC, as well as calculations for mean cell volume (MCV), to demonstrate that approximately one-half of a 30% increase in Hct following swimming to maximum sustainable swimming velocity (U_{crit}) in freshwater rainbow trout was due to cellular swelling (circulating catecholamines levels were also shown to be significantly elevated at U_{crit}) (see Fig. 2b).

Venous blood is predicted to have a higher Hct than arterial blood because of CO_2-mediated red blood cell swelling. However, the expected difference in Hct between blood sampled in the ventral aorta and that in the dorsal aorta may not be detectable (e.g., see Kiceniuk and Jones, 1977; Gallaugher, 1994) because of the variable amount of plasma skimming that occurs within the gill vasculature.

B. Plasma Skimming

Fish possess primary and secondary circulations (Bushnell et al., 1992), and Hct measurements involve blood samples from the primary circulation,

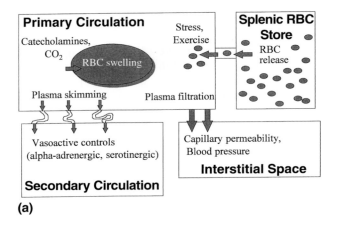

(a)

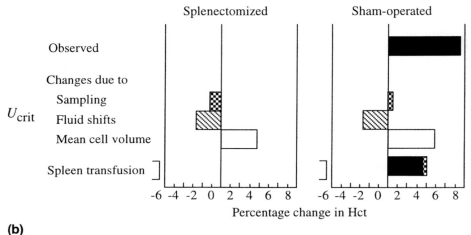

(b)

Fig. 2. (a) Theoretical mechanisms that can influence the measurement of Hct and [Hb] in the primary circulation in a short-term, transient manner. (b) An example of some of the calculated contributions to the changes in Hct observed in splenectomized and sham-operated rainbow trout at U_{crit}. Adapted from Gallaugher *et al.*, 1992.

which contains the branchial heart. The secondary circulation is connected to the arterial side of the primary circulation. The interconnecting vessels are narrow and tortuous, sometimes having hairlike projections that screen

Table 2

Hct, [Hb], and MCHC Values for Selected Species of Bony and Cartilaginous Fishes and Cyclostomes

Common name	Species name	Sampling	Variable	Hct (%)	[Hb](g/100 ml)	MCHC (g/L)	Ref.
Cyclostomes							
Pacific hagfish	*Eptatretus cirrhatus*	Acute		13.5			Forster *et al*., 1989
Pacific hagfish	*Eptatretus cirrhatus*	VA cannula		12.6	3.1	243	Wells and Forster, 1989
	Myxine glutinosa L.	Acute		15–26			Johansen *et al*., 1962
Lamprey	*Lampetra fluviatilis*	Acute	2°C		6.1	194	Nikinmaa and Weber, 1984
			15°C		9.5	284	
Rays							
Bluespotted fantail ray	*Taeniura lymna*	Acute		14.8	4.8	318	Baldwin and Wells, 1990
Shovelnosed ray	*Rhinbatos batillum*	Acute		10.5	3.8	373	
Sharks							
Dogfish	*Squalus acanthias*	Extracorporeal		14.8	3.1	208	Perry and Gilmour, 1996
Dogfish	*Squalus acanthias*	DA cannula		23			Opdyke *et al*., 1981
Epaulette shark	*Hemiscyllium ocellatum*	Acute		13.4	3.6	272	Baldwin and Wells, 1992
Blacktip shark	*Carcharhinus melanopterus*	Acute		17.1	4.1	243	
Lemon shark	*Negaprion acutiidens*	Acute		18.2	5.5	300	
Shovelnosed ray	*Rhinobatos batillum*	Acute		10.5	3.8	373	
Black-tipped reef shark	*Carcharhinus melanopterus*	Acute		17	4.1	243	Wells *et al*., 1992
Lemon shark	*Negaprion acutidens*	Acute		13	3.6	277	
Epaulette shark	*Hemiscyllium acellatum*	Acute		19.7	5.6	286	
Shovelnosed ray	*Rhinobatos batilum*	Acute		13.8	3.9	281	
Blue shark	*Prionace glauca*	Acute		11	2.9	264	Wells *et al*., 1986
	Dasyatis sabina	Acute		14			see Butler and Metcalfe, 1988
	Cephaloscyllium isabella	Acute		15			
	Squalus acanthias	Acute		11.5			
	Squalus suckleyi	Acute		20			
	Scyliorhinus stellaris	Cannula		16			
	Scyliorhinus canicula	Acute		20–22			
	Negaprion breviraostris	Acute		16			
Mako shark	*Isurus oxyrhinchus*	Acute		34	10.5	318	Wells *et al*., 1986

(continues)

Table 2 (*Continued*)

Common name	Species name	Sampling	Variable	Hct (%)	[Hb] (g/100 ml)	MCHC (g/L)	Ref.
Teleosts							
Scombrids							
Skipjact tuna	*Katsuwonus pelamis*	Cannulation		40.6	15		Brill and Bushnell, 1991
Yellowfin tuna	*Thunnus albacares*	Cannulation		35.1	12.5		
Skipjack tuna	*Katsuwonus pelamis*	Acute		83	20.4	247	Wells et al., 1986
Yellowfin tuna	*Thunnus albacares*	Acute		75	18.9	254	
Frigate mackerel	*Auxis rochei*	Acute			17.8–21.2		
Marlins							
Striped marlin	*Tetrapturus audax*	Acute		55	13.7	249	Davie, 1990
Blue marlin	*Makaira nigricans*	Acute		43.1	10.4	311	
Black marlin	*Makaira indica*	Acute		43	10.3	242	
Salmonids							
Atlantic salmon (farmed)	*Salmo salar*	Acute		44–49	8.9–10.4	194–217	Sandnes et al., 1988
Brown trout	*Salmo trutta*	DA cannula	15°C	21	5.8		Butler et al., 1992
			5°C	19	6		
Chinook salmon	*Oncorhynchus tshawytscha*	Acute		39.3	10.8	277	Brauner et al., 1993
Chinook salmon	*Oncorhynchus tshawytscha*	DA cannula		26.8	8.8	335	Thorarensen et al., 1993
		DA cannula	Rest	23.3	7.2	304	
Chinook salmon	*Oncorhynchus tshawytscha*	DA cannula	U_{crit}	24.5	7.2	297	Gallaugher, 1994
			Rest	29.7	11.4	385	
Coho salmon	*Oncorhynchus kisutch*	Acute	U_{crit}	29.1	11.5	399	Brauner et al., 1992
Coho salmon	*Oncorhynchus kisutch*	DA cannula		35			Smith, 1966
			Hatchery stock	20–33			
			Wild stock	30			
Steelhead trout	*Oncorhynchus mykiss*	DA cannula		21			Houston and DeWilde, 1972
Sockeye salmon	*Oncorhynchus nerka*	DA cannula		26			
Brook trout	*Salvelinus fontinalus*	Acute		36.6	8		
Coral reef fishes							
Lowly trevally	*Caranx ignobilis*	Acute		39.1	12.6	323	Wells and Baldwin, 1990
Gold-spotted trevally	*Carangoides fulvoguttotus*	Acute		35	11.5	328	

Common name	Species	Method	Condition				Reference
Spangled emporer	*Lethrinus nebulosus*	Acute		19.8	5.85	298	
Wire-netting cod	*Epinephelus merra*	Acute		21.9	5.73	153	
Striped seaperch	*Lutjanus carponotatus*	Acute		17.8	5.03	282	
Blue tuskfish	*Choerodon albigeria*	Acute		19.4	5.26	270	
Slender suckerfish	*Echeneis naucrates*	Acute		21	5.22	248	
Snook	*Centropomus undecimalis*	Acute	Freshwater	28.7	4.1		Perez-Pinzon and Lutz, 1991
			Seawater	31.1	2.8		
Temperate pelagic marine and freshwater							
Yellowtail	*Seriola quinqueradiata*	DA cannula	Rest	26.1	8.4		Yamamoto, 1991
			Exh. exercise	41.7	13.1		
Yellowtail	*Seriola quinqueradiata*	DA cannula	Rest	25.7	7.69		Yamamoto et al., 1980
			Exh. exercise	36.7	11.04		
			Rest	28			
			U_{crit}	41			
Killifish	*Fundulus heteroclitus*	Acute	Normoxia	34–48			Hannah and Pickford, 1981
Sheepshead minnow	*Cyprindon varregatus*	Acute	Normoxia	26			Peterson, 1990
			Hypoxia	33			
			Normoxia	25			
Sailfin molly	*Poecia latipinna*	Acute	Hypoxia	29			Garcia et al., 1992
European sea bass	*Dicentrarchus labrax*	Acute		36.1			Young and Cech, 1994
Striped bass	*Morone saxatilis*	Acute	Rest	35.8			
			Burst swim	56.3			
Striped bass	*Morone saxatilis*	Acute	Control	40			Young and Cech, 1993
			Stress	58			
Atlantic cod	*Gadus morhua*	Venous cannula	Rest	24.4	5.2	213	Gallaugher, Axelsson, Farrell and Nilsson, unpublished data
			U_{crit}	25.7	5.1	198	
			Normoxia	23.2	5.6	240	
			Hypoxia	28.7	6.1	214	
			Rest	23.1	5.6	225	
			Stress	27.5	5.3	190	
Atlantic cod	*Gadus morhua*	Venous cannula	Rest	14.3			Butler et al., 1989
			U_{crit}	18.5			
Atlantic cod	*Gadus morhua*	Venous cannula	Brackish water	25.6	4.6		Nelson et al., 1994
Atlantic cod	*Gadus morhua*	Acute	Seawater	23.2	4.6		Khan, 1977
			Nondiseased	24	4.8		
			Diseased	15	2.8		

(*continues*)

199

Table 2 (*Continued*)

Common name	Species name	Sampling	Variable	Hct (%)	[Hb][g/100 ml]	MCHC (g/L)	Ref.
Benthic marine and freshwater species							
Fourhorn sculpin	*Myoxocephalus quadricronis*	Acute	With MS222	25.9			Oikari and Soivio, 1975
			No MS222	23.3			
Shorthorn sculpin	*Myoxocephalus scorpius*	Acute	With MS222	15.8			
			No MS222	17.4			
Blenny	*Zoarces viviparus*	Acute	With MS222	11.9			
			No MS222	10			
Winter flounder	*Pseudopleuronectes americanus*	Acute	Aug–Mar	23–26	4.9–6.1	250–300	Bridges *et al.*, 1976
Starry flounder	*Platichthys stellatus*	Caudal A cannual	Apr–Jul	17–19	4.3–5.7	300–330	Milligan and Wood, 1987
			Rest	13	4	300	
Flounder	*Pleuronectes flesus*	Cannula	Exh. exercise	18	4.5	260	Alkindi *et al.*, 1996
			Crude oil exp.	16	4.5		
Turbot	*Scophthalmus maximus*	Acute		8	2.5		Quentel and Obach, 1992
Pinfish	*Lagodon rhomboides*	Acute		16.7	5.5	331	Cameron, 1970
Mullet	*Mugil cephalus*	Acute		32.9	5.9		
				21.9	6.1		
European eel	*Anguilla anguilla*	Acute		24.5–34.8	3.57– 4.8 mM		Andersen *et al.*, 1985
White sucker	*Catostomus commersoni*	Cannula		13.4			Wilkes *et al.*, 1981
Northern pike	*Esox lucius*	Acute	Brackish water	27.8	4.4	268	Soivio and Oikari, 1976
		DA cannula	Freshwater	18.3			
Northern pike	*Esox lucius*	DA cannula		25.2	6.1	241	Oikari and Soivio, 1975
		Acute +MS222		18.3			
		Acute, no MS222		22.8			
		DA cannula		20.2			
Channel catfish	*Ictalurus punctatus*	Acute		15			McKim *et al.*, 1994
				33			
Tench	*Tinca tinca*	Acute		24.1	6.8	330	Eddy, 1973

200

Eurythermal freshwater species

	Species		Condition				Reference
Carp	*Cyprinus carpio*	DA cannula	Control	20.2	6.21	307	Takeda, 1990
Carp	*Cyprinus carpio*	DA cannula	Nitrite exposure	23.5 / 15.5	2.3 / 1.8	186 / 180	Jensen, 1990
Carp	*Cyprinus carpio*	Acute		27.8	7.1		Houston and DeWilde, 1972
Carp	*Cyprinus carpio*	Acute	2°C	32.9	7.5	179	Houston and Smeda, 1979
			10°C	32.3	7.5	179	
			18°C	33.4	8.2	172	
Goldfish	*Carassius auratus*	Acute	10°C	16.1	4.6		Houston and Gingras-Bedard, 1994
			20°C	16.8	4		
			30°C	17.7	5.9		
			20°C+/−10	13.5	4		
			10–30°C	17.6	5.7		
Goldfish	*Carassius auratus*	Acute	7.5°C	35.8	7.03		Chudzik and Houston, 1983
Goldfish	*Carassius auratus*	Acute	30°C	25	6.01		Houston and Schrapp, 1994
			12°C	26.6			
Goldfish	*Carassius auratus*	Acute	28°C	24.9			Houston and Murad, 1992
			20°C+/−8cycle	30.7			
			12–28°C random	25.6			
			35°C	25	6.3	380	
			25°C	22	6.7	420	
			15°C	16	5.7	450	
Deep-sea fishes							
	Antimora rostrata	Acute		30.5	5.04	170	Graham et al., 1985
	Lycodes esmarkii			24.3	4.4	181	
	Macrurus berglax			27.8	5.41	198	
Airbreathing Species							
Florida gar	*Lepisosteus platyrhincus*	Acute	Fed	32; 38	5.6	295	McLeod et al., 1978
Kissing gourami	*Helostoma temmincki*	Acute	Starved	18.7 / 12.1	4.3	347	Weinberg et al., 1976
Amazonian species							
	Hoplosternum littorale	Acute	Normoxia	39.1	12.9		Val et al., 1992
			Hypoxia	43.7	16.1		
Curimata	*Pterygoplichthys multiraduatus*	Acute	Wild	36	10.9	311	Val, 1993

(continues)

Table 2 (*Continued*)

Common name	Species name	Sampling	Variable	Hct (%)	[Hb](g/100 ml)	MCHC (g/L)	Ref.
	Colossoma macropomum	Acute	Lab-normoxia	29.8	9.2	311	Monteiro *et al.*, 1987
			Lab-hypoxia	30.4	11.4	373	
	Piaractus brachypomum	Acute	Normoxia	22.7	6.9		
			Hypoxia	28.5	8.5		
			Normoxia	27.3	8.2		
			Hypoxia	31.2	8.6		
	Prochilodus cf. nigricans	Acute	July	39.9	11.1	281	
			October	36.7	10.2	269	
			November	39.8	10.3	260	
	Mylossoma duriventris	Acute	Feb–Aug	27.5–28.2	6.8–8.2	244–299	
			Oct–Nov	36.3–37.6	10.3–10.7	286	
Antarctic species							
Borch	*Pagothenia borchgrevinki*	Acute	Rest	14.8	3.3	226	Franklin *et al.*, 1993
			Exh. exercise	35	5.3	157	
Borch	*Pagothenia borchgrevinki*	Acute	Normoxia	14.4	2.9	199	Wells *et al.*, 1989
			Chron. hypoxia	27.8	4.7	171	
Borch	*Pagothenia borchgrevinki*	Cannulation		13	1.03	79	
	Pagothenia bernachii	Acute cannulation		15.4	3.2	205	
				7.6	1.8	242	
Nototheniids	*Notothenia neglecta*	DA cannula	Rest	18.6	5.6	316	Egginton *et al.*, 1991
			Exh. exercise	17.6	5.2	316	
	Notothenia rossi	DA cannula	Rest	6.4			
			Exh. exercise	11.3			
Icefish	*Chionodraco kathleenee*	Acute		1.23			Wells *et al.*, 1990
Icefish	*Gyodraco antarticus*	Acute		1.25			Tetens *et al.*, 1984
	Rhigophila dearborni	Acute		15	1.4	94	
	Trematomus bernachii	Acute		13.5	0.9	63	
	Trematomus loennbergi	Acute		8	0.6	81	
	Dissostichus mawsoni	Cannulation		17.5	0.9	125	
	Notothenia angustata	Acute		18.5	1.7	92	

the opening (Steffensen and Lomholt, 1992). Plasma, but not necessarily red blood cells, passes freely from the primary into the secondary circulation, i.e., plasma skimming. Furthermore, some of the connections, especially those in the gills, are under powerful vasoactive control, including α-adrenergic constrictory and serotonergic dilatory controls (Sundin and Nilsson, 1992). Although blood in the secondary circulation is characterized by a low Hct, the volume of the secondary circulation is appreciable. In rainbow trout, the secondary circulation has a volume similar to that of the primary circulation (Steffensen and Lomholt, 1992), and in hagfish the secondary circulation has the larger volume by far (Satchell, 1991). Thus, the secondary circulation represents a large plasma pool that is connected to the primary circulation.

The nature and extent of plasma flow between the primary and the secondary circulations will affect the measured Hct in the primary circulation. Olson (1984) used perfused gills to clearly demonstrate the phenomenon of plasma skimming in three fish species. Plasma skimming in the gills consistently resulted in the blood leaving the gills via the primary circulation having a higher Hct compared with that draining the secondary circulation. Randall (1985) suggested that, although plasma skimming in the gills tends to increase Hct in arterial blood compared with venous blood, such changes are offset *in vivo* by a decrease in Hct because of CO_2 excretion at the gill.

C. Changes in Plasma Volume

Swimming in fish is characterized by major changes in water balance. In freshwater, rainbow trout increased body mass due to fluid gain while swimming aerobically (Wood and Randall, 1973). In contrast, seawater chinook salmon decreased body mass by around 10% while swimming (Gallaugher, 1994). In addition to these whole body fluid changes, fluid is redistributed between the intracellular and extracellular compartments, and between plasma and the interstitial space. These types of changes can be detected by changes in plasma protein concentrations and osmolality and should alter Hct. Gallaugher et al. (1992) showed a significant hemodilution due to shift of fluid into the plasma compartment when rainbow trout were swum to U_{crit} in freshwater (see Fig. 2b). In contrast, when chinook salmon and rainbow trout were swum in seawater, changes in Hct were relatively small compared with those observed in the same species swimming to U_{crit} in freshwater even though plasma osmolality was increased significantly (Gallaugher, 1994). This situation may have resulted from a fluid shift out of the plasma, which in turn "masked" the expected increases in Hct due to red cell swelling and/or transfusion of stored red blood cells. Other stress factors have also been linked to elevations in plasma osmolality and therefore may influence Hct determinations. For example, plasma osmolality was significantly increased in Atlantic cod following hypoxia and exhaustive exercise (Gallaugher, Axelsson, Farrell and Nilsson, unpublished observations) and in dogfish following exhaustive exercise (Opdyke *et al.*, 1982). An additional factor that may influence plasma volume during swimming and stress is high blood pressure, which is likely to alter the Starling forces, favoring a decrease in plasma volume. Clearly, changes in plasma volume must be factored in when evaluating Hct changes.

D. Splenic Release of Red Blood Cells

The spleen is the main storage site for red blood cells in many species of fish. Elasmobranchs have very large spleens, and teleost species have spleens of varying size. Cyclostomes do not have a spleen (Fange and Nilsson, 1985). The best measure of relative spleen size is the spleen somatic index (SSI), ratio of spleen mass to body mass, and reductions in this index can be interpreted as evidence of spleen contraction.

The contraction of the fish spleen is known to be under adrenergic nervous and/or humoral control (Nilsson and Grove, 1974; see Nilsson, 1983). Stressed fish tend to deplete their splenic stores and have an elevated Hct. Elevations in Hct have been linked with transfusion of cells from the spleen in a number of different species under various stress conditions—for example, burst swimming in rainbow trout (e.g., Primmett *et al.*, 1986); aerobic swimming in yellowtail (Yamomoto *et al.*, 1980), Antarctic fishes, (Franklin *et al.*, 1993), and rainbow trout (Pearson and Stevens, 1991b; Gallaugher *et al.*, 1992); exhaustive exercise in rainbow trout (Pearson and Stevens, 1991a) and yellowtail (Yamamoto *et al.*, 1980); anesthetic in rainbow trout (Wells and Weber, 1990); acute hypoxia in rainbow trout (Wells and Weber, 1990), Antarctic fish (Wells *et al.*, 1989), and yellowtail (Yamamoto *et al.*, 1985); low environmental pH in rainbow trout (Milligan and Wood, 1982); and prolonged swimming, acute hypoxia, and exhaustive exercise in Atlantic cod (Gallaugher, Axelsson, Farrell and Nilsson, unpublished observations) (see also Tables 1 and 2). Increases in Hct were associated with reductions in the SSI, where this index was measured. Because stress can cause splenic contraction over a period of minutes and because a certain amount of stress is inevitable when the spleen is removed for weighing, measurement of SSI may underestimate the real value. To circumvent this problem, Pearson and Stevens (1991a) employed a "window" in the body wall to observe splenic contraction *in vivo*.

Obviously, the extent to which Hct increases is dependent on the degree of emptying of the spleen and the total blood volume of the spleen relative to the systemic blood volume and normal Hct. For example, Wells and

Weber (1990) have estimated that full splenic contraction (spleen Hct = 93%) in a 400-g rainbow trout (systemic Hct = 25%) yields about 1 g of red blood cells. Pearson and Stevens (1991a) have shown that the splenic reservoir in rainbow trout contains approximately 21% of total body Hb and more than 95% of this is released within minutes of exposure to air or induced exercise. Yamamoto *et al.* (1980) have estimated that splenic red blood cell stores compose about 14% of the total erythrocytes in yellowtail, and >80% of these are released during prolonged swimming. The important point to emerge from these studies is that splenic release of red blood cells can effect a major increase in Hct.

The degree to which the spleen contracts and its effect on Hct is extremely variable among species. For example, in two different species of chronically cannulated nototheniids (ones that do not show evidence of adrenergically mediated cell swelling), Hct increased by 50% in one species following exhaustive exercise, but did not change in the other (Egginton *et al.*, 1991). More examples of the variable effects of splenic contraction on Hct are provided in Section IVC. Seasonal changes in spleen size have long been known for salmonids (see Fange and Nilsson, 1985), but to what degree these changes might affect Hct is unclear.

E. Serial Blood Sampling

With fish blood volume being around 5% of body mass, a 1-kg fish will have around 50 ml of blood in the primary circulation. Theoretically, a 1-ml blood sample could cause a 2% decrease in blood volume. Physiological compensations to maintain Hct would involve either red blood cell release and appropriate fluid shifts to restore plasma volume or just venoconstriction. Fluid shifts alone would restore blood volume, but Hct would decrease. A serial blood sampling study with 0.5- to 0.7-kg rainbow trout showed that removal of 1.2 ml of blood had no effect on Hct, but removal of 3 ml of blood reduced Hct by 4% (Gallaugher *et al.*, 1992). In a serial sampling study with 0.5- to 0.6-kg Atlantic cod, removal of 1.8 ml of blood caused Hct to decrease from 26 to 23% (Gallaugher, Axelsson, Farrell, and Nilsson, unpublished observations). In most studies involving serial sampling, extracted blood is either not replaced or replaced with equal volumes of saline, and so progressive anemia is a possibility (e.g., Milligan and Wood, 1986; Stormer *et al.*, 1996). This sort of hemodilution can be avoided by replacing each sample drawn with an equal volume of blood from a cannulated donor fish (e.g., Gallaugher *et al.*, 1995). Cross-reactions between blood cells from different individuals do not appear to be a problem in certain salmonid species (Gallaugher, 1994).

Blood loss due to poor cannulation techniques will deplete splenic reserves first and therefore make fish more prone to hemodilution through serial sampling. In the short term, this could confound data interpretation with respect to factors that normally stimulate splenic release of red blood cells (i.e., there might be none left to release!). In the long term, hemodilution may stimulate erythropoiesis.

Whether or not a cannulated fish is anemic can be assessed using known Hct values for that species. In the case of cannulated rainbow trout, a Hct value of less than 20% seems suspect against a background of much higher values in numerous other studies with that species. Anemia can then lead to other physiological discrepancies. For example, a Hct value of 13.5% reported for rainbow trout (Wilson and Egginton, 1994) not only suggests anemia, but also may account for the relatively low U_{crit} values that were observed in the same study.

III. INTERSPECIFIC DIVERSITY IN HEMATOCRIT VALUES

Of the estimated 21,000 species of fishes (Satchell, 1991), Hct values are known for only a few percent (see Table 2). Thus, any generalization about interspecific diversity must be treated with some caution until there is a greater species representation. A further word of caution is that in many cases Hct values are reported only as incidental data, and so given the variety of factors that affect Hct within a species (see Sections II and IV), Hct data may not always be directly comparable between studies even though studies by the same investigator(s) tend to be internally consistent [e.g., see values of Houston and co-workers for carp and goldfish (acute sampling) and Thorarensen and Gallaugher for salmonids (chronic cannulations); Tables 1 and 2].

Tables 1 and 2 report selected Hct values for a wide range of fish species. Although most of the available information is for hatchery-reared rainbow trout (Table 1), a number of generalizations are possible from the more limited data on other species (Table 2). To assist species-specific comparisons, the data in Table 2 are grouped into broad phyletic and environmental categories. Many of the data contained in Table 2 are presented in Fig. 3 to illustrate the general relationship between Hct and Hb between these broad phyletic and environmental categories. From this relationship, it is clear that Hct is the major determinant of [Hb] because the correlation coefficient is 83%. None of the fish groupings deviate in a major way from this relationship even though data for Hct values from cannulated and noncannulated fish are included. (Red blood cell swelling would increase

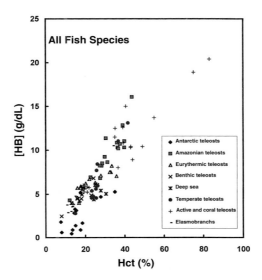

Fig. 3. The relationship between measured values of Hct and [Hb] in various phyletic and environmental groupings of fish species. Data obtained from Table 2. The regression is [Hb] = 0.26 Hct − 0.12 (R^2 = 0.83).

the variability around the relationship, but splenic release of red blood cells should not have any effect unless immature blood cells having a lower MCHC had been released from the spleen.)

From a phylogenetic perspective, Atlantic hagfishes have Hct values of 15–26% whereas Pacific hagfish have Hct values of 12–14%. Lampreys have Hct values ranging from 13 to 30% depending on temperature. Many elasmobranchs have Hct values less than 20%, values ranging from 10.5 to 19.7% (Table 2). A notable exception is the active mako shark, which has a Hct of 34% (noncannulated). Teleosts generally have Hct values greater than 20%, although there are important exceptions that are noted later. This evolutionary trend toward a higher Hct in teleosts could reflect a difference in metabolic scope between the teleosts and the cyclostomes and elasmobranchs, but such an assertion is presently lacking careful experimental validation.

Hct values in scombrids are characteristically the highest recorded among fishes. However, the highest values of 75–83% reported for skipjack and yellowfin tuna sampled via acute techniques (Wells *et al.*, 1986) undoubtedly reflect a highly stressed state with splenic red blood cell release and possibly red blood cell swelling. The Hct values of 40.6 and 35% reported for cannulated skipjack and yellowfin tuna, respectively, are probably a better reflection of the routine Hct (Brill and Bushnell, 1991). Caution should also be used with the Hct values reported for marlin. Stress and blood loss are factors that could contribute to the variability in these data because blood samples are usually from angled fish that are gaffed before sampling.

Hct values for salmonids as a group range from <20 to >50%. In contrast, the range for chronically cannulated fish is generally between 25 and 35% (Table 1; see Wells and Weber, 1991). Wells and Baldwin (1990) acutely sampled a number of coral reef fishes and found a broad range of Hct values from 17.8 to 39.8%. The most active of these species (the trevally) had the highest Hct value.

A large range of Hct values exists among the grouping for temperate freshwater and marine fishes (Table 2). One of the highest values is for striped bass. The lowest Hct values in this group are found among the benthic species such as starry flounder (13%), blenny (10%), and shorthorn sculpin (15.8%). However, a broad generalization that all benthic fishes have low Hct values does not hold since Hct values greater than 25% have been reported for both the winter flounder and the fourhorn sculpin (Table 2). The data for winter flounder suggest that seasonal variation in Hct (with higher values in the winter months) could be a confounding factor with such comparisons. Hct values for Atlantic cod are also quite variable, and here water quality (e.g., brackish versus seawater) and parasitic infections play a role in the variability (Table 2).

Among the Amazonian and airbreathing fish, Hct values (all obtained *via* acute sampling procedures) are typically higher than 30% (Table 2). Whether this high Hct reflects the higher water temperatures and/or the bouts of hypoxia experienced by these fish is unclear. There is also some seasonal variability in Hct, with higher values being associated with the dry season when hypoxia prevails (Monteiro *et al.,* 1987).

Antarctic fishes inhabit frigid waters and are characterized by having no, or few, red blood cells in their primary circulation. Icefish species (family Channichthyidae), such as *Chionodraco kathleenee* and *Gyodraco antarcticus,* have an Hct of around 1% (Wells *et al.,* 1990). These fishes are typically very slow moving. Some of the more active fishes (e.g., *Notothenia neglecta* and *N. rossii*) have higher Hct values, ranging from 6 to 18% (Egginton *et al.,* 1991). The more active of these species (*N. rossii*), however, had the lower Hct value. In addition, some Antarctic fish can double Hct when either stressed or exercised (Franklin *et al.,* 1993). A splenic release of red blood cells moves their Hct values into a range more typical of temperate water teleosts. The loss or reduction of red blood cells in icefish has been described as an adaptation to the greater blood viscosity that would result from these cold temperatures (see Wells *et al.,* 1990). However, other determinants such as a low metabolic rate and a poor pressure-generating capacity of the hearts could be equally important (see Section V).

Although there are numerous examples of active fishes having high Hct values, and a relationship between Hct and activity has been alluded to several times in the literature (Fange, 1992; Farrell, 1991, 1995), others have found no clear relationships between capacity for activity and Hct ([Hb]) in teleosts (Wells and Baldwin, 1990) and elasmobranchs (Baldwin and Wells, 1990). Similarly, there is no clear-cut relationship between Hct and activity/behavior for a number of mammals (e.g., see Birchard and Tenney, 1990; Jones and Lindstedt, 1993). The relationship between activity and V_{O_2} is more complex, and other factors are clearly involved in determining maximum V_{O_2}. Conversely, red blood cells have roles in addition to O_2 transport. An additional complicating factor in fishes is that temperature may be a major determinant of V_{O_2}, yet temperature has relatively small effects on Hct within a given species, as shown in the next section.

IV. INTRASPECIFIC REGULATION OF HEMATOCRIT

A survey of Table 1 points to considerable intraspecific variability beyond the differences due to sampling, as described earlier. This variability will result either from long-term influences on erythropoietic activity or from short-term physiological influences.

A. Erythropoiesis

The production of new red blood cells is termed erythropoiesis. The site of erythropoiesis in fish varies among species as follows: in hagfish, the major site is in the intestinal submucosa and pronephros (head kidney) as well as in the circulating blood; in elasmobranchs and some teleosts, the major site is the spleen; for most teleosts, the pronephros is the predominant site, but, as in the case of salmonids, sites between the kidney tubules may also be involved (Fange and Nilsson, 1985; Fange, 1986, 1992; Houston et al., 1996b). Erythropoietic tissue is also associated with the heart in sturgeon (Fange, 1986). Iuchi and Yamomoto (1983) showed that the site of major erythropoietic activity shifts from one organ to another during development in rainbow trout. The relative importance of different sites for erythropoiesis may also change in adult fish. For example, Houston et al. (1996a) demonstrated that in goldfish the erythropoietic response to maximal stress is much more significant in the pronephros than in the spleen.

Erythrocytes are released from hematopoietic tissue at an early stage in their development and continue development (including Hb synthesis) while being circulated (Fange, 1986). Thus, the ratio of immature/mature cells in circulation as well as Hb isomorph composition may be influenced by physiological and/or environmental conditions. Unfortunately, little is known regarding the ontogeny of erythropoiesis and the regulation of erythropoiesis in adult fish.

In mammals, the primary regulator of erythropoiesis, the hormone erythropoietin (Epo), is produced in the kidney and circulates in plasma at picomolar levels. The structure of this glycoprotein hormone is well known, and the gene has been cloned (Bauer and Kurtz, 1989). The mechanism for Epo production is not well understood, but it is known that regulation of Epo levels takes place at the transcriptional level rather than at the point at which Epo is released from the kidney. Epo synthesis most likely involves a feedback loop, including a sensor that detects changes in blood O_2 concentrations, the level of serum Epo, and the numbers of circulating erythrocytes (Bauer and Kurtz, 1989). Receptors for Epo located on erythroids have been characterized (and cloned) as belonging to the "cytokine" superfamily of receptors (Bauer and Kurtz, 1989).

Some of the physiological conditions known to stimulate Epo production in mammals include anemia, tissue hypoxemia, hypobaric hypoxia, reduced blood flow to the kidney, and changes in Hb–O_2 affinity (Bauer and Kurtz, 1989). To our knowledge, Epo has not been separated from fish plasma, and there is no information regarding the signal transduction for the hormone in fish.

B. Long-Term Influences on Hct

A number of environmental and physiological stimuli have been directly linked to increases in erythropoietic activity in fish. It is likely that fish encounter each of these stimuli under natural conditions in the wild.

ANEMIA AND STARVATION EFFECTS

A common response to anemia is a stimulation of erythropoiesis, as shown in pinfish (Cameron and Wohlschlag, 1969), Florida garr (McLeod *et al.,* 1978), rainbow trout (Byrne and Houston, 1988), and goldfish (Houston and Murad, 1995). This response is likely humorally based, possibly involving Epo, because plasma from anemic fish (or injection of human urinary Epo) stimulated both erythropoiesis and erythrocyte Hb synthesis in the kissing gourami (Weinberg *et al.,* 1976). Temperature may influence recovery from anemia (Chudzik and Houston, 1983). Recovery from anemia is slowed by splenectomy in rainbow trout (Lane, 1979), indicating an involvement of the spleen in the humoral control of erythropoiesis. Prolonged starvation can lead to anemia, which in turn stimulated erythropoiesis in both rainbow trout (Lane *et al.,* 1981) and kissing gourami (Weinberg *et al.,* 1976). Again, Epo may be involved in this response since human urinary Epostimulated erythropoiesis in starved, but not in fed, species of tropical teleosts (Pradhan *et al.,* 1989). Relationships reported between body mass and Hct and/or [Hb] found for plaice (Smith, 1977) and for rainbow trout (Martinez *et al.,* 1994) could, therefore, reflect individual variability in nutritional status. Whereas nutritional status can be (or should be) controlled in laboratory studies, it is more difficult to control nutritional status in fish sampled from the wild. Thus, the anemic state reported for certain species sampled from the wild (e.g., flounder; Wood *et al.,* 1979) could reflect nutritional deficiency.

TEMPERATURE EFFECTS

The response to temperature in fish is variable and, when present, rarely results in a large change in Hct. Temperature stimulated erythropoiesis in goldfish (Houston and Murad, 1992), but a cycling temperature had a stronger effect than a constant temperature. Temperature has also been positively correlated with Hct and [Hb] in winter flounder (Cech *et al.,* 1976). However, laboratory-based temperature acclimation studies with rainbow trout are equivocal. Some studies reported an increased Hct with temperature (Jones, 1971; Houston and Smeda, 1979; Marinsky *et al.,* 1990; Perry and Reid, 1994; Martinez *et al.,* 1994), whereas Barron *et al.* (1987) showed small or no changes in Hct and/or [Hb] over a temperature range from 5 to 20°C. In contrast, Nikinmaa *et al.* (1981), Wood *et al.* (1979), and Taylor *et al.* (1993) all found a negative correlation between Hct and/or [Hb] and temperature (4–20°C). Some of these differences could be a result of the sampling problems alluded to earlier in the chapter. Houston and Schrapp (1994) suggested that the temperature-induced shifts in the relative abundance of circulating mature and immature red blood cells and the relative proportions of Hb isomorphs were more prominent than the stimulation of erythropoiesis.

HYPOXIA AND PHOTOPERIOD EFFECTS

Evidence for long-term hypoxia as a stimulant for erythropoiesis in fishes is contradictory (see Murad *et al.,* 1990). Some studies suggested that hypoxia produced only minor increases in Hct and [Hb] independent of temperature and photoperiod, whereas photoperiod alone was a more powerful stimulant (see Tun and Houston, 1986; Houston *et al.,* 1996a). The interactive effects of these stimuli also significantly stimulated the erythrocyte maturation process (Marinsky *et al.,* 1990; Houston and Schrapp, 1994; Houston *et al.,* 1996a).

SEASONAL EFFECTS

Bridges *et al.* (1976) observed significantly lower values for both Hct and [Hb] in winter flounder under winter than under summer conditions. However, Andersen *et al.* (1985) were unable to show seasonal variation in either of these parameters for European eels acclimatized in the wild, and a similar lack of correlation between hematological parameters and season was observed by Sandnes *et al.* (1988) in farmed Atlantic salmon. Hct may also vary intraspecifically according to seasonal events such as reproduction; e.g., Hct was higher in mature male than in female rainbow trout (36% versus 29%) and this difference was associated with a threefold larger spleen in males (H. Thorarensen and P. Davie, unpublished observations).

HABITAT

Evidence for differences in Hct and [Hb] due to habitat differences is weak. Nelson *et al.* (1994) showed only slight differences in [Hb] between brackish and seawater Atlantic cod. Soivio and Oikari (1976) found slightly higher Hct and [Hb] levels in freshwater than in brackish Northern pike.

STOCK EFFECTS

A survey of the chronic cannulation data in Tables 1 and 2 for Hct in rainbow trout and Atlantic cod shows considerable variability between groups of investigators. For Atlantic cod, Butler *et al.* (1989) can be compared with Nelson *et al.* (1994), and for rainbow trout, Nikinmaa *et al.* (1981) and Thomas *et al.* (1987) can be compared with Wilson and Egginton (1994) and Sorensen and Weber (1995). Although these differences point to the possibility of genotypic and/or phenotypic variability, complications due to sampling and disease need to be eliminated before firm conclusions can be made.

EXERCISE-TRAINING (CONDITIONING) EFFECTS

A number of studies have examined long-term exercise-training or conditioning effects on O_2-transport capacity in fishes. The results are contradictory and may be related to life cycle. Hochachka (1961) and Zbanyszek and Smith (1984) both reported small increases in Hct and [Hb] in exercise-trained rainbow trout fingerlings and coho salmon smolts, respectively. In contrast, Woodward and Smith (1985) and Davie *et al.* (1986) found no change in Hct in trained adult rainbow trout. However, small increases in both Hct and [Hb] occurred following exercise training protocols with largemouth bass (Farlinger and Beamish, 1978) and adult chinook salmon (Thorarensen *et al.*, 1993; Gallaugher, 1994) (Table 2).

C. Short-Term Influences on Hct

Transient changes in Hct due to the combined effects of erythrocyte swelling, transfusion of stored erythrocytes from the spleen, and intercompartmental fluid shifts were described earlier. These types of mechanisms are likely involved in the short-term changes in Hct highlighted next and detailed in Tables 1 and 2.

Aerobic Swimming

Swimming to U_{crit} has been associated with a graded increase in Hct and/or [Hb] in several studies with rainbow trout (e.g., Thomas *et al.*, 1987; Nielsen and Lykkeboe, 1992; Gallaugher *et al.*, 1992; Thorarensen *et al.*, 1996), chinook salmon (Thorarensen *et al.*, 1993; Gallaugher, 1994), and yellowtail (Yamamoto *et al.*, 1980).

Using rainbow trout, Gallaugher *et al.* (1992) showed that splenectomy (used to prevent splenic release of red blood cells) prevented the graded release of red blood cells that normally increased Hct from 27.5 to 34.0% with increasing swimming velocity (see Fig. 2b). Interestingly, when the spleen was surgically tied off in this study, the spleen was already visibly contracted as a result of netting and anesthesia. Therefore, Hct was already elevated (at 32.5%) in splenectomized fish because the ligation prevented resequestration of the erythrocytes that had been released prior to surgery.

Swimming may induce much larger increases in Hct in other species. For example, Hct increased from 28% at rest to 41% at high cruising speeds in yellowtail (Yamamoto *et al.*, 1980). About 40% of this increase (i.e., a 20% change in Hct) was attributed to a graded release of erythrocytes from the spleen, and a further 35% to a measured decrease in plasma volume. Similar increases in Hct were reported for Atlantic cod swimming to U_{crit} (Butler *et al.*, 1989), and even larger increases in Hct were observed at maximal prolonged swimming velocities in an Antarctic nototheniid (Franklin *et al.*, 1993). In the latter case, more than 60% of a 136% increase in Hct was attributed to spleen transfusion, and the 30% reduction in MCHC at maximal swimming velocity indicated that a significant portion of the increase in Hct was also due to cellular swelling.

Swimming-induced increases in Hct and [Hb] occurred in some but not all studies. Increases occurred in freshwater and seawater coho (Zbanyszek and Smith, 1984) and in freshwater snook (Perez-Pinzon and Lutz, 1991), but not in seawater snook (Perez-Pinzon and Lutz, 1991), freshwater brown trout (Butler *et al.*, 1992), freshwater rainbow trout (Kiceniuk and Jones,

1977; Nielsen and Lykkeboe, 1992), freshwater coho (Brauner *et al.,* 1992), or seawater chinook salmon (Brauner *et al.,* 1993; Thorarensen *et al.,* 1993). The apparent absence of Hct changes in some of the seawater studies could be due to the observed increase in plasma osmolality and outward shift of fluid from the plasma masking the increase in Hct (as discussed in Section II).

EXHAUSTIVE EXERCISE AND BURST SWIMMING

Increases in Hct and [Hb] have been observed in a variety of species following short periods of exhaustive (induced) exercise and burst swimming. For example, exhaustive exercise in rainbow trout elevated plasma catecholamine levels and resulted in a 60% increase in Hct, a 25% increase in [Hb], a 20% decrease in MCHC, and a 28% shift of whole body extracellular fluid volume into the intracellular compartment (Milligan and Wood, 1986, 1987). Similar trends, though of lesser magnitude, were observed in starry flounder (Milligan and Wood, 1987). Pearson and Stevens (1991a) reported that exhaustive exercise in rainbow trout, in addition to increasing Hct and [Hb], decreased both SSI and spleen [Hb]. Increases in Hct and/or [Hb] following exhaustive exercise in yellowtail (Yamamoto *et al.,* 1980; Yamamoto, 1991) and Atlantic cod (Gallaugher, Axelsson, Farrell, and Nilsson, unpublished observations) were also associated with splenic transfusion, as indicated by a decrease in SSI. Other species that showed increases in Hct and [Hb] include largemouth bass (Farlinger and Beamish, 1978), striped bass (Young and Cech, 1994), and nototheniids (Egginton *et al.,* 1991).

HYPOXIA

Acute hypoxic exposure is reported to have variable effects on Hct and/or [Hb] in rainbow trout (Table 1). Small increases in Hct and/or [Hb] are reported in several studies (Tetens and Lykkeboe, 1981; Hughes and Kikuchi, 1984; Tun and Houston, 1986; Boutilier *et al.,* 1988; Wells and Weber, 1990, 1991), and no change is reported in several others (Bushnell *et al.,* 1984; Aota *et al.,* 1990; Perry and Reid, 1994; Houston *et al.,* 1996a; Perry and Gilmour, 1996). Marinsky *et al.* (1990) reported an increase in Hct but not in [Hb] with hypoxia. Similar variability exists for other species (see Table 2). Acute hypoxia produced small increases in Hct and/or [Hb] in sheepshead minnow and sailfin molly (Peterson, 1990), Antarctic borch (Wells *et al.,* 1989), some Amazonian fishes (Val *et al.,* 1992), yellowtail (Yamamoto *et al.,* 1985), and Atlantic cod (Gallaugher, Axelsson, Farrell, and Nilsson, unpublished observations). No Hct changes were observed for hypoxic goldfish (Murad *et al.,* 1990) or dogfish (Perry and Gilmour,

1996). Wherever hypoxia produced a significant increase in Hct and [Hb], cellular swelling was involved and sometimes splenic contraction.

HYPERCAPNIA AND HYPEROXIA

As with hypoxia, Hct responses to hypercapnia and hyperoxia are variable (Tables 1 and 2). For rainbow trout, Hct and [Hb] initially (after 20 min) increased with hypercapnia, but the response diminished after 3 h of exposure (Nielsen and Lykkeboe, 1992). In contrast, either Hct and [Hb] did not change or only Hct increased in studies by Perry and Gilmour (1996) and Ishimatsu et al. (1992), respectively. There was no Hct change in hypercapnic dogfish (Perry and Gilmour, 1996), and no hematological response to hyperoxia was evident in carp (Takeda, 1990).

EXPOSURE TO TOXICANTS AND LOW pH

Fish toxicologists routinely use a change in Hct as an indicator of sublethal stress. Elevated catecholamines are regarded as a primary stress response, and since this is associated with an increase in Hct, it is perhaps not surprising that exposure to a number of toxicants has resulted in increased Hct (Table 2). For example, a 3-day exposure to low pH in rainbow trout, elevated Hct and [Hb], caused red blood cell swelling, reduced plasma volume, caused splenic transfusion, and elevated blood viscosity (Milligan and Wood, 1982). Similarly, brown trout held at sublethal pH displayed an elevated Hct and [Hb] and an indication of an outward fluid shift from the plasma (Butler et al., 1992). Waterborne copper exposure in brown trout increased Hct and [Hb], caused erythrocyte swelling, and reduced plasma volume (Wilson and Taylor, 1993a). These later effects were somewhat ameliorated in full-strength seawater compared with freshwater or dilute seawater (Wilson and Taylor, 1993b). In contrast, anemia resulted from exposure of rainbow trout to waterborne thiocyanate (Lanno and Dixon, 1996), of flounder to crude oil (Alkindi et al., 1996), of rainbow trout to nitrite (Jensen, 1990; Stormer et al., 1996), and of Atlantic cod to parasites (Khan, 1977).

V. CRITIQUE OF THE OPTIMAL HEMATOCRIT THEORY

Based on the most consistent data (i.e., blood collected from unstressed fish via chronic cannulation; Tables 1 and 2), it is apparent that Hct is well regulated in most species and the range for normocythemia within a given species tends to be relatively narrow. Although, on an individual basis, Hct may vary in response to environmental or physiological stimuli, the extent of change in Hct is often not that large. The question posed here is: Does the optimal Hct theory apply to fish?

The optimal Hct hypothesis, originally based on studies with mammals (Richardson and Guyton, 1961; Crowell and Smith, 1967), suggests a balance (or tradeoff) between the inadequate concentration of O_2 in arterial blood that results from abnormally low Hct levels (anemia) and the viscosity-induced decreases in cardiac output that result from abnormally high Hct values (polycythemia). Although this hypothesis remains in some medical physiology textbooks, problems exist with its validity. Initial experiments were conducted with anesthetised mammals. However, more recent experimental work with conscious and exercising polycythemic mammals has provided strong evidence that Hct can be increased considerably without negative effects on cardiac output. This is because physiological adjustments can apparently compensate for the supposed detrimental effects of elevated viscosity associated with increased Hct. Further, $V_{O_2 max}$ and exercise performance can be improved with experimentally induced polycythemia in mammals even though one would have predicted viscosity-related constraints based on the optimal hematocrit theory (see Gallaugher *et al.,* 1995, for references). In fact, the observed increases in Hct, known to occur during exercise in many vertebrate species, including fishes, would argue against an optimal Hct based simply on a tradeoff between O_2 transport and blood viscosity.

As for fish, the optimal Hct theory, as originally put forward, appears to be too simplistic to account for our present state of knowledge. For example, Wells and Baldwin (1990) and Wells and Weber (1991) examined whether an optimal Hct applied to rainbow trout and to various tropical reef fishes using known values for viscosity and O_2 transport potential. This theoretical exercise clearly revealed discrete optima for Hct. However, these theoretical optimum Hct values rarely coincided with the measured Hct. Similar concerns about the optimal Hct theory were raised in physiological studies with rainbow trout. Using blood doping studies, Gallaugher *et al.* (1995) examined the influence of Hct on $V_{O_2 max}$ and swimming performance, with the idea that the proposed viscosity limitations on O_2 transport would be most likely revealed under a condition of maximum aerobic activity when the demands on the cardiovascular system are greatest. Blood doping extended Hct to polycythemic levels ($>55\%$) that were far in excess of increases normally related to environmental or physiological perturbations. However, cardiac performance, T_{O_2}, and U_{crit} were not compromised. Instead, there was a rather broad plateau for Hct between the normocythemic and polycythemic states over which $V_{O_2 max}$ and U_{crit} changed very little (Fig. 4). These results clearly argue against a blood viscosity-related limitation for Hct in rainbow trout except at very extreme levels of polycy-

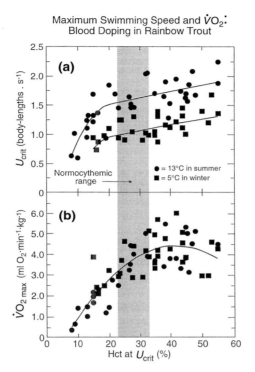

Fig. 4. Results of anemia and blood doping studies with rainbow trout to illustrate the relationships between Hct and V_{O_2max} and between Hct and U_{crit}. Adapted from Gallaugher *et al.,* 1995.

themia, a finding consistent with the recent mammalian literature in this area. If anything, V_{O_2max} would be better classed as relatively insensitive to changes in Hct within the range of normocythemia and moderate polycythemia. In contrast, V_{O_2max} falls off sharply with anemic states (Fig. 4).

Hemoglobin-free Antarctic icefish are often put forward as support for the optimal Hct hypothesis. The evolutionary loss of erythrocytes in these fish is viewed as an adaptation to reduce blood viscosity at the extremely cold water temperatures. However, if blood viscosity is such a constraint under these frigid conditions, why would some Antarctic fish double their Hct when exercising? Clearly, factors other than O_2 transport and blood viscosity must play important roles in setting Hct. Thus, although the optimal Hct hypothesis, as originally set out, correctly identifies the lower limit for Hct as being set by an anemia-related reduction in the capacities for

T_{O_2}, V_{O_2max}, and swimming performance, the factors setting the upper limit for Hct are less clear. Some of the possibilities are introduced here.

One possibility is that the inotropic capabilities of the heart have to be considered since they vary considerably between fish species. Trout, for example, are capable of homeometric regulation well above their normal blood pressure (Farrell and Jones, 1992). In contrast, hemoglobin-free icefishes cannot generate high arterial blood pressures (Tota et al., 1991). Thus, the ability of the heart to accommodate the additional workload associated with blood of a higher viscosity will vary considerably among fish species, and hemoglobin-free icefishes will not be representative of fishes as a whole.

A second possibility that has not been explored is the important role that Hb plays in buffering and CO_2 transport. These roles could be particularly important in exercise when Hct is normally increased. Improved V_{O_2} with elevated Hct could also benefit physiological functions other than swimming performance, such as osmoregulation and digestion during swimming and recovery from respiratory and metabolic acidoses following swimming (Thorarensen et al., 1993; Gallaugher et al., 1995).

Polycythemia may have constraints in addition to viscosity, such as the arterial hypoxemia revealed in the blood doping studies with rainbow trout (Gallaugher et al., 1995). Pa_{O_2} was characteristically decreased at U_{crit}, but in addition, the magnitude of the arterial hypoxemia was Hct-dependent. A Hct-dependent arterial hypoxemia during exercise may indicate that O_2 transfer at the gills is diffusion-limited under polycythemic states whereas with normocythemia it is perfusion-limited.

The metabolic costs associated with Hb synthesis and transport need to be considered. However, at this time, we have only preliminary information on the cost of circulation. For example, Farrell and Steffensen (1987) estimated that the cost of circulating blood is around 1–2% of V_{O_2}. Consequently, fish appear to "unnecessarily" expend almost 1% of V_{O_2} to circulate an appreciable venous O_2 store because venous blood is normally 60–70% saturated and tissue O_2 extraction can increase two- to threefold with exercise. A more cost-effective, long-term strategy might be to store the extra O_2-transport capacity in an enlarged spleen and keep routine O_2 extraction at a much higher level. Indeed, this type of strategy may be important in Antarctic fish where some of the largest exercise-induced increases in Hct occur. However, narrow cost–benefit analyses such as these can be misleading because the evolution of the cardiovascular system has been shaped by a far more complex series of selection pressures. For example, there are limits on venous P_{O_2} since many fish hearts rely solely on the O_2 contained in venous blood for their O_2 supply. Thus, Farrell (1995) referred to venous P_{O_2} as the "tail that wags the dog" in terms of supporting the cardiac metabolism. Second, although splenic release of red blood cells boosts O_2-carrying capacity, splenic release does not provide O_2 to the tissues as rapidly as increased tissue O_2 extraction. Therefore, the venous O_2 reserve serves to minimize the O_2 debt associated with sudden increases in tissue O_2 demand, unless, of course, the demand can be anticipated. Safeguards of this sort within the fish cardiovascular design are presumably "worth" around 1% of V_{O_2}.

VI. CONCLUSIONS

We have collated and described data on Hct in fishes from very diverse sources. Comparisons were often limited because of methodological complications. In particular, evidence was presented to suggest that Hct values may be elevated with acute blood sampling techniques as a result of stress-induced red blood cell swelling and splenic red blood cell release. We also advise caution when interpreting data obtained in experiments involving invasive surgery. The potential blood loss could obscure or exacerbate normal responses of the O_2-transport system in normocythemic fish. Interspecific patterns in Hct were difficult to discern with the present database, and intraspecific changes were generally quite small, with the noted exception in some species of large increases in Hct associated with splenic release of red blood cells under certain physiological conditions. Regardless of these difficulties, extremely strong correlations were found between Hct and [Hb] for both intraspecific and interspecific comparisons. Furthermore, it was clear that anemia in fish can compromise O_2 transport and exercise performance, and therefore it is highly likely that the needs of O_2 transport play a pivotal role in setting the lower limit for Hct in normocythemic fish. However, it was equally clear that undue emphasis has been placed on the role played by blood viscosity in setting the upper limit for Hct in normocythemic fish. Other factors need to be considered. Potential areas for future research include controlled studies of intraspecific diversity in Hct, splenic control of Hct, molecular understanding of the effects of environmental stimuli on erythropoiesis, and energy budgets for erythropoiesis and Hb transport. Moreover, virtually nothing is known about the embryology of hematology, effects of physiological and environmental influences during development on Hct regulation, or the links between Hct (Hb), nutrition, and growth.

REFERENCES

Alkindi, A. Y. A., Brown, J. A., Waring, C. P., and Collins, J. E. (1996). Endocrine, osmoregulatory, respiratory and haematological parameters in flounder exposed to the water soluble fraction of crude oil. *J. Fish Biol.* **49**, 1291–1305.

Andersen, N. A., Laursen, J. S., and Lykkeboe, G. (1985). Seasonal variations in hematocrit, red cell hemoglobin and nucleoside triphosphate concentrations, in the European eel *Anguilla anguilla. Comp. Biochem. Physiol. A* **81,** 87–92.

Aota, S., Holmgren, K. D., Gallaugher, P., and Randall, D. J. (1990). A possible role for catecholamines in the ventilatory responses associated with internal acidosis or external hypoxia in rainbow trout *Oncorhynchus mykiss. J. Exp. Biol.* **151,** 57–70.

Baldwin, J., and Wells, R. M. G. (1990). Oxygen-transport potential in tropical elasmobranchs from the Great Barrier Reef: Relationship between hematology and blood viscosity. *J. Exp. Mar. Biol. Ecol.* **144,** 145–155.

Barron, M. G., Tarr, B. D., and Hayton, W. L. (1987). Temperature-dependence of cardiac output and regional blood flow in rainbow trout, *Salmo gairdneri* (Richardson). *J. Fish Biology* **31,** 735–744.

Bauer, C., and Kurtz, A. (1989). Oxygen sensing in the kidney and its relation to erythropoietin production. *Ann. Rev. Physiol.* **51,** 845–856.

Birchard, G. F., and Tenney, S. M. (1990). Relationship between blood–oxygen affinity and blood volume. *Respir. Physiol.* **83,** 365–374.

Boutilier, R. G., Dobson, G., Hoeger, U., and Randall, D. J. (1988). Acute exposure to graded levels of hypoxia in rainbow trout (*Salmo gairdneri*): Metabolic and respiratory adaptations. *Respir. Physiol.* **71,** 69–82.

Brauner, C. J., Shrimpton, J. M., and Randall, D. J. (1992). Effect of short-duration seawater exposure on plasma ion concentrations and swimming performance of coho salmon (*Oncorhynchus kisutch*) parr. *Can. J. Fish. Aqua. Sci.* **49,** 2399–2405.

Brauner, C. J., Val, A. L., and Randall, D. J. (1993). The effect of graded methaemoglobin levels on the swimming performance of chinook salmon (*Oncorhynchus tshawytscha*). *J. Exp. Biol.* **185,** 121–135.

Bridges, D. W., Cech, J. J. Jr., and Pedro, D. N. (1976). Seasonal hematological changes in winter flounder *Pseudopleuronectes americanus. Trans. Am. Fish. Soc.* **105,** 596–600.

Brill, R. W., and Bushnell, P. G. (1991). Metabolic and cardiac scope of high energy demand teleosts, the tunas. *Can. J. Zool.* **69,** 2002–2009.

Bushnell, P. G., Steffensen, J. F., and Johansen, K. (1984). Oxygen consumption and swimming performance in hypoxia-acclimated rainbow trout *Salmo gairdneri. J. Exp. Biol.* **113,** 225–235.

Bushnell, P. G., Jones, D. R., and Farrell, A. P. (1992). The Arterial System. *In* "Fish Physiology" (W. S. Hoar, D. J. Randall, A. P. Farrell, eds.), Vol. XIIA. pp. 89–139. Academic Press, New York.

Butler, P. J., and Metcalfe, J. D. (1988). Cardiovascular and Respiratory Systems. *In* "Physiology of Elasmobranch Fishes" (T. J. Shuttleworth, ed.), pp. 1–48. Springer-Verlag, Berlin.

Butler, P. J., Day, N., and Namba, K. (1992). Interactive effects of seasonal temperature and low pH on resting oxygen uptake and swimming performance of adult brown trout *Salmo trutta. J. Exp. Biol.* **165,** 195–212.

Butler, P. J., Axelsson, M., Ehrenstrom, F., Metcalfe, J. D., and Nilsson, S. (1989). Circulating catecholamines and swimming performance in the Atlantic cod, *Gadus morhua. J. Exp. Biol.* **141,** 377–387.

Byrne, A. P., and Houston, A. H. (1988). Use of phenylhydrazine in the detection of responsive changes in hemoglobin isomorph abundances. *Can. J. Zool.* **66,** 758–762.

Cameron, J. N. (1970). The influence of environmental variables on the hematology of pinfish (*Lagondon rhomboides*) and striped mullet (*Mugil cephalus*). *Comp. Biochem. Physiol.* **23,** 175–192.

Cameron, J. N., and Wohlschlag, D. E. (1969). Respiratory response to experimentally induced anemia in the pinfish (*Lagodon rhomboides*). *J. Exp. Biol.* **50,** 307–317.

Cameron, J. N., and Davis, J. C. (1970). Gas exchange in rainbow trout (*Salmo gairdneri*) with varying blood oxygen capacity. *J. Fish. Res. Bd. Can.* **27**, 1069–1085.

Cech, J. J. Jr., Bridges, D. W., Rowell, D. M., and Balzer, P. J. (1976). Cardiovascular responses of winter flounder, *Pseudopleuronectes americanus* (Walbaum), to acute temperature increase. *Can. J. Zool.* **54**, 1383–1388.

Chudzik, J., and Houston, A. H. (1983). Temperature and erythropoiesis in goldfish. *Can. J. Zool.* **61**, 1322–1325.

Crowell, J. W., and Smith, E. E. (1967). Determinant of the optimal hematocrit. *J. Appl. Physiol.* **22**, 501–504.

Currie, S., and Tufts, B. (1993). An analysis of carbon dioxide transport in arterial and venous blood of the rainbow trout, *Oncorhynchus mykiss*, following exhaustive exercise. *Fish Physiol. Biochem.* **12**, 183–192.

Davie, P. S. (1990). "Pacific Marlins: Anatomy and Physiology." pp. 1–87. Massey Univ., Palmerston North, New Zealand.

Davie, P. S., Wells, R. M. G., and Tetens, V. (1986). Effects of sustained swimming on rainbow trout muscle structure, blood oxygen transport, and lactate dehydrogenase isozymes: Evidence for increased aerobic capacity of white muscle. *J. Exp. Zool.* **237**, 159–171.

Eddy, F. B. (1973). Oxygen dissociation curves of the blood of the tench, *Tinca tinca. J. Exp. Biol.* **58**, 281–293.

Egginton, S., Taylor, E. W., Wilson, R. W., Johnston, I. A., and Moon, T. W. (1991). Stress response in the Antarctic teleosts (*Notothenia neglecta* Nybelin and *N. rossii* Richardson). *J. Fish Biol.* **38**, 225–235.

Fange, R. (1986). Physiology of haemopoiesis. *In* "Fish Physiology: Recent Advances" (S. Nilsson and S. Holmgren, eds.), pp. 1–23. Croom Helm, London, Sydney.

Fange, R. (1992). Fish red blood cells. *In* "Fish Physiology" (W. S. Hoar, D. J. Randall, A. P. Farrell, eds.), XIIA. pp. 2–46. Academic Press, New York.

Fange, R., and Nilsson, S. (1985). The fish spleen: Structure and function. *Experientia* **41**, 52–157.

Farlinger, S., and Beamish, F. W. H. (1978). Changes in blood chemistry and critical swimming speed of largemouth bass, *Micropterus salmoides*, with physical conditioning. *Trans. Am. Fish. Soc.* **107**, 523–527.

Farrell, A. P. (1991). From hagfish to tuna: A perspective on cardiac function in fish. *Physiol. Zool.* **64**(5), 1137–1164.

Farrell, A. P. (1995). Cardiac output in fish: Regulation and limitations. *In* "The Vertebrate Gas Transport Cascade: Adaptations to Environment and Mode of Life" (J. E. Bicudo, ed.), pp. 208–214. CRC Press, Boca Raton, FL.

Farrell, A. P., and Jones, D. R. (1992). The heart. *In* "Fish Physiology" (W. S. Hoar, D. J. Randall, and A. P. Farrell, eds.), XIIA. pp. 1–88. Academic Press, New York.

Farrell, A. P., and Steffensen, J. F. (1987). An analysis of the energetic cost of the branchial and cardiac pumps during sustained swimming in trout. *Fish Physiol. Biochem.* **4**, 73–79.

Forster, M. E., Davison, W., Satchell, G. H., and Taylor, H. H. (1989). The subcutaneous sinus of the hagfish, *Eptatretus cirrhatus* and its relation to the central circulating blood volume. *Comp. Biochem. Physiol.* **93**, 607–612.

Franklin, C. E., Davison, W., and McKenzie, J. C. (1993). The role of the spleen during exercise in the Antarctic teleost, *Pagothenia borchgrevinki. J. Exp. Biol.* **174**, 381–386.

Gallaugher, P. (1994). The role of hematocrit in oxygen transport and swimming in salmonid fishes. 248 pp. Ph.D. thesis. Simon Fraser Univ., Burnaby, BC, Canada.

Gallaugher, P., Axelsson, M., and Farrell, A. P. (1992). Swimming performance and haematological variables in splenectomized rainbow trout, *Oncorhynchus mykiss. J. Exp. Biol.* **171**, 301–314.

Gallaugher, P., Thorarensen, H., and Farrell, A. P. (1995). Hematocrit in oxygen transport and swimming in rainbow trout (*Oncorhynchus mykiss*). *Respir. Physiol.* **102**, 279–292.

Garcia, M. P., Echevarria, G., Martinez, F. J., and Zamora, S. (1992). Influence of blood sample collection on the haematocrit value of two teleosts: Rainbow trout (*Oncorhynchus mykiss*) and European sea bass (*Dicentrarchus labrax* L.). *Comp. Biochem. Physiol. A* **101**, 733–736.

Graham, M. S., and Farrell, A. P. (1992). Environmental influences on cardiovascular variables in rainbow trout, *Oncorhynchus mykiss* (Walbaum). *J. Fish Biol.* **41**, 851–858.

Graham, M. S., and Fletcher, G. L. (1983). Blood plasma viscosity of winter flounder: Influence of temperature, red cell concentration, and shear rate. *Can. J. Zool.* **61**, 2344–2350.

Graham, M. S., Haedrich, R. L., and Fletcher, G. L. (1985). Hematology of three deep-sea fishes: A reflection of low metabolic rates. *Comp. Biochem. Physiol. A* **80**, 79–84.

Hannah, G. S., and Pickford, G. E. (1981). Diurnal peaks in the hematocrit and in serum sodium in the killifish, *Fundulus heteroclitus,* and their absence in serum potassium and chloride: A lack of correlation. *Comp. Biochem. Physiol. A* **70**, 157–159.

Hochachka, P. W. (1961). The effect of physical training on oxygen debt and glycogen reserves in trout. *Can. J. Zool.* **39**, 767–776.

Houston, A. H., and DeWilde, M. A. (1972). Some observations upon the relationship of microhaematocrit values to haemoglobin concentrations and erythrocyte numbers in the carp *Cyprinus carpio* L. and brook trout *Salvelinus fontinalis* (Mitchill). *J. Fish Biol.* **4**, 109–115.

Houston, A. H., and Gingras-Bedard, J. H. (1994). Variable versus constant temperature acclimation regimes: Effects on hemoglobin isomorph profile in goldfish, *Carassius auratus.* *Fish Physiol. Biochem.* **13**, 445–450.

Houston, A. H., and Koss, T. F. (1984). Plasma and red cell ionic composition in rainbow trout exposed to progressive temperature increases. *J. Exp. Biol.* **110**, 53–67.

Houston, A. H., and Murad, A. (1992). Erythrodynamics in goldfish, *Carassius auratus* L.: Temperature effects. *Physiol. Zool.* **65**, 55–76.

Houston, A. H., and Murad, A. (1995). Erythrodynamics in fish: Recovery of the goldfish *Carassius auratus* from acute anemia. *Can. J. Zool.* **73**, 411–418.

Houston, A. H., and Schrapp, M. P. (1994). Thermoacclimatory hematological response: Have we been using appropriate conditions and assessment methods? *Can. J. Zool.* **72**, 1238–1242.

Houston, A. H., and Smeda, J. S. (1979). Thermoacclimatory changes in the ionic microenvironment of haemoglobin in the stenothermal rainbow trout (*Salmo gairdneri*) and eurythermal carp (*Cyprinus carpio*). *J. Exp. Biol.* **80**, 317–340.

Houston, A. H., Dobric N., and Kahurananga, R. (1996a). The nature of hematological response in fish. Studies on rainbow trout *Oncorhynchus mykiss* exposed to simulated winter, spring and summer conditions. *Fish Physiol. Biochem.* **15**, 339–347.

Houston, A. H., Roberts, W. C., and Kennington, J. A. (1996b). Hematological response in fish: Pronephric and splenic involvements in the goldfish, *Carassius auratus* L. *Fish Physiol. Biochem.* **15**, 481–489.

Hughes, G. M., and Kikuchi, Y. (1984). Effects of *in vivo* and *in vitro* changes in P_{O_2} on the deformability of red blood cells of rainbow trout (*Salmo gairdneri* R.). *J. Exp. Biol.* **111**, 253–257.

Ishimatsu, A., Iwama, G. K., Bentley, T. B., and Heisler, N. (1992). Contribution of the secondary circulatory system to acid–base regulation during hypercapnia in rainbow trout (*Oncorhynchus mykiss*). *J. Exp. Biol.* **170**, 43–56.

Iuchi, I., and Yamamoto, M. (1983). Erythropoiesis in the developing rainbow trout, *Salmo gairdneri*: Histochemical and immunochemical detection of eythropoietic organs. *J. Exp. Zool.,* **226**, 409–417.

Jensen, F. B. (1990). Nitrite and red cell function in carp: Control factors for nitrite entry, membrane potassium ion permeation, oxygen affinity and methaemoglobin formation. *J. Exp. Biol.* **152,** 149–166.

Johansen, K., Fange, R., and Johannessen, M. W. (1962). Relations between blood, sinus fluid, and lymph in *Myxine glutinosa* L. *Comp. Biochem. Physiol.* **7,** 23–28.

Jones, D. R. (1971). The effect of hypoxia and anemia on the swimming performance of rainbow trout (*Salmo gairdneri*). *J. Exp. Biol.* **55,** 541–551.

Jones, J. H., and Lindstedt, S. L. (1993). Limits to maximal performance. *Annu. Rev. Physiol.* **55,** 547–569.

Khan, R. (1977). Blood changes in Atlantic Cod (*Gadus morhua*) infected with *Trypanosoma murmanensis. J. Fish. Res. Board Can.* **34,** 2193–2196.

Kiceniuk, J. W., and Jones, D. R. (1977). The oxygen transport system in trout (*Salmo gairdneri*) during sustained exercise. *J. Exp. Biol.* **69,** 247–260.

Kikuchi, Y., Hughes, G. M., and Duthie, G. G. (1985). Effects of moderate and severe exercise in rainbow trout on some properties of arterial blood, including red blood cell deformability. *Jap. J. Ichthyol.* **31,** 422–426.

Korcock, D. E., Houston, A. H., and Gray, J. D. (1988). Effects of sampling conditions on selected blood variables of rainbow trout, *Salmo gairdneri* Richardson. *J. Fish Biol.* **33,** 319–330.

Lane, H. C. (1979). Some haematological responses of normal and splenectomized rainbow trout (*Salmo gaidneri*) to a 12% blood loss. *J. Fish Biol.* **14,** 159–164.

Lane, H. C., Rolfe, A. E., and Nelson, J. R. (1981). Changes in the nucleotide triphosphate/haemoglobin and nucleotide triphosphate/red cell ratios of rainbow trout, *Salmo gairdneri* Richardson, subjected to prolonged starvation and bleeding. *J. Fish Biol.* **18,** 661–668.

Lanno, R. P., and Dixon, D. G. (1996). Chronic toxicity of waterborne thiocyanate to rainbow trout (*Oncorhynchus mykiss*). *Can. J. Fish. Aquat. Sci.* **53,** 2137–2146.

McKim, J. M., Nichols, J. W., Lien, G. J., and Bertelsen, S. L. (1994). Respiratory-cardiovascular physiology and chloroethane gill flux in the channel catfish, *Ictalurus punctatus. J. Fish Biol.* **44,** 527–547.

McLeay, D. J., and Gordon, M. R. (1977). Leucocrit: A simple hematological technique for measuring acute stress in salmonid fish, including stressful concentrations of pulpmill effluent. *J. Fish Res. Board Can.* **34,** 2164–2175.

McLeod, T. F., Sigel, M. M., and Yunis, A. A. (1978). Regulation of erythropoiesis in the Florida gar, *Lepisosteus platyrhincus. Comp. Biochem. Physiol. A* **60,** 145–150.

Marinsky, C. A., Houston, A. H., and Jurad, A. (1990). Effects of hypoxia on the hemoglobin isomorph abundances in rainbow trout, *Salmo gairdneri. Can. J. Zool.* **68,** 884–888.

Martinez, F. J., Garcia-Riera, M. P., Canteras, M., De Costa, J., and Zamora, S. (1994). Blood parameters in rainbow trout (*Oncorhynchus mykiss*): Simultaneous influence of various factors. *Comp. Biochem. Physiol. A* **107,** 95–100.

Milligan, C. L., and Wood, C. M. (1982). Disturbances in haematology, fluid volume distribution and circulatory function associated with low environmental pH in the rainbow trout, *Salmo gairdneri. J. Exp. Biol.* **99,** 397–415.

Milligan, C. L., and Wood, C. M. (1986). Intracellular and extracellular acid–base status and H^+ exchange with the environment after exhaustive exercise in the rainbow trout. *J. Exp. Biol.* **123,** 93–121.

Milligan, C. L., and Wood, C. M. (1987). Regulation of blood oxygen transport and red cell pH_i after exhaustive activity in rainbow trout (*Salmo gairdneri*) and starry flounder (*Platichthys stellatus*). *J. Exp. Biol.* **133,** 263–282.

Monteiro, P. J., Val, A. L., and De Almeida-Val, M. F. (1987). Biological aspects of Amazonian fishes. Hemoglobin, hematology, intraerythrocytic phosphates, and whole blood Bohr effect of *Mylossoma duriventris. Can. J. Zool.* **65,** 1805–1811.

Murad, A., Houston, A. H., and Samson, L. (1990). Haematological response to reduced oxygen-carrying capacity, increased temperature and hypoxia in goldfish, *Carassius auratus* L. *J. Fish Biol.* **36**, 289–305.

Nelson, J. A., Tang, Y., and Boutilier, R. G. (1994). Differences in exercise physiology between two Atlantic cod (*Gadus morhua*) populations from different environments. *Physiol. Zool.* **67**, 330–354.

Nielsen, O. B., and Lykkeboe, G. (1992). Changes in plasma and erythrocyte K^+ during hypercapnia and different grades of exercise in trout. *J. Appl. Physiol.* **72** (4) 1285–1290.

Nikinmaa, M., Soivio, A., and Railo, E. (1981). Blood volume of *Salmo gairdneri*: Influence of ambient temperature. *Comp. Biochem. Physiol. A* **69**, 767–769.

Nikinmaa, M., and Weber, R. (1984). Hypoxic acclimation in the lamprey, *Lampetra fluviatilis*: Organismic and erythrocytic responses. *J. Exp. Biol.* **109**, 109–119.

Nilsson, S. 1983. "Autonomic Nerve Function in the Vertebrates." 253 pp. Springer-Verlag, Berlin/Heidelberg/New York.

Nilsson, S., and Grove, D. J. (1974). Adrenergic and cholinergic innervation of the spleen of the cod: *Gadus morhua*. *Eur. J. Pharmac.* **28**, 135–143.

Oikari, K., and Soivio, A. (1975). Influence of sampling methods and anaesthetization on various haematological parameters of several teleosts. *Aquaculture* **6**, 171–180.

Olson, K. R. (1984). Distribution of flow and plasma skimming in isolated perfused gills of three teleosts. *J. Exp. Biol.* **109**, 97–108.

Opdyke, D. F., Carroll, R. G., and Keller, N. E. (1982). Catecholamine release and blood pressure changes induced by exercise in dogfish. *Am. J. Physiol.* **242**, R306–R310.

Pearson, M. P., and Stevens, E. D. (1991a). Size and hematological impact of the splenic erythrocyte reservoir in rainbow trout, *Oncorhynchus mykiss*. *Fish Physiol. Biochem.* **9**, 39–50.

Pearson, M. P., and Stevens, E. D. (1991b). Splenectomy impairs aerobic swim performance in trout. *Can. J. Zool.* **69**, 2089–2092.

Perez-Pinzon, M. A., and Lutz, P. L. (1991). Activity related cost of osmoregulation in the juvenile snook (*Centropomus undecimalis*). *Bull. Mar. Sci.* **48**, 58–66.

Perry, S. F., and Gilmour, K. M. (1996). Consequences of catecholamine release on ventilation and blood oxygen transport during hypoxia and hypercapnia in an elasmobranch (*Squalus acanthias*) and a teleost (*Oncorhynchus mykiss*). *J. Exp. Biol.* **199**, 2105–2118.

Perry, S. F., and Kinkead, R. (1989). The role of catecholamines in regulating arterial oxygen content during acute hypercapnic acidosis in rainbow trout (*Salmo gairdneri*). *Respir. Physiol.* **77**, 365–378.

Perry, S. F., and Reid, S. D. (1992). Relationship between blood O_2 content and catecholamine levels during hypoxia in rainbow trout and American eel. *Am. J. Physiol.* **263**, R240–R249.

Perry, S. F., and Reid, S. G. (1994). The effects of acclimation temperature on the dynamics of catecholamine release during acute hypoxia in the rainbow trout *Oncorhynchus mykiss*. *J. Exp. Biol.* **186**, 289–307.

Perry, S. F., Kinkead, R., Gallaugher, P., and Randall, D. J. (1989). Evidence that hypoxemia promotes catecholamine release during hypercapnic acidosis in rainbow trout (*Salmo gairdneri*). *Respir. Physiol.* **77**, 351–364.

Peterson, M. S. (1990). Hypoxia-induced physiological changes in two mangrove swamp fishes: Sheepshead minnow, *Cyprinodon variegatus* (Lacepede) and sailfin molly, *Poecilia latipinna* (Leseur). *Comp. Biochem. Physiol. A* **97**, 17–21.

Pradhan, R. K., Saini, S. K., Biswas, J., and Pati, A. K. (1989). Influence of human urinary erythropoietin and L-thyroxine on blood morphology and energy reserves in two tropical species of fed and starved teleosts. *Gen. Comp. Endocrin.* **76**, 382–389.

Primmett, D. R. N., Randall, D. J., Mazeaud, M., and Boutilier, R. G. (1986). The role of catecholamines in erythrocyte pH regulation and oxygen transport in rainbow trout (*Salmo gairdneri*) during exercise. *J. Exp. Biol.* **122**, 139–148.

Quentel, C., and Obach, A. (1992). The cellular composition of the blood and haematopoietic organs of turbot *Scophthalmus maximus* L. *J. Fish Biol.* **41**, 709–716.

Railo, E., Nikinmaa, M., and Soivio, A. (1985). Effects of sampling on blood parameters in the rainbow trout, *Salmo gairdneri* Richardson. *J. Fish Biol.* **26**, 725–732.

Randall, D. J. (1985). Shunts in fish gills. *In* "Cardiovascular Shunts: Phylogenetic, Ontogenetic and Clinical Aspects." (K. Johansen and W. Burggren, eds.), pp. 71–82. Alfred Benzon Symposium 21. Munksgaard, Copenhagen.

Richardson, T. Q., and Guyton, A. C. (1961). Effects of polycythemia and anemia on cardiac output and other circulatory factors. *Am. J. Physiol.* **197** (6), 1167–1170.

Sandnes, K., Lie, O., and Waagbo, R. (1988). Normal ranges of some blood chemistry parameters in adult farmed Atlantic salmon, *Salmo salar*. *J. Fish Biol.* **32**, 129–136.

Satchell, G. H. (1991). "Physiology and Form of Fish Circulation." Cambridge Univ. Press, Cambridge.

Smith, J. C. (1977). Body weight and the haematology of the American plaice *Hipposglossoides platessoides*. *J. Exp. Biol.* **67**, 17–28.

Smith, L. S. (1966). Blood volumes of three salmonids. *J. Fish. Res. Bd. Can.* **23**, 1439–1446.

Soivio, A., and Oikari, A. (1976). Haematological effects of stress on a teleost, *Esox lucius* L. *J. Fish Biol.* **8**, 397–411.

Sorensen, B., and Weber, R. E. (1995). Effects of oxygenation and the stress hormones adrenaline and cortisol on the viscosity of blood from the trout *Oncorhynchus mykiss*. *J. Exp. Biol.* **198**, 953–959.

Steffensen, J. F., and Lomholt, J. P. (1992). The secondary vascular system. *In* "Fish Physiology" (W. S. Hoar, D. J. Randall, A. P. Farrell, eds.), Vol. XIIA. pp. 185–218. Academic Press, New York.

Stormer, J., Jensen, F. B., and Rankin, J. C. (1996). Uptake of nitrite, nitrate, and bromide in rainbow trout, *Oncorhynchus mykiss*: Effects on ionic balance. *Can. J. Fish. Aquat. Sci.* **53**, 1943–1950.

Sundin, L., and Nilsson, S. (1992). Arterio–venous branchial blood flow in the Atlantic cod, *Gadus morhua*. *J. Exp. Biol.* **163**, 73–84.

Takeda, T. (1990). Ventilation, cardiac output and blood respiratory parameters in the carp, *Cyprinus carpio*, during hyperoxia. *Respir. Physiol.* **81**, 227–240.

Taylor, S. E., Egginton, S., and Taylor, E. W. (1993). Respiratory and cardiovascular responses in rainbow trout (*Oncorhynchus mykiss*) to aerobic exercise over a range of acclimation temperatures. *J. Physiol.* **459**, 19P.

Tetens, V., and Lykkeboe, G. (1981). Blood respiratory properties of rainbow trout, *Salmo gairdneri*: Responses to hypoxia acclimation and anoxic incubation of blood in vitro. *J. Comp. Physiol.* **145**, 117–125.

Tetens, V., and Lykkeboe, G. (1985). Acute exposure of rainbow trout to mild and deep hypoxia: O_2 affinity and O_2 capacitance of arterial blood. *Respir. Physiol.* **61**, 221–235.

Tetens, V., Wells, R. M. G., and DeVries, A. L. (1984). Antarctic fish blood: Respiratory properties and the effects of thermal acclimation. *J. Exp. Biol.* **109**, 265–279.

Thomas, S., Poupin, J., Lykkeboe, G., and Johansen, K. (1987). Effects of graded exercise on blood gas tensions and acid–base characteristics of rainbow trout. *Respir. Physiol.* **68**, 85–97.

Thorarensen, H. (1994). Gastrointestinal oxygen transport in fed and unfed chinook salmon, *Oncorhynchus tshawytscha*. Ph.D. thesis. Simon Fraser Univ., BC, Canada.

Thorarensen, H., Gallaugher, P. E., Kiessling, A. K., and Farrell, A. P. (1993). Intestinal blood flow in swimming chinook salmon *Oncorhynchus tshawytscha* and the effects of haematocrit on blood flow distribution. *J. Exp. Biol.* **179**, 115–129.

Thorarensen, H., Gallaugher, P., and Farrell, A. P. (1996). Cardiac output in swimming rainbow trout, *Oncorhynchus mykiss,* acclimated to seawater. *Physiol. Zool.* **69**, 139–153.

Tota, B., Acierno, R., and Agnisola, C. (1991). Mechanical performance of the isolated and perfused heart of the haemoglobinless Antarctic icefish *Chionodraco hamatus* (Lonnberg): Effects of loading conditions and temperature. *Phil. Trans. R. Soc. Lond. B* **332**, 191–198.

Tun, N., and Houston, A. H. (1986). Temperature, oxygen, photoperiod, and the hemoglobin system of the rainbow trout, *Salmo gairdneri. Can. J. Zool.* **64**, 1883–1888.

Val, A. L. (1993). Adaptations of fishes to extreme conditions in fresh waters. *In* "The Vertebrate Gas Transport Cascade: Adaptations to Environment and Mode of Life" (J. E. Bicudo, ed.), pp. 43–53. CRC Press, Boca Raton.

Val, A. L., de Almeida-Val, V. M., and Affonso, E. G. (1990). Adaptive features of Amazon fishes: Hemoglobins, hematology, intraerythrocytic phosphates and whole blood Bohr effect of *Pterygophlichthys multiradiatus* (Siluriformes). *Comp. Biochem. Physiol. B* **97**, 435–440.

Val, A. L., Affonso, E. G., and de Almeida-Val, V. M. (1992). Adaptive features of Amazon fishes: Blood characteristics of Curimata (*Prochilodus* cf. *nigricans,* Osteichthyes). *Physiol. Zool.* **65**, 832–843.

Val, A. L., Mazur, C. G., de Salvo-Souza, R. H., and Iwama, G. K. (1994). Effects of experimental anaemia on intra-erythrocytic phosphate levels in rainbow trout, *Oncorhynchus mykiss. J. Fish Biol.* **45**, 269–277.

Vorger, P., and Ristori, M.-T. (1985). Effects of experimental anemia on the ATP content and the oxygen affinity of the blood in the rainbow trout (*Salmo gairdneri*). *Comp. Biochem. Physiol. A* **82**, 221–224.

Weinberg, S. R., LoBue, J., Siegel, C. D., and Gordon, A. S. (1976). Hematopoiesis of the kissing gourami (*Helostoma temmincki*). Effects of starvation, bleeding, and plasma-stimulating factors on its erythropoiesis. *Can. J. Zool.* **54**, 1115–1127.

Wells, R. M. G., and Baldwin, J. (1990). Oxygen transport potential in tropical reef fish with special reference to blood viscosity and hematocrit. *J. Exp. Mar. Biol. Ecol.* **141**, 131–143.

Wells, R. M. G., and Forster, M. E. (1989). Dependence of blood viscosity on haematocrit and shear rate in a primitive vertebrate. *J. Exp. Biol.* **145**, 483–487.

Wells, R. M. G., and Weber, R. E. (1990). The spleen in hypoxic and exercised rainbow trout. *J. Exp. Biol.* **150**, 461–466.

Wells, R. M. G., and Weber, R. E. (1991). Is there an optimal haematocrit for rainbow trout, *Oncorhynchus mykiss* (Walbaum)? An interpretation of recent data based on blood viscosity measurements? *J. Fish Biol.* **38**, 53–65.

Wells, R. M. G., Baldwin, J., and Ryder, J. M. (1992). Respiratory function and nucleotide composition of erythrocytes from tropical elasmobranchs. *Comp. Biochem. Physiol. A* **103**, 157–162.

Wells, R. M. G., Macdonald, J. A., and diPrisco, G. (1990). Thin-blooded Antarctic fishes: A rheological comparison of the haemoglobin-free icefishes, *Chionodraco kathleenae* and *Cryodraco antarcticus,* with a red-blooded nototheniid, *Pagothenia bernacchii. J. Fish Biol.* **36**, 595–609.

Wells, R. M. G., McIntyre, R. H., Morgan, A. K., and Davie, P. S. (1986). Physiological stress responses in big gamefish after capture: Observations on plasma chemistry and blood factors. *Comp. Biochem. Physiol. A* **84**, 565–571.

Wells, R. M. G., Grigg, G. C., Beard, L. A., and Summers, G. (1989). Hypoxic responses in a fish from a stable environment: Blood oxygen transport in the Antarctic fish *Pagothenia borchgrevinki. J. Exp. Biol.* **141**, 97–111.

Wilkes, P. R. H., Walker, R. L., McDonald, D. G., and Wood, C. M. (1981). Respiratory, ventilatory, acid–base and ionoregulatory physiology of the white sucker *Catostomus commersoni*: The influence of hyperoxia. *J. Exp. Biol.* **91**, 239–254.

Wilson, R., and Egginton, S. (1994). Assessment of maximum sustainable swimming performance in rainbow trout (*Oncorhynchus mykiss*). *J. Exp. Biol.* **192**, 299–305.

Wilson, R. W., and Taylor, E. W. (1993a). The physiological responses of freshwater rainbow trout, *Oncorhynchus mykiss,* during acutely lethal copper exposure. *J. Comp. Physiol* **163**, 38–47.

Wilson, R. W., and Taylor, E. W. (1993b). Differential responses to copper in rainbow trout (*Oncorhynchus mykiss*) acclimated to sea water and brackish water. *J. Comp. Physiol.* **163**, 239–246.

Wilson, R. W., and Egginton, S. (1994). Assessment of maximum sustainable swimming performance in rainbow trout (*Oncorhynchus mykiss*). *J. Exp. Biol.* **192**, 299–305.

Wood, C.M., and Randall, D.J. (1973). The influence of swimming activity on water balance in the rainbow trout (*Salmo gairdneri*). *J. Comp. Physiol.* **82**, 257–276.

Wood, C. M., McMahon, B. R., and McDonald, D. G. (1979). Respiratory, ventilatory, and cardiovascular responses to experimental anaemia in the starry flounder, *Platichthys stellatus*. *J. Exp. Biol.* **82**, 139–162.

Woodward, J. J., and Smith, L. S. (1985). Exercise training and the stress response in rainbow trout, *Salmo gairdneri* Richardson. *J. Fish Biol.* **26**, 435–447.

Yamamoto, K. I. (1991). Increase of arterial O_2 content in exercised yellowtail (*Seriola quinqueradiata*). *Comp. Biochem. Physiol. A* **98**, 43–46.

Yamamoto, K. I., Itazawa, Y, and Kobayashi, H. (1980). Supply of erythrocytes into the circulating blood from the spleen of exercised fish. *Comp. Biochem. Physiol. A* **65**, 5–11.

Yamamoto, K. I., Itazawa, Y., and Kobayashi, H. (1985). Direct observation of fish spleen by an abdominal window method and its application to exercised and hypoxic yellowtail. *Jap. J. Ichthyol.,* **31**, 427–433.

Young, P. S., and Cech, J. J., Jr. (1993). Effects of exercise conditioning on stress responses and recovery in cultured and wild young-of-the-year striped bass, *Morone saxatilis. Can. J. Fish. Aquat. Sci.* **50**, 2094–2099.

Young, P. S., and Cech, J. J., Jr. (1994). Effects of different exercise conditioning velocities on the energy reserves and swimming stress responses in young-of-the-year striped bass (*Morone saxatilis*). *Can. J. Fish. Aquat. Sci.* **51**, 1528–1534.

Zbanyszek, R., and Smith, L. (1984). Changes in carbonic anhydrase activity in coho salmon smolts resulting from physical training and transfer into seawater. *Comp. Biochem. Physiol. A* **79**, 229–233.

7

CARBON DIOXIDE TRANSPORT AND EXCRETION

BRUCE TUFTS AND STEVE F. PERRY

I. Introduction
II. Carriage of CO_2 in Blood
III. Carbon Dioxide Transport and Excretion
    A. Steady-State Conditions in Teleosts
    B. Non-steady-state Conditions in Teleosts
    C. Elasmobranchs
    D. Steady-State Conditions in Agnathans
    E. Non-steady-state Conditions in Agnathans
IV. Future Directions
    References

I. INTRODUCTION

Metabolism produces CO_2 at variable rates that are dictated by aerobic metabolic requirements. In aqueous solutions, CO_2 acts as a weak acid, and consequently the processes of CO_2 transport/excretion and acid–base balance are closely linked. To avoid acid–base imbalances, CO_2 production is matched by CO_2 excretion under steady-state conditions. In most fish species that have been examined, the majority of CO_2 is excreted across the gill into the water as O_2 is absorbed across the gill into the blood. The processes of O_2 uptake and CO_2 excretion (collectively termed "respiratory gas transfer") share common pathways, are governed by several mutual principles, and are intricately related (see Chapter 8 by Brauner and Randall in this volume). In many respects, however, CO_2 transport/excretion is considerably more complex than O_2 uptake/transport, and therefore the two components of respiratory gas transfer are often reviewed separately.

Several comprehensive reviews on gas transfer in fish have been written in the past 20 years, and many of these contain extensive sections on blood carbon dioxide transport and excretion (Cameron, 1979; Randall and Daxboeck, 1984; Randall *et al.*, 1982; Wood and Perry, 1985; Perry, 1986,

Fish Physiology, Volume 17:
FISH RESPIRATION

1997a; Butler and Metcalfe, 1988; Perry and Wood, 1989; Piiper, 1989; Randall, 1990; Perry and Laurent, 1990; Swenson, 1990; Jensen, 1991; Nikinmaa, 1992; Thomas and Perry, 1992; Perry and McDonald, 1993; Walsh and Henry, 1991; Brauner and Randall, 1996). For the most part, these reviews have focused on teleost fishes (in particular the rainbow trout) and have dealt largely with CO_2 transport and excretion under steady-state conditions. In recent years, however, there have been significant new developments concerning CO_2 transport in the more "primitive" elasmobranchs and agnathans. Additional information has also now been collected from some species during non-steady-state conditions (e.g., during exercise or after adrenergic stimulation). Thus, the intent of this chapter is to review and revise the basic principles of CO_2 transport and excretion in teleost, elasmobranch, and agnathan fish under both steady- and non-steady-state conditions. The reader is also referred to Chapter 4 by Henry and Heming, Chapter 8 by Brauner and Randall, and Chapter 9 by Gilmour in this volume, which cover topics related to CO_2 transport and excretion.

II. CARRIAGE OF CO_2 IN BLOOD

Carbon dioxide is carried in the blood as physically dissolved CO_2, bicarbonate (HCO_3^-), and carbamino CO_2. The solubility of molecular CO_2 in plasma is low (Boutilier et al., 1984), and thus physically dissolved CO_2 (the product of solubility and partial pressure) constitutes only a small percentage (<5%) of the total carbon dioxide pool in the blood. Generally, HCO_3^- constitutes >90% of the carbon dioxide within the blood compartment. Carbamino CO_2 is less important in teleost fish than in mammals owing to the acetylation of the terminal amino groups of hemoglobin (Riggs, 1970), which disables these groups as sites of CO_2 binding. CO_2 attachment to hemoglobin is confined to amino groups of the β chains, where its binding is constrained by competition with organic phosphates. Therefore, the contribution of carbamino CO_2 to overall carbon dioxide carriage in teleost fish is minor (Heming et al., 1986). In elasmobranchs, carbamino CO_2 may be more important for overall CO_2 carriage because the terminal amino groups of hemoglobin chains are not acetylated (F. Jensen, personal communication). Because metabolism produces molecular CO_2 at rates far exceeding the carrying capacity of the blood for physically dissolved CO_2, the ability of the blood to transport adequate quantities of carbon dioxide relies on a continual conversion of CO_2 to HCO_3^-. Thus, the high capacitance of the blood for carbon dioxide essentially reflects the capacity for HCO_3^- carriage.

The capacitance of the blood for carbon dioxide is determined experimentally by construction of CO_2 dissociation (combining) curves that relate P_{CO_2} and total CO_2 (or HCO_3^- since it constitutes >90% of the total). At the low P_{CO_2}'s, which are typical of fish blood, such CO_2 dissociation curves are steep and nonlinear, and thus the capacitance may vary markedly as a function of slight P_{CO_2} fluctuations (Albers, 1970; Baumgarten-Schumann and Piiper, 1968). The CO_2 capacitance of the blood increases as P_{CO_2} decreases (e.g., see Fig. 1) and thus fish with low P_{CO_2}'s, such as the elasmobranchs (Butler and Metcalfe, 1988), generally have high blood CO_2 capacitance. Typically, the capacitance of fish blood for CO_2 is higher than in that mammalian blood; this is a result of the low P_{CO_2} of fish blood rather than intrinsic differences in the combining properties of the blood. Indeed, at PCO_2's typical of fish blood, mammalian blood exhibits a considerably higher CO_2 capacitance owing to its greater buffering capacity (Albers and Pleschka, 1967). At physiological partial pressures, the CO_2 capacitance far exceeds the O_2 capacitance of the blood. For this reason, arterial–venous P_{CO_2} differences are exceedingly small in relation to arterial–venous P_{CO_2} differences. Table 1 summarizes arterial and venous P_{CO_2} and P_{O_2} values in representative teleosts, elasmobranchs, and agnathans. Generally, arterial–venous P_{CO_2} differences in teleosts are less than 3 torr; whereas arterial–venous P_{O_2} differences may exceed 80 torr.

In teleost fish, the high CO_2 capacitance of true plasma predominantly reflects the buffering power of the red blood cell (RBC; see Table 2). Thus, CO_2 added to the blood diffuses into the RBC where extensive formation

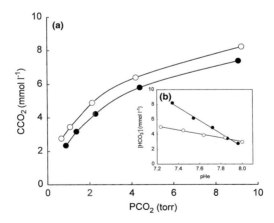

Fig. 1. *In vitro* (a) CO_2 combining and (b) buffer curves for true (solid circles) and separated (unfilled circles) plasma for dogfish, *Scyliorhinus stellaris*. Previously unpublished data.

Table 1

Gas Transfer Rates (in mmol kg^{-1} h^{-1}) and Blood Gas Partial Pressures of O_2 and CO_2 in Selected Teleost, Elasmobranch, and Agnathan Species

Species	$MCO_2{}^a$	$MO_2{}^b$	Pa_{CO_2} (torr)	Pv_{CO_2} (torr)	Pa_{O_2} (torr)	Pv_{O_2} (torr)	Ref.
Teleosts							
O. mykiss	2.53	2.28	2.0	—	102	—	Greco et al., 1995
O. mykiss	1.89	2.56	—	—	—	—	Gilmour and Perry, 1994
O. mykiss	2.72	3.12	—	—	—	—	Playle et al., 1990
O. mykiss	5.65	7.61	—	—	—	—	Steffensen et al., 1987
O. mykiss	—	—	2.3	3.2	—	—	Currie and Tufts, 1993
O. mykiss	—	—	2.0	3.3	117	31	Eddy et al., 1977
O. mykiss	—	—	2.3	5.7	85	19	Stevens and Randall, 1967
O. mykiss	—	—	1.7	2.2	93	24	Thomas et al., 1994
O. mykiss	2.70^c	—	3.2	4.2	91.6	17.3	Eddy, 1976
T. tinca	1.07^c	—	3.3	5	35.8	7	Eddy, 1974
P. stellatus	0.98^c	—	2.5	3.0	34.9	13.4	Wood et al., 1979
A. anguilla	1.0	—	3.4	—	42.8	—	P-Waitzenegger and Soulier, 1989
Elasmobranchs							
S. stellaris	1.37^c	—	2.0	2.6	49	10	B.-Schumann and Piiper, 1968
S. cannicula	—	—	1.0	—	98.1	—	Wood et al., 1994; Perry et al., 1996
			1.1		93.0		
S. stellaris	2.09^c	—	—	—	—	—	Piiper et al., 1970; Piiper et al., 1977
	2.57^c						
S. suckleyi	—	—	2.3	3.4	68.8	8.1	Lenfant and Johansen, 1966
S. cannicula	2.08^c	—	—	—	—	—	Metcalfe and Butler, 1988
S. cannicula	—	—	0.6	—	—	—	Truchot et al., 1980
R. ocellata	0.65^c	—	0.8	—	100	—	Graham et al., 1990
S. acanthias	1.20^c	—	1.7	—	111	—	Swenson and Maren, 1987
Agnathans							
P. marinus	—	—	1.7	1.9	120	42	Tufts et al., 1992

[a] Carbon dioxide excretion rate.

[b] Oxygen uptake rate.

[c] Calculated assuming a respiratory exchange ratio value of 0.8.

of $HCO_3{}^-$ is facilitated by end-product removal. Specifically, the H^+ and $HCO_3{}^-$ produced via the hydration of CO_2 are removed by hemoglobin buffering and transmembrane $Cl^-/HCO_3{}^-$ exchange, respectively. Consequently, in teleosts, the CO_2 capacitance of separated plasma (limited by plasma protein buffering) is much lower than that in whole blood or true plasma. A strikingly different pattern, however, is exhibited by elasmobranchs and agnathans. In the elasmobranchs *Scyliorhinus stellaris* (Fig. 1) and *Squalus acanthias* (Lenfant and Johansen, 1966), the CO_2 capacitances of

Table 2
Buffer Values in Representative Fish Species

Species	Whole blood	Red blood cells	Separated plasma	Ref.
Teleosts				
K. pelamis	−8.0 (25)	—	−3.1	Perry et al., 1985
O. mykiss	−9.7 (25)	—	−2.6	Wood et al., 1982
P. stellatus	−7.0 (25)	—	−2.9	Wood et al., 1982
I. punctatus	−14.3 (25)	—	−5.8	Cameron and Kormanik, 1982
Elasmobranchs				
S. suckleyi	−9.0	—	−6.5	Lenfant and Johansen, 1966
S. stellaris	−8.8 (18)	—	−2.6	S. F. Perry, unpublished
R. ocellata	−11.0 (13)	−70.8[a]	−6.6	Graham et al., 1990; Wood et al., 1990
Agnathans				
P. marinus	−0.1	−37.0	—	Tufts and Boutilier, 1989
P. marinus	−1.6 (17)	−48.3	—	Ferguson et al., 1992
L. fluviatalis	—	−43.0	—	Nikinmaa and Mattsoff, 1992
M. glutinosa	−8.3	−9.7	−5.2 −8.4 (true plasma)	

Note. Haematocrits are indicated in parentheses.
[a] Expressed per unit intracellular water.

separated and true plasma are equal. Although in some elasmobranch species, the buffering capacity of separated plasma is high (see Table 2), this cannot be the sole explanation for the high (relative to true plasma) CO_2 capacitance of separated plasma. Further, the low buffer value of the separated plasma of S. stellaris (Fig.1; Table 2) is inconsistent with the high CO_2 capacitance. An additional possible explanation for the relatively high CO_2 capacitance of dogfish separated plasma is that the plasma proteins participate in significant carbamino CO_2 formation.

Lampreys exhibit a unique pattern of CO_2 transport among vertebrates in which the RBCs display a markedly high CO_2 capacitance in comparison to true plasma (Tufts and Boutilier, 1989). This is due to the absence of rapid anionic exchange in their RBCs (see Section IIID).

The noncarbonic acid (also termed "nonbicarbonate") buffer capacity of blood is generally determined by titration with increasing levels of CO_2. The resultant changes in pH and $[HCO_3^-]$ are measured, and the noncarbonic acid buffer value (denoted as β) is determined as $\Delta[HCO_3^-]/\Delta pH$. The primary factor influencing the buffer value of blood is the concentration of the principal buffer, hemoglobin. Thus, changes in hematocrit will markedly affect blood buffer values. For this reason, it is useful to construct relationships between hematocrit (or [hemoglobin]) and β (e.g., Wood *et al.*, 1982). Table 2 summarizes blood buffer values from representative teleosts, elasmobranchs, and agnathans. In teleosts, the buffering value of the plasma tends to be low and accounts for about 20–40% of the whole blood buffering. Interestingly, in certain elasmobranchs, the contribution of plasma to whole blood buffering may be considerably more important. For example, the plasma buffer values of *S. suckleyi* (-6.5 mmol L^{-1}/pH unit) and *R. ocellata* (-6.6 mmol L^{-1}/pH unit) are higher than those in the teleosts and constitute 72 and 60% of whole blood buffering, respectively. The relatively large contribution of the plasma to whole blood buffering in these species cannot be explained by unusually high plasma protein levels (Graham *et al.*, 1990) but may instead reflect unusual buffering properties of these proteins. The elevated buffering capacity of the plasma is not a universal feature among the elasmobranchs because *S. stellaris* displays the typical teleost pattern (Table 2). The lampreys are particularly interesting because of the enormous differences between whole blood and RBC buffer values (Table 2). In these fish, hemoglobin acts as an effective buffer but because of the absence of appreciable anionic exchange activity (see Section IIID), the HCO_3^- formed as the result of buffering is retained within the RBC. Thus, with respect to buffering, the plasma and RBCs are effectively uncoupled. Nevertheless, the very low buffer values of the whole blood (see Table 2) suggest unusually poor plasma buffering.

In whole blood, carbon dioxide is partitioned between the plasma and RBCs. The relative contributions of the plasma and RBCs to overall whole blood CO_2 transport varies among the species (see Table 3). In teleosts and elasmobranchs that have been examined, the majority (generally about 90%) of carbon dioxide is carried within the plasma as HCO_3^- (Heming *et al.*, 1986; Currie and Tufts, 1993; Tufts *et al.*, 1997). Thus, the RBC with its large complement of noncarbonic acid buffers is largely responsible for the high carbon dioxide capacitance of the blood yet carries only a small fraction of the total (Table 3). This is because the HCO_3^- formed within the cell via CO_2 hydration is rapidly transferred to the plasma by a membrane-associated band 3 anionic exchanger to maintain Donnan equilibrium. In lampreys, a comparable RBC anion exchange mechanism is believed to be either absent or functionally restricted (Tufts and Boutilier, 1989; Ferguson *et al.*, 1992; Nikinmaa and Mattsoff, 1992; Cameron and Tufts, 1994; Cameron *et al.*, 1996). This feature, coupled with a large Haldane effect (Ferguson *et al.*, 1992; Nikinmaa and Mattsoff, 1992) and an elevated RBC intracellular pH (Nikinmaa and Mattsoff, 1992), allows these agnathans to transport considerable quantities of carbon dioxide within their RBCs (Tufts *et al.*, 1992; Ferguson *et al.*, 1992). Hagfish also lack appreciable RBC anionic

Table 3

In Vivo and *In Vitro* Partitioning of Carbon Dioxide between Whole Blood and Plasma in Selected Fish Species

Species	Pa_{CO_2} (torr)	% CO_2 carried in RBC	Ref.
In vivo			
O. mykiss (8)	2.93 ± 0.53	9.7 ± 2.2	Perry *et al.*, 1996b
A. anguilla (5)	2.40 ± 0.60	11.6 ± 1.6	Perry *et al.*, 1996b
S. maximus (5)	2.48 ± 0.23	8.0 ± 2.0	Perry *et al.*, 1996b
S. cannicula (8)	1.05 ± 0.15	13.0 ± 2.6	Perry *et al.*, 1996b
P. marinus (6)	Arterial 1.65	Arterial—16	Tufts *et al.*, 1992
	Venous 1.90	Venous—22	
In vitro			
O. mykiss (5–7)	2.4 ± 0.02	8	Heming *et al.*, 1986
O. mykiss (10)	1.5	4	Tufts *et al.*, 1997
M. glutinosa (5)	1.5	5	Tufts *et al.*, 1997
P. marinus (6)	Oxy—1.5	Oxy—17	Ferguson *et al.*, 1992
	Deoxy—1.5	Deoxy—26	

exchange activity (Ellory *et al.*, 1987). Unlike the lampreys, however, hagfish transport a smaller proportion of CO_2 in their RBCs (Tufts and Boutilier, 1990a) and thus more closely resemble teleost fish with respect to the partitioning of carbon dioxide between the plasma and RBCs (Table 3). Differences between the two groups of agnathans may reflect, at least in part, the smaller Haldane effect in hagfish blood (Tufts *et al.*, 1997).

The relative contribution of the RBC to CO_2 carriage in blood *in vivo* is somewhat difficult to compare among the species (e.g., see Table 1) because it is highly dependent on P_{CO_2}, oxygenation status, intracellular pH, and carbamino CO_2 formation. The proportion of carbon dioxide carried within the RBC is markedly increased as P_{CO_2} is elevated or as the blood is deoxygenated (Heming *et al.*, 1986; Ferguson *et al.*, 1992; Tufts *et al.*, 1997). Consequently, the RBCs in the venous circulation carry a greater proportion of the total CO_2 than in arterial blood (Tufts *et al.*, 1992). Because blood P_{CO_2} and oxygenation status vary among the species, these differences may mask interspecific variations in RBC CO_2-carrying capacities. For example, in the arterial blood of the dogfish *Scyliorhinus stellaris*, the RBCs carry 13% of the total CO_2 (Table 3), and thus the dogfish appears to be similar to the teleosts with respect to the partitioning of CO_2 between RBCs and plasma. However, because the arterial blood P_{CO_2} in this species of dogfish is substantially lower than in the comparable teleosts (Table 1), it is possible that the CO_2-carrying capacity of the dogfish RBCs is higher than that of the teleosts. Indeed, it appears that the dogfish RBCs may more closely resemble those of the lamprey in having a high intrinsic capacity to transport CO_2.

The partitioning of CO_2 between the plasma and RBCs is influenced by exercise and adrenergic stimulation; these factors are discussed in detail in Section IIIB.

III. CARBON DIOXIDE TRANSPORT AND EXCRETION

Under steady-state conditions, carbon dioxide excretion matches the production at the tissues and consequently respiratory acid–base disturbances are avoided. Excluding some of the air-breathers, the gill is the principal site of carbon dioxide excretion in fish. However, the mechanisms of excretion may differ both qualitatively and quantitatively among the species. Under non-steady-state conditions, there may be substantial imbalances between CO_2 production and excretion leading to acid–base disturbances and altered patterns of excretion at the gill. The following sections will cover CO_2 transport and excretion under both steady- and non-steady-state conditions in the teleosts, elasmobranchs, and agnathans. The reader is referred to Randall *et al.* (1981) for a detailed account of carbon dioxide excretion in air-breathing fishes.

A. Steady-State Conditions in Teleosts

As blood flows through the tissue capillaries, metabolic CO_2 enters a series of chemical pathways that ultimately result in the carriage of large quantities of HCO_3^- within the plasma; these chemical reactions are summarized in Fig. 2. The initial step is the diffusion of molecular CO_2 from tissue to plasma. A potential impediment to rapid and sustained diffusion of CO_2 into the plasma is accumulation of CO_2 within the interstitial fluid separating the tissue and plasma. Recently, however, Henry *et al.* (1997a) demonstrated the presence of extracellular carbonic anhydrase (CA) in rainbow trout skeletal muscle, at least some of which was associated with sarcolemmal membranes. Thus, in trout skeletal muscle and perhaps other tissues, CO_2 diffusion into the plasma is facilitated by the catalyzed hydration of CO_2 within the interstitium (Enns, 1967; see Fig. 2). Henry *et al.* (1997a) also showed that the production of H^+ within the interstitium via this pathway aided NH_4^+ formation, thereby facilitating NH_3 diffusion from tissue to plasma by sustaining the PNH_3 gradient.

In trout and presumably other teleosts, there is no CA accessible to catalyze CO_2–HCO_3^-–H^+ reactions within the tissue capillaries or venous

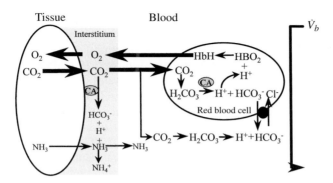

Fig. 2. A schematic model depicting the movements of carbon dioxide, oxygen, and ammonia between the tissue and blood compartments in a typical teleost fish. The stippled area represents the interstitial compartment separating the tissue and blood spaces. CA, carbonic anhydrase; $\dot{V}_b$, blood flow.

plasma (Perry *et al.,* 1997). Thus, HCO_3^- formation within the blood occurs almost exclusively within the RBC via catalyzed CO_2 hydration. The H^+ ions formed from the hydration of CO_2 are buffered by hemoglobin; and a portion of the bound H^+ (Bohr protons) contribute to O_2 off-loading via the Bohr effect (Fig. 2). The HCO_3^- ions formed from the hydration of CO_2 are removed from the RBC by electroneutral Cl^-/HCO_3^- exchange.

As blood arrives at the gill, it contains carbon dioxide predominantly in the form of HCO_3^- dissolved in the plasma. Within the transit time through the gill vasculature (approximately 0.5–2.5 s; Cameron and Polhemus, 1974), sufficient HCO_3^- is converted (via dehydration) to molecular CO_2 and excreted at a rate that matches production at the tissues. In a single passage through the gill, approximately 12–35% of blood total CO_2 is excreted (Perry, 1986). Relatively few studies have reported pre- and postbranchial blood respiratory and acid–base data, and even fewer have combined these with measurements of CO_2 excretion. It is clear, however, that the excretion of CO_2 is associated with a slight decrease in the P_{CO_2} of the blood (Table 1). Although it is generally thought that arterial blood pH is higher than venous pH, the direction and magnitude of the arterial–venous pH difference may vary markedly and are highly dependent upon environmental factors (see Chapter 8).

The rate of the uncatalyzed dehydration of HCO_3^- is too slow to permit significant excretion in the absence of catalysis (Edsall, 1969). Thus, the CA enzyme (for reviews see Maren, 1967; and Chapter 4) plays a vital role in allowing rapid dehydration of plasma HCO_3^- to CO_2 within the gill transit time. Unlike the endothelia of mammalian pulmonary capillaries that contain CA that is oriented toward the plasma (Effros *et al.*, 1978; Klocke, 1978), the blood channels (incorrectly referred to as "capillaries" in some literature) of the teleost gill lack plasma-facing CA (Henry *et al.*, 1988, 1993; Rahim *et al.*, 1988; Perry and Laurent, 1990; Perry *et al.*, 1997). For this reason, the RBCs are the exclusive site of catalyzed HCO_3^- dehydration in teleosts. The accessibility of plasma HCO_3^- to RBC CA is governed by a membrane-associated anionic exchanger that facilitates the electroneutral exchange of HCO_3^- for Cl^-, a process termed the "chloride shift" (Cameron, 1978; Romano and Passow, 1984; Jensen and Brahm, 1996). The trout RBC exchanger has been cloned and sequenced (Hubner *et al.*, 1992) and is similar to the well-characterized mammalian erythrocyte band 3 (AE1) protein (Alper, 1991). The velocity of the chloride shift is probably the rate-limiting process in the overall CO_2 excretion pathway because it is relatively slow in comparison to the rate of CO_2 diffusion or catalyzed HCO_3^- dehydration.

Molecular CO_2 formed within the RBC enters the plasma and traverses the gill epithelium by diffusion (Fig. 3). CO_2 entering the water is removed physically by ventilatory convection and chemically by hydration to HCO_3^- and H^+ within a boundary layer adjacent to the gill epithelium (Wright *et al.*, 1986, 1989; Randall and Wright, 1989; Randall *et al.*, 1991). The latter process may be catalyzed by CA associated with the mucous (Wright *et al.*, 1986) or the external surface of apical membrane microplice (Rahim *et al.*, 1988). The physical and chemical removal of CO_2 from the ventilatory water serves to maintain P_{CO_2} diffusion gradients as blood flows through the gill.

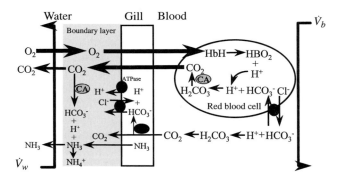

Fig. 3. A schematic model depicting the movements of carbon dioxide, oxygen, and ammonia between the blood and water in a typical teleost fish. The stippled area represents the boundary layer separating the gill epithelium and the bulk water flow. CA, carbonic anhydrase; V_b, blood flow; V_w, water flow.

Carbonic anhydrase is also found within the cytosol of all gill epithelial cells (Dimberg et al., 1981; Rahim et al., 1988). Consequently, a portion of the CO_2 crossing the epithelium is hydrated to HCO_3^- and H^+ within the pavement cells and chloride cells of the gill lamellae (Fig. 3). The H^+ ions are used as a substrate for an apical membrane vacuolar-type H^+-ATPase or "proton pump" (Lin and Randall, 1991, 1993, 1995; Lin et al., 1994; Sullivan et al., 1995, 1996; Perry and Fryer, 1997; Goss et al., 1996), whereas the HCO_3^- provides a counterion for an apical Cl^-/HCO_3^- exchange mechanism (Perry, 1997b). The direct excretion of HCO_3^- via this mechanism, though important for ionic and acid–base regulation (see Goss et al., 1992, 1995; Evans, 1993; Heisler, 1993; Marshall, 1995), likely accounts for less than 5% of total carbon dioxide excretion (Perry et al., 1981).

According to the model of teleost carbon dioxide excretion depicted in Fig. 3, the overall process is a complex pathway consisting of numerous steps within the plasma, RBC, gill cells, and external boundary layer. Thus, the control of excretion could conceivably be exerted at multiple sites by various means. The potential factors controlling carbon dioxide excretion in teleosts are discussed later.

AVAILABILITY OF CARBONIC ANHYDRASE

In the circulatory system of teleosts, CA is restricted to the interior of the RBC (Rahim et al., 1988; Henry et al., 1988, 1993; Perry and Laurent, 1990; Perry et al., 1997). It is generally accepted that the catalytic capacity of RBC CA to dehydrate plasma HCO_3^- far exceeds the actual rates of CO_2 excretion in vivo (e.g., Swenson and Maren, 1987). This suggests that CO_2 excretion in vivo is constrained by the availability of plasma HCO_3^- to RBC CA, which is governed by the rate of Cl^-/HCO_3^- exchange. Using an isotopic CO_2 excretion assay (Wood and Perry, 1991), Perry and Gilmour (1993) demonstrated that lysis of RBCs in vitro markedly increased the rate of production of CO_2 from plasma HCO_3^- and thus provided direct evidence that accessibility of plasma HCO_3^- to RBC CA normally limits CO_2 production in vitro. It was suggested that the relatively slow rate of Cl^-/CO_3^- exchange was insufficient to match the catalytic capacity of RBC CA to dehydrate HCO_3^-. Recently, Wood and Munger (1994) provided clear evidence that CO_2 excretion in rainbow trout in vivo is restricted by the accessibility of plasma HCO_3^- to RBC CA. In that study, addition of bovine CA to the plasma via intravascular injection caused a pronounced respiratory alkalosis (25% reduction in arterial PCO_2) in resting fish. Other studies (e.g., Gilmour et al., 1994b) that have failed to detect changes in arterial P_{CO_2} after injection of CA likely did not monitor blood respiratory status for sufficient time since the effects are gradual (Wood and Munger, 1994). Clearly, CO_2 excretion, at least in trout, is limited by the slow rate of RBC Cl^-/HCO_3^- exchange that constrains the access of plasma HCO_3^- to CA. This, in part, explains the apparent diffusion limitations for carbon dioxide excretion in fish (Cameron and Polhemus, 1974; Malte and Weber, 1985; Perry, 1986; Piiper, 1989) despite the high intrinsic permeation coefficient of molecular CO_2. Indeed, there is accumulating evidence that alterations of the blood-to-water diffusion distance cause greater impact on arterial PCO_2 than arterial PO_2 (Bindon et al., 1994; Greco et al., 1995; Powell and Perry, 1996; Perry et al., 1996). Such results are expected if carbon dioxide excretion across the gill was diffusion-limited while O_2 uptake was perfusion-limited.

The RBC has a vital role in CO_2 excretion owing to its abundance of CA, and thus CO_2 excretion is positively correlated with hematocrit both *in vitro* (Tufts et al., 1988; Wood and Perry, 1991; Perry and Gilmour, 1993) and *in vivo* (Haswell and Randall, 1978; Wood et al., 1982). However, during conditions of anemia *in vivo*, several physiological adjustments may serve to maintain CO_2 excretion despite the loss of RBC numbers. In particular, increased cardiac output during anemia (Cameron and Davis, 1970; Wood and Shelton, 1980) maintains CO_2 excretion at normal levels in rainbow trout until a critical hematocrit of 5–10% is reached (Wood et al., 1982). When cardiac output is experimentally maintained (Perry et al., 1982), CO_2 excretion is related to hematocrit in a linear manner. When trout experience chronic severe anemia (~5% hematocrit), CO_2 excretion is impaired initially but then recovers to resting levels after 2 to 3 weeks (Gilmour and Perry, 1996). The mechanisms by which CO_2 excretion is restored to resting levels during chronic anemia are unknown although adaptations at the level of the RBC are not apparently involved (Gilmour and Perry, 1996). Several fish species, such as certain flatfishes (*Platichthys stellatus;* Wood et al., 1979; *Scopthalmus maximus;* Perry et al., 1996b), display low natural hematocrits. These species have adapted to this condition by increasing their cardiac output (Wood et al., 1979), by exploiting a large Haldane effect to enhance CO_2 excretion during oxygenation (Perry et al., 1996b), and possibly by an enhanced rate of RBC Cl^-/HCO_3^- exchange (Perry et al., 1996b). The latter two adaptations are particularly important considering that the increased cardiac output will markedly reduce the transit time of blood within the gill and thus constrain CO_2 excretion. Under such conditions of reduced transit time, there is an obvious benefit of accelerated Cl^-/HCO_3^- exchange. Severe anemia, though restricting availability to CA, might also reduce the total amount of CA in the blood to limiting levels. For example, Perry and Gilmour (1993) demonstrated that the addition of bovine CA to lysed blood caused an increase in the rate of HCO_3^- dehydration *in vitro* but only at hematocrits lower than 5%. A most extreme example of natural anemia is found in certain species of

Antarctic fishes that lack hemoglobin (e.g., Holeton, 1970). Virtually nothing is known about the ability of these fish to excrete CO_2 although the mechanisms (e.g., high cardiac output) known to aid O_2 uptake (Holeton, 1970) presumably also assist CO_2 excretion.

A second site of CO_2 hydration–dehydration reactions in the excretion pathway is the gill epithelium (Fig. 3). Although previous studies suggested that branchial CA played a key role in dehydrating plasma HCO_3^- (Haswell and Randall, 1978; Haswell et al., 1980), subsequent research has shown that it has only a minor role in CO_2 excretion (Perry et al., 1982). A role for branchial CA in dehydrating plasma HCO_3^- is precluded because the basal (plasma-facing) membranes of gill epithelial cells are relatively impermeable to HCO_3^- (Perry et al., 1982, 1984), and thus the HCO_3^- in the plasma cannot gain access to the abundant CA within the gill epithelial cells.

The final site of CO_2 hydration–dehydration reactions in the excretion pathway (Fig. 3) is the boundary layer adjacent to the gill epithelium. Currently, the existence of boundary layer CA is debated (see Chapter 4). However, if present, theory predicts an important role for boundary layer CA in facilitating CO_2 excretion because the catalyzed hydration of excreted CO_2 will lower the P_{CO_2} in the boundary layer and thus sustain the CO_2 diffusion gradient during gill transit. Although it has been shown experimentally that acidification of the boundary layer by enzyme-catalyzed hydration of CO_2 facilitates ammonia diffusion (Wright et al., 1989), no experimental data support a role for boundary layer CA in CO_2 excretion. This in an area of research that merits further attention.

RBC Cl$^-$/HCO$_3^-$ EXCHANGE

The Cl^-/HCO_3^- exchanger of the RBC is a member of an anion exchanger (AE) gene family comprising at least three members (AE1-3; Kopito, 1990). The RBC exchanger is a product of the AE1 gene and is generally referred to as band 3 protein. The band 3 protein contains both membrane-spanning and cytoplasmic domains. It is the C-terminal membrane-spanning region that functions as an electroneutral Cl^-/HCO_3^- exchanger. Cloning of the trout band 3 cDNA (Hubner et al., 1992) has revealed a deduced protein composed of 918 amino acid residues with a molecular mass of 102 kDa. These values are similar to those obtained from mammalian (mouse, human) and avian (chicken) band 3 proteins (Hubner et al., 1992).

Although trout and mammalian band 3 proteins appear to be similar in structure, there are notable differences in function between fish and mammalian Cl^-/HCO_3^- exchangers. For example, at similar temperatures, the rate of Cl^-/HCO_3^- exchange in RBCs of four teleost species is higher than that in human blood (Jensen and Brahm, 1996). This feature, in addition to blunted temperature sensitivity (Obaid *et al.*, 1979; Jensen and Brahm, 1996), may be advantageous to fish that often function at considerably colder (and variable) temperatures than mammals and that possess blood of lower hematocrits. It is generally accepted that a universal feature of band 3 protein is an inhibition of Cl^-/HCO_3^- exchange function in the presence of disulfonic stilbene derivatives such as SITS or DIDS. Although sensitivity of RBC Cl^-/HCO_3^- exchange to SITS or DIDS has clearly been demonstrated in trout blood (Perry *et al.*, 1982; Tufts *et al.*, 1988; Wood and Perry, 1991; Perry and Gilmour, 1993; Cameron *et al.*, 1996; Jensen and Brahm, 1996), this may not be true of all teleosts. Recently, Jensen and Brahm (1996) provided evidence for rapid Cl^-/HCO_3^- exchange in carp (*Cyprinus carpio*) RBCs that was largely unaffected by 0.2 mmol L^{-1} DIDS.

Although time constants (the time required to complete 63% of response) for teleost RBC Cl^-/HCO_3^- exchange ($\sim$100 to $\sim$400 ms at 15°C) are markedly lower than previous estimates (e.g., see Romano and Passow, 1984; Cameron, 1978), they do, however, approximate lower estimates of gill transit time. Regardless of whether RBC Cl^-/HCO_3^- exchange is completed within gill transit, it is nevertheless clear that the relatively slow rate of this process will restrict the access of plasma HCO_3^- to RBC CA during a considerable portion of the time that blood resides within the gill. Consequently, the effective time for carbon dioxide diffusion across the gill is shortened by the time required to form molecular CO_2 from plasma HCO_3^-. As discussed earlier, there is little doubt that RBC Cl^-/HCO_3^- exchange is the rate-limiting process in CO_2 excretion and that it imposes "effective diffusion limitations" on the transfer of CO_2 across the gill.

The rate of RBC Cl^-/HCO_3^- exchange is sensitive to oxygenation status (Perry and Gilmour, 1993; Wood and Simmons, 1994; Perry *et al.*, 1996b). Under steady-state conditions *in vitro*, the rate of HCO_3^- dehydration in whole blood is enhanced under deoxygenated conditions indicating an acceleration of Cl^-/HCO_3^- exchange. It has been suggested that this effect may reflect oxygenation-linked changes in intracellular pH (Perry and Gilmour, 1993; Perry *et al.*, 1996b) or allosteric interactions between the band 3 protein and hemoglobin (Wood and Simmons, 1994). Regardless of the mechanism, such an effect would benefit CO_2 excretion across the gill during environmental hypoxia and during exercise. Interestingly, two recent studies failed to observe changes in the rate of Cl^-/HCO_3^- exchange (Jensen and Brahm, 1996) or RBC HCO_3^- dehydration (Brauner *et al.*, 1996) during steady-state changes in oxygenation status.

Although relatively few species have been examined under comparable conditions, there appear to be marked interspecific differences in the intrinsic rate of RBC Cl^-/HCO_3^- exchange among the teleosts. Owing to the sensitivity of the Cl^-/HCO_3^- exchanger to temperature (Jensen and Brahm, 1996), and HCO_3^- dehydration to oxygenation status (Perry and Gilmour, 1993; Wood and Simmons, 1994; Perry et al., 1996b) and $[HCO_3^-]$ (Wood and Perry, 1991; Perry and Gilmour, 1993), such comparisons must be made under identical conditions. Recently, two studies have examined absolute rates of Cl^-/HCO_3^- exchange (Jensen and Brahm, 1996) or RBC HCO_3^- dehydration (an index of Cl^-/HCO_3^- exchange rate; Perry et al., 1996b) in blood of several teleost species using identical assay conditions. The former study (Jensen and Brahm, 1996) revealed the following order of Cl^-/HCO_3^- exchange rates within the four species that were examined: Atlantic cod (*Gadus morhua*) > rainbow trout > carp > European eel (*Anguilla anguilla*). The latter study (Perry et al., 1996b) revealed the following sequence of RBC HCO_3^- dehydration rates among three teleost species: turbot > European eel = rainbow trout. Perry et al. (1996b) suggested that a higher intrinsic rate of Cl^-/HCO_3^- exchange in turbot blood would be advantageous to counteract the low hematocrit in this species and allow adequate HCO_3^- dehydration during the short gill transit time imposed by unusually high cardiac outputs in flatfishes.

THE SUPPLY OF H^+ AS A LIMITING FACTOR

Equimolar quantities of H^+ are required to dehydrate plasma HCO_3^-. These H^+ are provided from the dissociation of noncarbonic acid buffers (Bidani and Heming, 1991) and also originate as Bohr protons derived from hemoglobin during the oxygenation process (the Haldane effect). The buffering process, which is largely independent of oxygenation status, may be less important in fish than in mammals because of the lower buffering capacities of fish hemoglobins (Jensen, 1989; for reviews, see Jensen, 1991; Weber and Jensen, 1988). Thus, the oxylabile component of the overall H^+ supply may be more important in teleost fish (Jensen, 1989); in support of this idea, teleost fish are known to possess large Haldane effects (Jensen, 1986, 1989, 1991). Recently, several studies have shown the importance of the Haldane effect to CO_2 excretion (Perry and Gilmour, 1993; Perry et al., 1996b; Brauner et al., 1996) by demonstrating a marked increase in the rate of RBC HCO_3^- dehydration during rapid oxygenation. The reader is referred to Chapter 8 by Brauner and Randall for a detailed analysis of the linkages between oxygenation and CO_2 excretion.

B. Non-steady-state Conditions in Teleosts

Under most circumstances, the rate of CO_2 production by a fish is equivalent to the rate of CO_2 excretion, CO_2 stores are relatively constant, and the gas exchange ratio (MCO_2/MO_2) typically ranges between 0.7 and

1.0. There are times, however, when rates of CO_2 excretion no longer match rates of CO_2 production. These non-steady-state imbalances may result in acid–base disturbances, changes in the forms and distribution of CO_2 within the body, and altered patterns of excretion at the gill. Since CO_2 stores in fish are large compared to oxygen stores (Randall and Daxboeck, 1984), non-steady-state disturbances may also have a large impact on the gas exchange ratio. A number of non-steady-state conditions, including exercise, hypoxia, hyperoxia, hypercapnia, and anemia, will disrupt the normal patterns of CO_2 transport and excretion in fish. In recent years, however, the vast majority of studies have focused on the effects of exhaustive exercise on CO_2 transport and excretion in teleosts and on the role of catecholamines under these conditions. The following section, therefore, will be primarily restricted to recent developments in these areas.

METHODOLOGICAL CONSIDERATIONS

Most attempts to examine the blood CO_2 transport properties and acid–base status of fish under non-steady-state conditions involve measurements of the appropriate variables after the CO_2 reactions in the blood sample have come to equilibrium. Furthermore, variables such as P_{CO_2}, which are difficult to measure in fish, are often calculated using measurements of other acid–base variables and appropriate constants. These calculations also assume equilibrium conditions for the CO_2 reactions in the blood. Although these approaches have been used to describe relative differences in these variables under a wide variety of conditions, it has recently been demonstrated that the CO_2 reactions in the blood of trout are not in equilibrium (Gilmour et al., 1994; Perry et al., 1997). Moreover, the magnitude of the disequilibrium is increased under non-steady-state conditions such as during metabolic acidoses or when circulating catecholamines are elevated (Gilmour and Perry, 1996b). Thus, the commonly measured equilibrium values for CO_2 and acid–base variables in the blood of fish will be subject to some degree of error, and the magnitude of this error will likely be increased under the non-steady-state conditions that occur following exhaustive exercise. One of the major challenges for future studies within this area will therefore be to obtain more accurate measurements of nonequilibrium values for CO_2 and acid–base variables under these non-steady-state conditions.

EFFECTS OF EXERCISE ON BLOOD CO_2 CARRIAGE

A number of studies have examined the CO_2 transport characteristics and acid–base status of arterial blood following exhaustive exercise (see also reviews by Wood and Perry, 1985; Wood, 1991). Following the exercise bout, there is usually a reduction in plasma HCO_3^- levels and a significant increase in arterial P_{CO_2} tension. There is also a large increase in the metabolic proton load of the blood arising from anerobic metabolism in the white muscle as well as from protons extruded from pH regulatory mechanisms of the RBCs. The combined result of this metabolic and respiratory acidosis is typically a large decrease in the plasma pH that persists for several hours during the recovery period.

In addition to the changes in plasma CO_2 transport properties, which have been well documented, there are likely to be significant changes in the CO_2 transport properties of the RBCs following exhaustive exercise, although this aspect of blood CO_2 transport in fish has received much less attention. Since (1) exhaustive exercise results in an increase in arterial P_{CO_2}, (2) the nonbicarbonate buffer value of the RBCs is normally much greater than that of the true plasma, and (3) the RBC pH of many fish species is well maintained following exhaustive exercise via adrenergic mechanisms (Nikinmaa, 1990, 1998), the amount of CO_2 carried within the RBC of many fishes should increase after exhaustive exercise. The only experimental evidence currently available seems to confirm this prediction since the calculated CO_2 content of RBCs in the arterial blood of rainbow trout is increased from 0.5 mmol L^{-1} at rest to 2.3 mmol L^{-1} after exercise (Currie and Tufts, 1993). These postexercise changes in RBC CO_2 content sharply contrast those occurring in the plasma, but their relative importance is probably insignificant in postbranchial blood.

In marked contrast to the numerous studies that have focused on the changes in CO_2 transport variables and acid–base status of arterial blood in fish following exercise, relatively few studies have examined the postexercise changes in these variables in venous blood. In several respects, the relative changes occurring in the CO_2 variables and acid–base status of venous blood are similar to those observed in arterial blood (Nikinmaa and Jensen, 1986; Currie and Tufts, 1993). This is expected since disturbances such as the profound metabolic acidosis after exercise will have a similar impact on these variables in both the arterial and venous compartments. Since CO_2 production by the tissues is greatly elevated during recovery from exercise, the increase in the P_{CO_2} of venous blood is much greater than that in arterial blood (Nikinmaa and Jensen, 1986; Currie and Tufts, 1993). As in arterial blood, titration of bicarbonate stores in venous blood by protons also contributes to a relative reduction in the plasma bicarbonate concentration during recovery despite the increase in venous P_{CO_2} (Nikinmaa and Jensen, 1986).

Venous RBC CO_2 stores are much greater than those in arterial blood due to both the Haldane effect and the increased P_{CO_2}. Interestingly, the calculated CO_2 content of the RBC in venous blood also increases significantly following exhaustive exercise (Currie and Tufts, 1993). In contrast to the situation in arterial blood, these changes in venous RBC CO_2 stores may have significant consequences for CO_2 excretion after exercise since plasma bicarbonate's access to RBC CA is thought to be a significant rate-limiting step in the CO_2 excretion process and may even be further reduced by the presence of elevated catecholamines after exercise. Thus, a postexercise increase in the relative amount of CO_2 carried within the RBC where CA is readily accessible may be an important factor facilitating CO_2 excretion at the instant that venous blood first reaches the gill. Although RBC CO_2 stores in most fishes are typically much lower than those in plasma, the relative importance of the RBCs in this regard will also be increased due to the elevated hematocrit following exhaustive exercise.

Measurements of the total CO_2 content in pre- and postbranchial blood indicate that the amount of CO_2 removed during passage of blood through the gill is significantly increased after exercise (Currie and Tufts, 1993). This increase probably results from a postexercise increase in the relative amount of CO_2 removed from both the plasma and RBCs during branchial blood transit. Postexercise increases in the arterio–venous differences for plasma HCO_3^- (Nikinmaa and Jensen, 1986), CO_2 content (Currie and Tufts, 1993), and RBC Cl^- concentrations (Nikinmaa and Jensen, 1986) all confirm an increase in the relative amount of CO_2 excreted from the plasma under these conditions. Since the magnitude of the CO_2 disequilibria in the blood of teleosts should be vastly increased after exercise (Gilmour and Perry, 1996b), large cumulative errors will likely arise during any attempt to calculate the contribution of the RBCs toward the arterio–venous differences in CO_2 content under these conditions. Thus, the relative importance of changes in RBC versus plasma CO_2 stores toward CO_2 excretion after exercise cannot be easily determined.

SIGNIFICANCE OF THE ARTERIAL P_{CO_2} INCREASE AFTER EXERCISE

The magnitude of the arterial P_{CO_2} rise in fish following exhaustive exercise varies according to species and may range anywhere from 50 to 400% (Wood and Perry, 1985). Injection of carbonic anhydrase into the circulation prior to exercise attenuates the postexercise increase in arterial P_{CO_2} (Wood and Munger, 1994; Currie et al., 1995). Titration of plasma HCO_3^- by protons extruded from muscle as well as from adrenergically stimulated RBCs is undoubtedly an important factor contributing to this P_{CO_2} increase since this reaction will occur in plasma at the uncatalyzed rate and will therefore continue in the postbranchial blood. It has also been suggested that adrenergic inhibition of HCO_3^- flux through the RBCs has a significant role in creating this arterial P_{CO_2} increase (Wood and Perry,

1985; Perry *et al.*, 1991). The latter phenomenon, often termed the "CO_2 retention theory," will be addressed later.

Wood and Perry (1985) proposed that the postexercise elevation in P_{CO_2} may have an important role in maintaining hyperventilation during recovery from exhaustive exercise, thereby facilitating correction of the O_2 debt. Subsequent experiments by Wood and Munger (1994) provide evidence to support this view. In these experiments, injection of CA (10 mg kg^{-1}) into rainbow trout prior to exhaustive exercise resulted in about a 50% reduction in the postexercise arterial P_{CO_2} increase as compared to saline-injected control animals (Fig. 4). CA injection also attenuated the

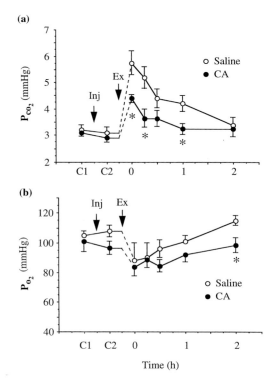

Fig. 4. The effects of 6 min of exhaustive exercise (dotted lines) on the arterial (a) carbon dioxide tension (P_{co_2}) and (b) oxygen tension (P_{o_2}) in the blood of rainbow trout injected intravascularly with either saline (open circles, $N = 8$) or carbonic anhydrase (10 mg kg^{-1}, filled circles, $N = 11$). Injections (arrow) were given immediately after the first control sample (C1) and the second control sample (C2) was taken 1 h later. Values are means ± 1 SEM. Asterisks indicate significant differences ($P \le 0.05$) between the saline and carbonic anhydrase treatments at the same sample time. Redrawn from Wood and Munger (1994).

hyperventilatory response after exercise and significantly reduced oxygen consumption during the recovery period. Based on this evidence, the authors concluded that their results provide functional significance for the phenomenon of arterial P_{CO_2} elevation following exercise.

ADRENERGIC INVOLVEMENT IN CO_2 TRANSPORT AND EXCRETION AFTER EXERCISE

Exhaustive exercise results in a large increase in circulating catecholamine levels in fish (Primmett *et al.,* 1986). This elevation in blood catecholamines stimulates a wide range of physiological responses during recovery from exercise, including activation of a sodium/proton (Na^+/H^+) exchange mechanism within the RBC membrane. The role of this exchanger in regulating RBC pH and hemoglobin–oxygen carriage after stresses such as exhaustive exercise has been intensively studied in recent years, and the mechanisms involved in these aspects of the RBC adrenergic response are well described elsewhere (Nikinmaa and Tufts, 1988; Nikinmaa, 1990, 1992, 1998). Another interesting feature of this RBC adrenergic response, however, is its potential impact on the CO_2 transport properties of the blood in fish. The relative importance of adrenergic involvement in blood CO_2 transport and excretion in fish has been the focus of numerous recent studies.

According to the current model of the RBC adrenergic response in fish (Nikinmaa, 1990, 1998), activation of the Na^+/H^+ exchange mechanism on the RBC membrane by catecholamines should shift the reaction H^+ + HCO_3^- ↔ (CO_2 + H_2O to the left, thereby increasing the RBC HCO_3^- concentration. In addition, titration of plasma HCO_3^- by protons extruded from the RBC will contribute to a P_{CO_2} rise in the blood and a further reduction in the HCO_3^- concentration gradient between the plasma and the RBCs. Perry *et al.* (1991) proposed that this reduction or reversal of the chemical gradient for the inward movement of HCO_3^- from plasma to RBC may reduce the flux of plasma HCO_3^- through the adrenergically stimulated RBC during the process of CO_2 excretion. Two other mechanisms by which adrenergic activation of RBC Na^+/H^+ exchange might cause inhibition of the net conversion rate of plasma HCO_3^- to CO_2 have also been proposed. Perry *et al.* (1991) further suggested that the normal P_{CO_2} gradient from RBC to plasma will be reversed following adrenergic stimulation due to "CO_2 recycling" (Motais *et al.,* 1989; Thomas and Perry 1992), which should also cause a reduction in the net dehydration rate of plasma HCO_3^-. Finally, it has also been proposed that adrenergic activation of RBC Na^+/H^+ exchange could inhibit the bicarbonate dehydration reaction by simply reducing the availability of protons that are necessary for the reaction to proceed (Randall and Brauner, 1991; Wood and Simmons, 1994).

Wood and Perry (1985) first reported unpublished experimental evidence collected by Perry and Heming showing that catecholamines inhibited HCO_3^- flux through rainbow trout RBCs *in vitro* in a dose-dependent fashion. Although Tufts *et al.* (1988) were unable to replicate these findings using a modification of the assay used by Perry and Heming, a different *in vitro* assay developed by Wood and Perry (1991) has confirmed the adrenergic inhibition of HCO_3^- flux through salmonid RBCs in a number of subsequent studies. Using this assay, it was found that the inhibitory effect was not due to any direct inhibition of the RBC Cl^-/HCO_3^- exchanger, but that the inhibitory effect was strongly linked to the adrenergic activation of the RBC Na^+/H^+ exchanger (Fig. 5; Perry *et al.*, 1991). Using the same assay, Wood and Simmons (1994) determined that the adrenergic inhibition of plasma HCO_3^- dehydration is not a result of the reduced availability of intracellular protons since alterations in proton availibility by manipulation of blood oxygenations status did not significantly affect the observed responses. As proposed by Perry *et al.* (1991), it therefore seems that the inhibition results mainly from the decrease in the plasma to RBC HCO_3^- gradient and perhaps also from a temporary reversal of the RBC to plasma P_{CO_2} gradient caused by CO_2 recycling after adrenergic stimulation.

Despite the evidence that elevated catecholamines inhibit the net rate of plasma HCO_3^- dehydration by trout blood *in vitro*, the relative importance of this inhibitory effect *in vivo* has been more difficult to determine.

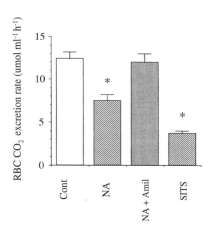

Fig. 5. Red blood cell (RBC) CO_2 excretion rate under control conditions or in the presence of either noradrenaline (NA, 1000 nmol L^{-1}), NA plus amiloride (10^{-4} mol L^{-1}), or SITS (10^{-4} mol L^{-1}). Values shown are means + 1 SEM.; $N = 6$ in all cases. An asterisk indicates a significant difference ($P < 0.05$) from the control value. Redrawn from Perry *et al.* (1991).

As mentioned, exhaustive exercise *in vivo* results in a relative increase in the arterio–venous difference in blood total CO_2 content (Nikinmaa and Jensen, 1987; Currie and Tufts, 1993) and an overall increase in whole animal CO_2 excretion (Steffensen *et al.*, 1987; Scarabello *et al.*, 1992). Moreover, CO_2 excretion by resting rainbow trout is either unchanged (Steffensen *et al.*, 1987) or increased (Playle *et al.*, 1990) following infusion of catecholamines. Finally, measurements of the CO_2 content in the pre- and postbranchial blood of rainbow trout indicate that the carbon dioxide clearance coefficient (i.e., the efficiency of carbon dioxide removal) across the gills is not significantly altered by catecholamine infusion (Nikinmaa and Vihersaari, 1993). Thus, the results of numerous *in vivo* experiments initially appear to conflict with the *in vitro* evidence for adrenergic inhibition of plasma HCO_3^- dehydration. More recent experiments by Wood (1994), however, appear to resolve much of the apparent discrepancy between these *in vitro* and *in vivo* observations. These experiments indicate that the dehydration rate of plasma HCO_3^- measured *in vitro* is actually increased by about 50% for blood samples removed from rainbow trout following exhaustive exercise *in vivo*. Interestingly, this increased rate of net HCO_3^- dehydration is eliminated if the P_{CO_2} of the blood sample is returned to that of resting fish prior to running the *in vitro* assay. Thus, as demonstrated *in vitro* by Wood and Perry (1991), the net rate of plasma HCO_3^- dehydration *in vivo* appears to be increased by factors such as the increase in blood P_{CO_2} following exercise (Wood, 1994). If these stimulatory effects are eliminated (for example, by reducing the blood P_{CO_2} prior to the *in vitro* assay), the relative inhibition of net plasma HCO_3^- dehydration rate by catecholamines predicted from *in vitro* studies can also be detected in blood samples removed from fish exhaustively exercised *in vivo* (Wood, 1994). According to Wood (1994), these *in vivo* observations therefore support the idea first proposed by Wood and Perry (1985) that a "relative" inhibition of the net plasma HCO_3^- dehydration rate caused by elevated catecholamines may be a significant factor contributing to the rise in arterial P_{CO_2} following exhaustive exercise in fish (i.e., CO_2 retention theory).

Although many of the apparent discrepancies in the literature in this area may now be resolved, a number of important factors must still be considered when evaluating the effects of catecholamines on the CO_2 transport properties of fish blood. For example, confirmation of the adrenergic inhibition of net plasma HCO_3^- dehydration by RBCs using an approach that more closely resembles branchial blood transit times (i.e., a few seconds) is probably warranted. It is also extremely difficult to experimentally determine the relative importance of any potential inhibition of HCO_3^- dehydration by RBCs *in vivo* since titration of plasma HCO_3^- stores by protons arising from tissues such as muscle and RBCs will also contribute to the arterial P_{CO_2} rise in fish. Finally, since other factors such as increased RBC CO_2 carriage in venous blood, increased hematocrit (i.e., relative increase in anion exchangers per ml of blood), and a rise in venous P_{CO_2} contribute to a measurable increase in CO_2 excretion in fish following exhaustive exercise, it is questionable whether any relative effects of cate-cholamines on plasma HCO_3^- dehydration can truly be described as "CO_2 retention" when all the changes in the dynamics of CO_2 transport and excretion after exercise are considered.

C. Elasmobranchs

The pattern of carbon dioxide excretion in the elasmobranchs differs from the teleosts in several important respects. Notably, elasmobranch blood does not exhibit a Haldane effect (reviewed by Butler and Metcalfe, 1988), and thus there is no oxylabile supply of H^+ in these fish. The absence of a Haldane effect in elasmobranchs is compensated for, at least in part, by a higher buffering capacity of hemoglobin (Jensen, 1989). On the other hand, the intrinsic rate of RBC Cl^-/HCO_3^- exchange does not appear to be increased in elasmobranchs (Perry *et al.*, 1996b).

A fascinating feature of elasmobranch blood is the presence of CA within the plasma (Wood *et al.*, 1994; Gilmour *et al.*, 1996; Henry *et al.*, 1997b) in addition to the RBC (Swenson and Maren, 1987). This naturally occurring extracellular CA likely explains the relatively large contribution of the plasma to whole blood HCO_3^- dehydration in isotopic *in vitro* assays (Wood *et al.*, 1994; Perry *et al.*, 1996b). The arterial blood of elasmobranchs is characterized by low P_{CO_2}'s (e.g., see Table 1) suggesting highly efficient CO_2 transfer across the gill. It is tempting to speculate that such efficient CO_2 transfer, especially in the absence of a Haldane effect, reflects an important contribution of plasma CA to HCO_3^- dehydration *in vivo*. How-ever, specific inhibition of the plasma CA with benzolamide did not elevate arterial P_{CO_2} in the dogfish *Squalus acanthias* (Swenson and Maren, 1987; Gilmour *et al.*, 1996). These data do not support a significant role for extracellular CA in CO_2 excretion at rest although it is possible that other mechanisms were activated (such as increased ventilation) to compensate for the absence of extracellular CA activity. It is possible that the role of extracellular CA is more important during exercise when CO_2 production is increased and gill transit time is reduced. Future research should focus on the role of extracellular CA in elasmobranch CO_2 excretion and acid–base regulation both at rest and during exercise. The reader is referred to Chapter 9 by Gilmour for a detailed account of the role of extracellular CA in accelerating CO_2–HCO_3^-–H^+ reactions within the plasma of elasmo-branchs.

In a recent study, Gilmour *et al.* (1996) successfully inhibited RBC $Cl^-/$
HCO_3^- exchange in dogfish *in vivo* by intra-arterial injection of DIDS.
Although the capacity of the blood to dehydrate HCO_3^- was reduced by
~40% (as assayed *in vitro*), DIDS treatment did not affect arterial P_{CO_2} *in*
vivo. This result suggests a CO_2 excretory system in dogfish with considerable reserve capacity in which extracellular CA and red cell internal HCO_3^-
stores may become more important during inhibition of Cl^-/HCO_3^- exchange.

D. Steady-State Conditions in Agnathans

The agnathans represent an ancient lineage of the vertebrates that is
thought to have separated from the rest of the vertebrates as early as the
Cambrian period, more than 500 million years ago (Hildebrand, 1988).
Recent studies examining the respiratory physiology of agnathans are therefore becoming the basis for an important comparison with the phylogenetically more recent fishes and may lead to a much better understanding of
respiratory processes in the ancestral fishes.

The agnathans consist of two major groups, the hagfish and the lamprey.
Although these groups are similar in many aspects of their respiratory
physiology, there are also a number of important differences between the
hagfish and the lampreys regarding their strategies for the transport of
respiratory gases. The following sections will therefore describe what is
known about carbon dioxide transport and excretion in these two groups
prior to providing more general insights and conclusions from the Agnatha.

Evidence of Anion Exchange Limitations in Lamprey Red Blood Cells

In contrast to the situation in more recent fishes, and most other vertebrates examined to date, chloride/bicarbonate exchange across the RBC
membrane of lamprey appears to be extremely limited, if not entirely
absent, in some species. The first account of these peculiar RBC properties
was provided by Ohnishi and Asai (1985), who were unable to detect
any evidence of anion exchange in RBCs of the lamprey (*Entosphenus*
japonicus). It is difficult to draw definite conclusions from this initial study,
however, since the authors stated that their experiments also failed to detect
anion exchange proteins in RBCs from teleost species such as carp, which
are known to possess significant anion exchange activity (Jensen and
Brahm, 1996). Shortly thereafter, Nikinmaa and Railo (1987) found that
the half-time for $^{36}Cl^-$ equilibration across the RBC membranes of the
river lamprey was about 2.5 h at 20°C (Fig. 6). In contrast, the half-time
for $^{36}Cl^-$ equilibration across the membranes of rainbow trout, which pos-

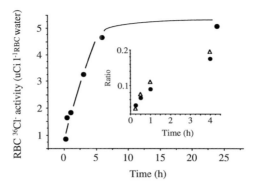

Fig. 6. Equilibration of chloride-36 across lamprey red blood cell membrane. Points are means of 5–10 experiments. The inset illustrates the ratio of $^{36}Cl^-$ across the RBC membrane over time for control RBCs (filled circles) and for RBCs treated with 10^{-4} M DIDS (open triangles). Values in inset are means of seven experiments. Redrawn from Nikinmaa and Railo (1987).

sess rapid anion exchange, is 0.8 s at 15°C (Romano and Passow, 1984). Interestingly, the $^{36}Cl^-$ equilibration value obtained by Nikinmaa and Railo (1987) for river lampreys was not significantly affected by the anion exchange inhibitor DIDS (Fig. 6) and is similar to that for lipid bilayers at the same temperature. Additional experiments by Nikinmaa and Railo (1987) and others (Tufts and Boutilier, 1989, 1990b; Cameron and Tufts, 1994; Cameron et al., 1996; Gilmour and Tufts, unpublished) have also failed to detect any DIDS-sensitive movements of bicarbonate across the RBCs of lampreys. Taken together, the existing physiological data would therefore seem to suggest that the anion exchange activity in lamprey RBCs is at best minimal, and perhaps even nonexistent in some species.

In contrast to earlier physiological studies, recent investigations involving biochemical and molecular approaches indicate that an anion exchangerlike protein may be present in the RBCs of at least some lamprey species (Brill et al., 1992; Kay et al., 1995; Cameron et al., 1996). This apparent conflict between physiological studies that have been unable to detect functional anion exchange and evidence from more recent biochemical/molecular studies that indicate that an anion exchangerlike protein is present in lamprey RBCs clearly indicates that further study is required. At present, however, these findings could be explained in several ways. The simplest explanation is that extremely low levels of anion exchange in lamprey RBCs are detectable using biochemical and/or molecular approaches, but are beyond the limits of resolution of most physiological methods. Another intriguing possibility is that the lamprey protein that has been detected in biochemical/molecular studies may lack the ability to carry out the anion transport function. In higher vertebrates, it is well known that the RBC anion exchange protein (AE1) has a number of important cellular functions in addition to anion transport (Salhany, 1990). Thus, it is conceivable that the protein in lamprey may be involved in other important cellular functions, but not anion transport. This latter possibility would certainly have interesting implications in terms of the evolution of structure and function of the anion exchanger. Further studies in this area may soon provide much needed information regarding the presence or absence and possible structure (i.e., amino acid sequence) of any anion exchangerlike proteins in lamprey RBCs. Once available, this molecular information may also form the basis for some very interesting comparisons of structure and function between the RBC anion exchangers of different vertebrates.

CO$_2$ CARRIAGE IN LAMPREY BLOOD

Reports that anion exchange was limited or possibly even absent in lamprey RBCs have been followed by *in vitro* and *in vivo* investigations to determine the implications of these unique RBC properties on the CO$_2$ transport processes in these primitive vertebrates. Although not fully complete, research in this area indicates that the strategy for CO$_2$ transport in agnathans such as the lampreys may indeed represent an interesting alternative strategy for CO$_2$ transport and excretion that does not rely heavily on RBC anion exchange.

In most fishes, much of the HCO$_3$$^-$ formed by the CO$_2$ hydration reaction within the erythrocyte is exchanged for plasma chloride via the anion exchange mechanism on the RBC membrane. Thus, at any given P_{CO_2}, the total CO$_2$ concentration of true plasma (i.e., plasma with RBCs present) is typically greater than that of either the whole blood or RBCs. In lamprey blood, however, the vast majority of CO$_2$ added to the blood *in vitro* remains within the RBC (Tufts and Boutilier, 1989). As P_{CO_2} is increased in lamprey blood, the total CO$_2$ concentration of the RBC therefore becomes higher than that of the true plasma (Fig. 4a). The elevated CO$_2$ content of the RBCs is probably not due to extensive formation of carbamino compounds since the amino terminal of the major hemoglobin component in lamprey is formylated (Fujiki *et al.,* 1970), thereby precluding carbamino formation (Nikinmaa and Matsoff, 1992). Addition of the artificial Cl$^-$/OH$^-$ ionophore, tri-*n*-propyl tin chloride to the blood of the sea lamprey also shifts the *in vitro* CO$_2$ dissociation properties toward the standard vertebrate pattern by markedly reducing the total CO$_2$ concentration of the RBCs and elevating the total CO$_2$ concentration of the true plasma (Fig. 7; Tufts and Boutilier, 1990b). Carbamino CO$_2$ would not be expected to leave the cell and enter the plasma when the RBC membrane is made

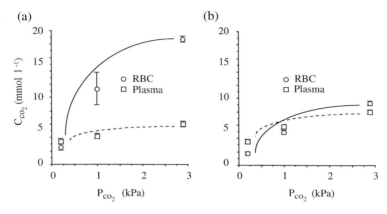

Fig. 7. Carbon dioxide dissociation curves for RBCs (circles) and true plasma (squares) of the lamprey (*Petromyzon marinus*) under (a) control conditions ($N = 4$) and (b) in the presence of tri-*n*-propyl tin chloride (5×10^{-6} M; $N = 8$). All values are means ± SE. Redrawn from Tufts and Boutilier, 1990b.

permeable to anions. These results therefore indicate that, prior to the addition of the ionophore, HCO_3^- is not passively distributed across the RBC membrane in lamprey blood. Thus, RBC anion exchange limitations are clearly an important factor contributing to the unique blood CO_2 transport properties of lampreys.

The Role of the Haldane Effect in Blood CO_2 Carriage in Lamprey

The Haldane effect is another important factor in the blood CO_2 transport processes of lamprey. The *in vitro* CO_2 carrying capacity of *P. marinus* whole bood, for example, increases by 15–27% upon deoxygenation at P_{CO_2}'s between 0.2 and 2.9 kPa (Ferguson *et al.*, 1992). A relatively large Haldane effect in lamprey blood is not surprising since the magnitude of the Bohr factor in lamprey is also relatively large when properly based on RBC pH (Nikinmaa and Matsoff, 1992; Ferguson *et al.*, 1992) and because the concept of linked functions (Wyman, 1964) implies that a large Bohr effect will be associated with a large Haldane effect.

In most vertebrates, including teleost fish, much of the additional HCO_3^- resulting from the increased proton-binding capacity of deoxygenated hemoglobin is redistributed to the plasma via the RBC anion exchange mechanism. Thus, the increased blood CO_2 carrying capacity mediated by the Haldane effect is typically associated with an increase in the total CO_2 concentration within both the plasma and RBCs as well as an increase in

RBC chloride concentration after the transfer of HCO_3^- to the plasma via the anion exchanger. An increase in rbc water content also normally occurs upon deoxygenation because of the increased RBC chloride concentration. In sharp contrast to this scenario, there is no significant increase in the total CO_2 concentration of lamprey plasma upon deoxygenation, and all the additional HCO_3^- resulting from the Haldane effect is retained within the RBC (Ferguson et al., 1992; Nikinmaa and Matsoff, 1992). Notably, deoxygenation of lamprey blood also has no effect on the RBC chloride or water content (Ferguson et al., 1992; Nikinmaa and Matsoff, 1992). Thus, the distribution of the CO_2 added to lamprey blood upon deoxygenation is again determined by the unique characteristics of the RBCs that lack significant anion exchange.

ARTERIO–VENOUS DIFFERENCES IN CO_2 CONTENT IN LAMPREY BLOOD

Analysis of the *in vitro* CO_2 transport properties of oxygenated and deoxygenated blood indicates that the strategy for CO_2 excretion across the gills of lamprey is also very different from that in more recent fishes. For example, when the CO_2 distribution in deoxygenated blood at an elevated CO_2 tension is compared to that of oxygenated blood at a low CO_2 tension (0.2 kPa), it is apparent that 71% of the difference in whole blood C_{CO_2} is attributable to changes in RBC C_{CO_2} (Ferguson et al., 1992). Unlike most vertebrates in which changes in plasma CO_2 content account for most of the CO_2 excreted across the respiratory organ, these *in vitro* data indicate that arterio–venous differences in RBC C_{CO_2} account for the majority of the CO_2 excreted across the gills of the lamprey. As mentioned earlier, *in vivo* attempts to partition CO_2 concentrations between plasma and RBCs should be viewed with caution since the measurements and calculations involved in these predictions assume equilibrium of CO_2 reactions in blood, and this may not be the case, particularly in postbranchial blood (Gilmour et al., 1994). Nonetheless, it is interesting to note that *in vivo* analysis of pre- and postbranchial lamprey blood (Table 4) supports the *in vitro* predictions and indicates that changes in RBC CO_2 content account for the majority (62%) of the arterio–venous difference in whole blood CO_2 content (Tufts et al., 1992). In most vertebrates, significant changes in RBC chloride concentration occur during passage of the blood through the respiratory organ since plasma HCO_3^- enters the RBC (where it will be dehydrated to CO_2) in exchange for chloride. Further evidence that RBC anion exchange and arterio–venous changes in plasma HCO_3^- are less important for CO_2 excretion in lamprey therefore arises from the fact that arterio–venous changes in RBC chloride concentrations that are

Table 4
Comparison of Resting Arterio–Venous Differences in the Sea Lamprey and the
Rainbow Trout

	Sea lamprey	Rainbow trout
Total arterio–venous difference in CO_2 (mmol L^{-1} blood)	1.13	0.99
Percentage of total arterio–venous difference in CO_2 from plasma	38	85[a]
Percentage of total arterio–venous difference in CO_2 from red blood cells	62	15[a]
RBC arterio–venous $[Cl^-]$ (meq L^{-1})	2.4	15.0[b]
plasma arterio–venous $[Cl^-]$ (meq L^{-1})	2.3	2.0[b]
Hematocrit (%)	27.4	23.2[b]

[a] Data from Heming (1984).
[b] Data from Nikinmaa and Jensen (1986). Used by permission from Tufts *et al.* (1992).

substantial (15 mM) in species such as trout (Nikinmaaa and Jensen, 1986)
are insignificant in lamprey (Tufts *et al.,* 1992; Table 4).

The CO_2 content of lamprey RBCs is influenced by both changes in
P_{CO_2} and changes in blood oxygenation since lamprey hemoglobin displays
a significant Haldane effect. The *in vitro* CO_2 dissociation properties of
lamprey blood indicate, however, that oxygenation-dependent changes in
RBC CO_2 content have a much greater impact on the loading and unloading
of CO_2 from lamprey RBCs. Based on *in vitro* CO_2 dissociation data for
Lampetra fluviatilis obtained by Nikinmaa and Matsoff (1992) under conditions resembling those in pre- and postbranchial blood, Nikinmaa *et al.*
(1995) calculated that more than 80% of the change in the total carbon
dioxide content of the RBCs between simulated arterial and venous conditions should be attributable to the oxygenation-dependent decrease in total
carbon dioxide content and less than 20% to the decrease in CO_2 tension.
Thus, although the absence of RBC anion exchange precludes efficient
utilization of the plasma for CO_2 carriage in lamprey, an extremely large
Haldane effect allows these animals to carry large amounts of CO_2 within
the RBC and to achieve arterio–venous differences in CO_2 content that
are surprisingly similar to those of more recent fishes (Table 4).

RELATIONSHIP BETWEEN RBC pH_i AND THE STRATEGY
FOR CO_2 CARRIAGE IN LAMPREY BLOOD

Several lines of evidence indicate that the lamprey RBC membrane
contains a Na^+/H^+ exchange mechanism (Nikinmaa, 1986; Nikinmaa *et al.,*
1986, 1993; Tufts, 1992; Virrki and Nikinmaa, 1994). This exchanger seems

to have an important role in maintaining an elevated RBC pH in lamprey even under steady-state conditions (Nikinmaa, 1986; Nikinmaa *et al.*, 1986; Tufts, 1992). The activity of this mechanism may also be increased following increases in ambient CO_2 tension (Tufts, 1992) or intracellular acidification (Nikinmaa *et al.*, 1993; Virrki and Nikinmaa, 1994) or upon deoxygenation (Ferguson *et al.*, 1992). Unlike the Na^+/H^+ exchange mechanism on the RBC membrane of teleost fish, however, the activity of the Na^+/H^+ exchanger on lamprey RBCs is not highly sensitive to catecholamines (Tufts, 1991; Virrki and Nikinmaa, 1994).

As in the RBCs of teleost fish following adrenergic stimulation, extrusion of protons from the RBC in lamprey should increase the apparent nonbicarbonate buffer value of the RBC and shift the CO_2 hydration reaction toward the formation of HCO_3^-. The extruded protons should also titrate plasma HCO_3^- and increase blood P_{CO_2} levels. The Na^+/H^+ exchanger on the RBC membrane in lamprey might therefore be expected to have an important influence on the distribution of CO_2 between plasma and RBCs. Experimental evidence indicating that this is in indeed the case has been obtained for the sea lamprey, *P. marinus*. If the Na^+/H^+ exchanger is disabled in these animals by removing extracellular sodium, there is a significant decrease in the RBC total CO_2 concentration and a significant increase in the extracellular total CO_2 concetration (Tufts, 1992; Cameron and Tufts, 1994). Thus, RBC Na^+/H^+ exchange and anion exchange limitations both appear to contribute to the unusual CO_2 transport properties of lamprey blood.

MODEL OF CO_2 TRANSPORT IN LAMPREY

The transport and excretion of carbon dioxide in lamprey do not conform to any of the standard models for CO_2 transport in vertebrates since these models all involve rapid anion exchange across the RBC membrane. A simple model of CO_2 transport and excretion in lampreys that does not involve rapid RBC anion exchange was proposed by Tufts and Boutilier (1989) and is illustrated in Fig. 8. Unlike other vertebrates, the majority of the CO_2 loaded into the blood of lamprey is carried within the RBC to the gill. Similar to most vertebrate RBCs, there are significant levels of CA within lamprey RBCs to catalyze the intracelluar CO_2 reactions (Nikinmaa *et al.*, 1986; Henry *et al.*, 1993). A small fraction of the CO_2 entering the plasma will also be hydrated to form HCO_3^- and a proton. To date, there has been no attempt to determine whether any intravascular CA is present within lamprey muscle that could facilitate CO_2 uptake into the blood. It has been shown, however, that there is no CA present within the plasma or on the endothelial membrane of the gill vasculature that could catalyze the CO_2 reactions in lamprey plasma (Henry *et al.*, 1993). Thus,

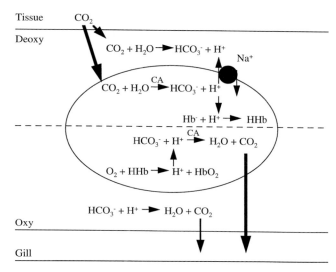

Fig. 8. Model of carbon dioxide transport in the blood of the lamprey. In the tissues (Deoxy), the majority of the CO_2 loaded into the blood diffuses into the RBC where it is hydrated to form bicarbonate (HCO_3^-) and a proton (H^+) that is bound to hemoglobin (Hb) or transferred to the plasma in exchange for sodium (Na^+). The resulting bicarbonate is not transported to the plasma, but is carried within the RBC to the gill. At the gill (Oxy), oxygenation of hemoglobin liberates a proton that combines with the HCO_3^-, and CO_2 is evolved. Carbonic anhydrase (CA) catalyzes the CO_2 hydration–dehydration reactions within the RBC, but is not available to catalyze these reactions in the plasma. Modified from Tufts and Boutilier (1989).

at the gill, the dehydration of plasma HCO_3^- to CO_2 will proceed at the uncatalyzed rate and will make only a minor contribution to overall CO_2 excretion.

The experimental evidence collected to date is basically consistent with this relatively simple model. At this time, the most important issue that currently remains unresolved is to directly assess whether the lamprey RBC membrane is at all permeable to HCO_3^- on a physiologically relevant time scale. Since the transit time for passage of blood through the respiratory organ (of aquatic vertebrates) is thought to range between 0.5 and 3 s (Cameron and Polhemus 1974; Randall and Daxboeck, 1984), HCO_3^- movements across the RBC membrane would have to occur within this time frame in order for plasma HCO_3^- to make a significant contribution to CO_2 excretion. At present, the possibility exists that even a relatively small number of anion exchangers or some other HCO_3^- transport mechanism within the lamprey RBC may still provide plasma HCO_3^- with some degree of access to RBC CA during the blood's transit through the gills. Earlier studies suggest that the HCO_3^- permeability of the lamprey RBC membrane is probably extremely low, but the possibility that these indirect measurements were influenced by other ion transport mechansims (e.g., Na^+/H^+ exchange) cannot be excluded. Results from more recent studies suggest that lamprey RBCs may be somewhat permeable to HCO_3^- (Tufts, 1992; Cameron and Tufts, 1994; Cameron et al., 1996), but none of these experiments have involved a time scale that is physiologically relevant for CO_2 excretion. Thus, further experiments that directly assess the HCO_3^- permeability of lamprey RBCs on a time scale that is physiologically relevant for CO_2 excretion, such as during the O^{18} exchange studies involving mass spectrometry, are clearly required to fully resolve this issue.

Since (1) anion exchange limitations that occur in lamprey RBCs may effectively isolate plasma CO_2 stores from RBC CA and (2) equilibration of acid–base equivalents between RBCs and plasma may be very slow (Nikinmaa and Railo, 1987), it might be expected that significant CO_2 disequilibria exist in lamprey plasma. Preliminary data (Gilmour, Tufts, and Perry, unpublished) indicate, however, that the magnitude of the CO_2 disequilibrium in postbranchial blood of the lamprey is in fact much smaller than that in species such as trout (Gilmour et al., 1994) that possess rapid RBC anion exchange. To date, however, the magnitude of the disequilibrium in venous blood of the lamprey has yet to be determined.

EVIDENCE FOR ANION EXCHANGE LIMITATIONS IN HAGFISH RBCs

Ellory et al. (1987) used a number of experimental approaches to quantify anion exchange activity in hagfish (Eptatretus stouti) RBCs. Assuming that the turnover rates of the apparent anion exchanger in hagfish and that in human RBCs at 11°C are comparable, these authors concluded that the observed H_2DIDS-sensitive $^{36}Cl^-$ flux activity would be accounted for by about 20 copies per cell in hagfish RBCs as compared to 1.2×10^6 copies per cell in human RBCs. More recently, Brill et al. (1992) provided further evidence that low, but detectable, anion exchange activity is present in the RBC of another species of hagfish, Myxine glutinosa. SDS–PAGE of RBC membranes from the hagfish (E. stouti) reveals low levels of a protein that co-migrates with the leading edge of the human anion exchanger (Ellory et al., 1987). These biochemical results are consistent with the extremely low physiological activity of RBC anion exchange for hagfish documented within the same study. To date, however, immunological confirmation of the identity of this hagfish protein as an early vertebrate band-3-like molecule is lacking, as is any information on its genetic sequence. Clearly, such information is required before any definite conclusions can be made with regard to the identity of this protein and before any attempt to compare this protein with the anion exchanger of more recent fishes. Nonetheless, the available evidence from the hagfish is very similar to that collected to date for lamprey in indicating that RBC anion exchange activity in the hagfish is extremely reduced compared to more recent fishes.

CO₂ CARRIAGE IN HAGFISH BLOOD

Interestingly, the *in vitro* CO_2 transport properties of the hagfish, *Myxine glutinosa* (Tufts and Boutilier, 1990a), are markedly different from those of the lamprey (Fig. 9). In *M. glutinosa,* the total CO_2 content of the true plasma is considerably greater than that of the RBCs throughout the physiological P_{CO_2} range. Moreover, a small, but significant, amount of HCO_3^- does appear to be transferred from RBCs to the plasma in hagfish blood incubated under steady-state conditions *in vitro* since the HCO_3^- content of true plasma is significantly greater than that of separated plasma. Unlike most other vertebrates, however, the HCO_3^- movement between the RBCs and plasma in hagfish is not significantly affected by the presence of the anion exchange inhibitor, DIDS.

The presence of an ionophore for anions, tri-*n*-propyl tin chloride, does not significantly alter the distribution of HCO_3^- between the RBCs and plasma in hagfish (Tufts and Boutilier, 1990b). In marked contrast to the situation in lamprey, these findings suggest that HCO_3^- may be passively distributed across the RBC membrane in hagfish, at least under steady-state conditions *in vitro*.

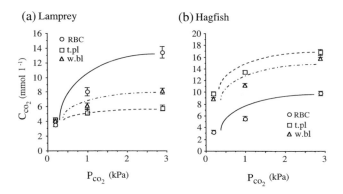

Fig. 9. Carbon dioxide dissociation curves for red blood cells (RBC, circles), true plasma (t.pl, squares) and whole blood (w.bl) of (a) Lamprey (*Petromyzon marinus; N* = 8; redrawn from Tufts and Boutilier, 1989) and (b) Hagfish (*Myxine glutinosa; N* = 6; redrawn from Tufts and Boutilier, 1990). All values are means ± SE.

The relative importance of the Haldane effect toward CO_2 carriage in *M. glutinosa* is also very different from that in lamprey. In *M. glutinosa,* deoxygenation of the blood at 0.96 kPa P_{CO_2} increases the CO_2 content of the RBCs by only about 4 mmol L^{-1} (Tufts *et al.,* 1997), whereas in the lamprey, the CO_2 content of the rbcs is increased by almost 20 mmol L^{-1} (Ferguson *et al.,* 1992). Since deoxygenation of blood does not result in any significant additional increase in the CO_2 content of the plasma, the overall magnitude of the Haldane effect in whole blood of *M. glutinosa* is therefore extremely small (Tufts *et al.,* 1997).

In apparent contrast to the findings for *M. glutinosa,* Wells *et al.* (1986) observed a much larger Haldane effect in the whole blood of another hagfish species, *Eptatretus cirrhatus.* Taking into consideration the low hematocrit (8.5%) observed in *E. cirrhatus,* the magnitude of the Haldane effect for the hemoglobin in this hagfish species approaches that observed for the lamprey (Ferguson *et al.,* 1992; Nikinmaa and Matsoff, 1992). Thus, there may be substantial differences in the blood CO_2 transport properties between different species of hagfish. Unfortunately, no data are currently available that partition the carriage of CO_2 between plasma and RBCs under oxygenated and deoxygenated conditions in *E. cirrhatus.* At present, it is therefore impossible to determine whether most of the CO_2 arising from the Haldane effect in *E. cirrhatus* is retained within the RBC as in the lamprey.

Taken together, the results from the *in vitro* studies carried out to date indicate that there are substantial differences in the strategy for blood CO_2 carriage within the agnathans. These differences undoubtedly result from several important differences in RBC characteristics between the lamprey and some species of hagfish (e.g., *M. glutinosa*). For example, the nonbicarbonate buffer value of the RBC (mainly reflective of Hb buffering) in the lamprey is much greater than that in *M. glutinosa* (Table 2*).* As mentioned earlier, RBC Na^+/H^+ exchange may also contribute to a further increase in the apparent nonbicarbonate buffer value of the lamprey RBC. In contrast, hagfish RBCs do not appear to possess significant Na^+/H^+ exchange (Nikinmaa *et al.,* 1993). Thus, as blood P_{CO_2}'s increase, much more HCO_3^- will be formed within the lamprey RBC than within the hagfish RBC, particularly under deoxygenated conditions. Interestingly, this results in a large gradient for HCO_3^- between the RBCs and plasma at elevated CO_2 levels in lamprey, but this is not the case in hagfish (Tufts and Boutilier, 1990b; Tufts *et al.,* 1997). The vastly different effects observed between lamprey and hagfish blood when the ionophore for anions, tri-*n*-propyl tin chloride, is present *in vitro* may therefore be largely explained by the fact that there is a large gradient established for HCO_3^- to move from the RBCs to the plasma in lamprey blood at elevated P_{CO_2}'s, whereas similar HCO_3^- gradients are not present across the RBC membrane of the hagfish. In *M. glutinosa*, RBC HCO_3^- levels never exceed those in the plasma within the physiological P_{CO_2} range (Tufts and Boutilier, 1990b; Tufts *et al.*, 1997). At least in *M. glutinosa*, the relative magnitude of the Haldane effect is also much less than that in the lamprey species examined to date (Tufts *et al.*, 1997). This does not appear to be the case for the hemoglobin of all species of hagfish, however, since the relative magnitude of the Haldane effect appears to be much greater in another species of hagfish, *E. cirrhatus* (Wells *et al.*, 1986).

At present, it is not known whether the HCO_3^- permeability of hagfish RBCs is significantly different from that of lamprey. It is noteworthy, however, that some evidence for very low levels of functional anion exchange has been obtained for hagfish (Ellory *et al.*, 1987; Brill *et al.*, 1992), whereas similar findings have not been obtained for lamprey. Thus, it is possible that relative differences in HCO_3^- permeability between the RBCs of hagfish and lamprey may also contribute to the observed differences in the *in vitro* blood CO_2 transport properties between these agnathans. To date, however, this possibility remains to be thoroughly examined.

ARTERIO–VENOUS DIFFERENCES IN CO_2 CONTENT IN HAGFISH BLOOD

Information regarding arterio–venous differences in CO_2 levels in hagfish is relatively limited. Based on the CO_2 dissociation properties of oxygenated and deoxygenated blood, Tufts *et al.* (1997) calculated that the amount of CO_2 arising from arterio–venous differences in RBC CO_2 content in the hagfish *M. glutinosa* is probably much less than that in lamprey. If plasma HCO_3^- does not make a significant contribution to CO_2 excretion in hagfish due to anion exchange limitations, the amount of CO_2 that could potentially be excreted across the gills of hagfish per unit blood volume would therefore be expected to be much lower than that in lamprey. These potential limitations for CO_2 excretion may be partially offset by the high blood volume in these animals relative to other vertebrates. Such a system is likely more than adequate for animals such as hagfish with very low metabolic rates (Smith and Hessler, 1974). In *E. cirrhatus*, arterio–venous differences in blood CO_2 content would be expected to be significantly greater than those in *M. glutinosa* since the Haldane effect is much larger in *E. cirrhatus* (Wells *et al.*, 1986). As stated previously, however, the relative importance of the plasma and RBCs toward CO_2 carriage in oxygenated and deoxygenated blood of *E. cirrhatus* has not been examined. Thus, the potential contribution of the plasma and RBCs toward arterio–venous differences in blood CO_2 content and CO_2 excretion in this species is unknown.

If plasma HCO_3^- does not have significant access to CA in hagfish during the blood's transit through the gill because of the limited anion permeability of hagfish RBCs, it is likely that the basic strategy for CO_2 transport in hagfish is similar to that in lamprey. It has been shown that hagfish RBCs possess significant levels of carbonic anhydrase (Maren *et al.*, 1980; Carlsson *et al.*, 1980). As in lamprey, the RBCs would therefore have the major role in CO_2 transport and excretion in hagfish if RBC CA is indeed unavailable to plasma HCO_3^-. Since hagfish species such as *Myxine glutinosa* lack an appreciable Haldane effect and possess a relatively low RBC nonbicarbonate buffering capacity, however, the amount of CO_2 that could be excreted across the gill per unit blood volume using such a strategy would probably be very reduced compared to the lamprey. In species such as *E. cirrhatus*, the potential for CO_2 removal using this strategy may be greatly increased due to the larger Haldane effect and may even approach that of the lamprey.

At present, additional information is required to obtain a complete understanding of CO_2 transport and excretion in hagfish. As in the lamprey, the major issue that needs to be fully resolved for the hagfish is whether plasma HCO_3^- has significant access to RBC CA during the short time that the blood resides in the gill. In the hagfish, it is also not known whether plasma bicarbonate has access to any other intravascular CA that could potentially catalyze the CO_2 reactions in this compartment. Although technically challenging, *in vitro* predictions of the strategy for CO_2 transport in hagfish will also require verification in *in vivo* studies. Finally, it appears that significant interspecific differences in CO_2 transport properties may occur in different species of hagfish, but this issue probably also warrants further study.

POTENTIAL INSIGHTS INTO THE EVOLUTION OF CO₂ TRANSPORT FROM AGNATHANS

Ellory *et al.* (1987) questioned whether the extremely low anion permeability that they observed in hagfish RBCs was truly a primitive vertebrate characteristic or whether it represented a secondary loss of RBC anion exchange activity. These authors speculated that the presence of CA in hagfish RBCs (Carlsson *et al.*, 1980) implied the earlier presence of a functional Jacobs–Stewart cycle in this species and therefore their results seemed to favor a secondary loss of RBC anion exchange activity. According to the model proposed by Tufts and Boutlier (1989), however, CA could still have a central role in the blood CO_2 transport process even in the absence of significant RBC anion exchange. Moreover, anion exchange limitations have been observed in the other primitive agnathan group, the lampreys. Thus, although further study in this area is certainly necessary, it is conceivable that low anion permeability may indeed be a characteristic of the RBCs of ancestral fishes. Additional information on the anion exchange properties of invertebrate RBCs might also provide valuable insights into this issue.

It is interesting to compare the strategies for CO_2 transport between the two groups of agnathans and the more recent fishes. Assuming plasma HCO_3^- is essentially isolated from RBC CA during CO_2 excretion in agnathans due to anion exchange limitations, an interesting pattern emerges within this comparison in terms of the amount of HCO_3^- having access to RBC CA during passage of blood through the gills. In the relatively sluggish hagfish species such as *M. glutinosa,* the amount of HCO_3^- that could be converted to CO_2 per liter of blood passed through the gills is relatively small since it will be restricted to arterio–venous differences in RBC CO_2 concentrations that are predicted to be minimal (based on *in vitro* CO_2 dissociation curves) due to the absence of a significant Haldane effect and a low RBC nonbicarbonate buffer value. In the more active lampreys, a much greater RBC nonbicarbonate buffer value, a large Haldane effect, and perhaps the incorporation of RBC Na^+/H^+ exchange greatly increase the amount of HCO_3^- carried within the RBC that could be converted to CO_2 during the blood's transit through the gill. Incorporation of rapid anion exchange in more recent fishes further increases the amount of blood HCO_3^- with access to RBC CA since this now includes both plasma and RBC HCO_3^- stores. It is clear that living members of the agnathans have continued to evolve in many ways since their divergence from a common ancestor with the remaining vertebrates more than 500 millions of years ago. Thus, it is likely that comparisons between the agnathans and more recent fishes such as the trout may not represent a true evolutionary sequence for the strategy of blood CO_2 transport in vertebrates. Nonetheless, this comparison does provide an indication of the potential amount of CO_2 that can be excreted from the blood as it passes through the gills via three very different strategies used by fishes for blood CO_2 transport. In addition, this comparison may provide some insight into the selective pressure for the evolution of rapid anion exchange in vertebrates. Although it seems that some selective advantage must be gained by the presence of rapid RBC anion exchange in terms of the potential amount of CO_2 that can be removed from the blood as it passes through the respiratory organ, it should also be considered that other important advantages may also be derived from an increased HCO_3^- permeability of the RBC. For example, changes in RBC pH are minimal in lamprey over a broad range of P_{CO_2}'s when

HCO_3^- is retained within the RBC, but much larger changes in RBC pH are observed when the HCO_3^- permeability of the RBC is increased using an artificial ionophore (Tufts and Boutilier, 1990b). Thus, another important consequence of increased RBC anion exchange will be to increase the magnitude of RBC pH changes during the loading of CO_2 into the blood, thereby providing a greater coupling of oxygen and carbon dioxide transport via the Bohr effect. In addition to an improved capacity for CO_2 carriage and excretion, it is therefore likely that other factors may have played an equally important role in the evolution of RBC anion exchange in more recent fishes and other vertebrates. Regardless of whether the present-day agnathans truly represent the ancestral vertebrate condition in their lack of significant quantities of RBC anion exchangers, these animals have certainly provided an interesting animal model to examine the potential selective pressures influencing the evolution of CO_2 transport strategies in fishes and other vertebrates. Further insights into some of these issues may also arise from the following section, which examines CO_2 transport and excretion in agnathans under non-steady-state conditions.

E. Non-steady-state Conditions in Agnathans

The hagfish are generally considered to be benthic animals with relatively low metabolic rates (Smith and Hessler, 1974). Field observations reported by Wells *et al.* (1986) indicate that there may be some exceptions, but most hagfish probably do not engage in protracted bouts of aerobic or anaerobic swimming activity. In contrast to the hagfish, a number of lamprey species are anadromous. Anadromous lamprey species such as *P. marinus* spend the parasitic phase of their life history in the ocean, or large bodies of freshwater (in the case of landlocked forms), and then migrate into freshwater rivers and streams prior to spawning. In many cases, lamprey spawning streams are the fast-flowing type co-habited by salmonids such as *Salmo salar*. Hence, the spawning migrations of lampreys may often involve significant bouts of intense aerobic and/or anaerobic swimming activity during their ascent to the spawning site.

The physiology of exercise in gnathostome fishes such as the salmonids has been intensively studied. Until recently, however, much less was known about the exercise physiology of agnathans such as the lamprey. Recent investigations examining carbon dioxide transport and acid–base regulation under non-steady-state conditions in lampreys, such as during recovery from intense swimming activity, have therefore begun to fill a significant gap in the comparative literature.

The following section reviews the recent literature focusing on CO_2 transport and acid–base regulation in agnathans under non-steady-state conditions. Several of these studies investigate recovery from anaerobic exercise in active lamprey species such as *P. marinus.* Other than a single study by Wells *et al.* (1986), which is also discussed, little information is available regarding similar non-steady-state disturbances in hagfish. Although relatively few studies examining non-steady-state conditions arising from other perturbations to the CO_2 transport system in agnathans are available, these are also included since they provide further insights into the agnathan strategy for CO_2 transport and excretion.

EFFECTS OF EXERCISE ON CO_2 TRANSPORT AND ACID–BASE REGULATION IN LAMPREY

The effects of exhaustive exercise on blood CO_2 transport properties and acid–base status in lamprey have now been examined in several studies (Tufts, 1991; Tufts *et al.,* 1992, 1996). As in other fishes, a relatively short bout (5–10 min) of exhaustive exercise in sea lampreys causes significant decreases in the total CO_2 content of the plasma in both arterial and venous blood due to the titration of plasma HCO_3^- by protons arising from the production of lactic acid during the exercise bout. This reaction also causes an increase in blood P_{CO_2} levels. A postexercise increase in CO_2 production by the tissues contributes to a greater elevation in the P_{CO_2} of venous blood, as compared to arterial blood.

Changes in the CO_2 content of lamprey RBCs following exercise are markedly different from those in the plasma. In both arterial and venous blood, exercise causes a significant increase in the RBC CO_2 content (Tufts, 1991; Tufts *et al.,* 1992). In venous blood, the total CO_2 concentration within the RBCs reaches levels that are more than double those at rest (Tufts *et al.,* 1992). Increases in the RBC CO_2 content in this situation are not surprising since the RBC pH is extremely well maintained in these animals at a time when blood P_{CO_2} is also increased. In combination with the relatively steep nonbicarbonate buffer slope of lamprey RBCs (Table 2), these factors largely explain the postexercise increase in RBC HCO_3^- levels. There are also significant decreases in blood PO_2 levels and hemoglobin–oxygen carriage in the arterial and venous blood of lampreys immediately after exercise (Tufts, 1991). Since lamprey hemoglobin has a large Haldane effect, these changes in blood oxygenation status contribute further to the rise in RBC HCO_3^- levels, particularly in venous blood.

It is interesting that lamprey RBC pH is so well maintained after exhaustive exercise since there is a large postexercise decrease in pH_e at this time (Tufts, 1991; Tufts *et al.,* 1992). In venous blood, there may even be a reversal of the pH gradient across the RBC membrane for the first few minutes after exercise (Tufts *et al.,* 1992). The relatively high nonbicarbonate buffer value of lamprey RBCs undoubtedly has an important role in the maintenance of RBC pH in these animals (Table 2). In contrast to more recent teleost fishes, lampreys do not regulate RBC pH via adrenergic mechanisms (Tufts, 1991; Virkki and Nikinmaa, 1994). The RBC Na^+/H^+ exchanger present in lamprey RBCs may be stimulated by increases in P_{CO_2} and/or decreases in pH and blood oxygenation (Ferguson et al. 1992; Nikinmaa et al. 1993; Virkki and Nikinmaa, 1994) although the maximal activity of this mechanism appears to be much less than that in teleost fish (Virkki and Nikinmaa, 1994). In lamprey, postexercise reductions in hemoglobin oxygenation (Tufts et al., 1992) will further contribute to an elevated RBC pH because of the large Haldane effect (Ferguson et al., 1992; Nikinmaa and Matsoff, 1992). As explained by Heisler (1986) and also predicted the Henderson–Hasselbach equation, the absence of HCO_3^- transfer to the plasma (via anion exchange) would also be expected to result in a relative increase in RBC pH at any given blood P_{CO_2} level. Thus, a number of factors, including the absence of functional anion exchange may be viewed as contributing to the maintenance of RBC pH in lamprey following exercise.

The increases in RBC CO_2 concentrations and decreases in plasma CO_2 concentrations observed in lampreys following exercise result in an increase in the relative contribution of the RBCs toward CO_2 carriage in both arterial and venous blood (Tufts et al., 1992). In addition, the relative importance of changes in RBC CO_2 content toward the arterio–venous differences in blood CO_2 levels (i.e., CO_2 excretion) is also affected by exercise in lampreys. Immediately after an exercise bout, changes in $[C_{CO_2}]_{rbc}$ accounted for 78% of the total CO_2 difference between arterial and venous blood of the sea lamprey as compared to 62% under resting conditions (Tufts et al., 1992). Thus, although these in vivo calculations will be subject to some degree of error, the relative importance of the RBC in CO_2 transport and excretion in lamprey appears to be even greater following exercise. Since the anion permeability of lamprey RBCs is low and blood transit time through the gills will likely be even shorter after exercise, these changes are probably not surprising and likely represent one of the only routes available to cope with the increased levels of CO_2 production in these animals during recovery from exercise. It is interesting that the magnitude of the arterio–venous difference in whole blood CO_2 levels can be almost doubled following exercise using this strategy in lamprey (Tufts et al., 1992). The magnitude of this postexercise increase in the arterio–venous difference in whole blood CO_2 levels in the lamprey is almost identical to that observed in exhaustively exercised rainbow trout (Currie and Tufts, 1993).

In phylogenetically more recent fishes, arterial P_{CO_2} increases by 50–400% immediately after exercise (Wood and Perry, 1985). The postexercise increase in arterial P_{CO_2} determined for sea lamprey (78%; Tufts, 1991) falls into the lower portion of this range. As explained previously, titration of plasma HCO_3^- by protons arising from anaerobic metabolism in muscle will contribute to the observed rise in arterial P_{CO_2} since this reaction will proceed at the uncatalyzed rate in plasma and will therefore continue after the blood has left the gill. In teleost fish, and possibly in lamprey, protons extruded from the RBC via Na^+/H^+ exchange will also contribute to this reaction. In addition, it has been suggested that temporary inhibition of HCO_3^- flux through the RBC may contribute to the P_{CO_2} rise in the arterial blood of more recent fishes after exercise (Wood and Perry, 1985). The relatively low anion permeability of lamprey RBCs may therefore be another factor contributing to the observed changes in arterial P_{CO_2}. If reduced anion permeability is indeed an important factor contributing to the postexercise rise in arterial P_{CO_2} in most fishes, however, it is somewhat surprising that the rise in arterial P_{CO_2} in lamprey is not far greater than that observed in more recent teleost fishes possessing RBCs with much greater anion permeability. The relatively small rise in arterial P_{CO_2} seen after exercise in lamprey would suggest that (1) the bicarbonate permeability of lamprey RBCs is much greater than previously anticipated or (2) factors other than changes in RBC HCO_3^- permeability probably have a more important role in creating the observed postexercise rise in arterial P_{CO_2}. Unfortunately, these arguments are currently complicated by the fact that most reported blood P_{CO_2} values in fish, including those of lamprey, are calculated values and assume equilibrium conditions for the CO_2 reactions in the blood, which is undoubtedly not the case following exercise (Gilmour and Perry, 1996). Thus, lamprey could provide an interesting addition to the literature discussed in Section III.B regarding the relationship between the postexercise increases in arterial P_{CO_2} and the rate of RBC HCO_3^- flux in teleost fish. Further experiments, including those involving instantaneous measurements of arterial P_{CO_2} after exercise in lamprey, will need to be carried out, however, before the importance of lamprey in these discussions can be clarified.

Since the transport and excretion of carbon dioxide is intimately linked to the regulation of acid–base status, it is interesting that the recovery of blood acid–base variables following exhaustive exercise in lamprey occurs as quickly as, if not more quickly than, in most teleost fish (Tufts, 1991). The relative importance of the lamprey gill in correcting the postexercise acid–base disturbance is roughly similar to that in teleosts (Wilkie et al., 1998). The rapid recovery of extracellular acid–base variables in lamprey does not therefore reflect compensation for blood CO_2 transport limitations by an extremely effective branchial regulation of extracellular H^+ and HCO_3^- levels. Thus, it seems that the unique strategy for the transport of carbon dioxide in lamprey blood does not have any obvious negative consequences for the regulation of blood acid–base status following bouts of burst activity.

Boutilier *et al.* (1993) determined the P_{CO_2} levels and acid–base status of locomotory muscle in lamprey following recovery from exhaustive exercise. The total proton and lactate load measured after exercise in lamprey was found to be much lower than that in rainbow trout white muscle (Tang and Boutilier, 1991). Although these results indicate that anaerobic capacity is lower in the lamprey, the increase in muscle P_{CO_2} levels was larger, and more prolonged in lamprey than that calculated for rainbow trout after a similar exercise regime (Tang and Boutilier, 1991). As suggested by Boutilier *et al.* (1993), these results could indicate that the blood CO_2 transport system in lamprey represents a significant limitation for the removal of CO_2 from muscle following exercise. These authors also point out that additional factors such as (1) a restricted ability of the muscle itself to clear the intracellular pool of CO_2, (2) a persistently high CO_2 production rate in muscle following exercise, or (3) the availability of muscle CA to facilitate CO_2 transport from muscle to blood may also explain the persistent elevation in intracellular P_{CO_2} in lamprey after exercise. It is therefore intriguing to speculate that RBC anion exchange limitations in lamprey could represent an important limitation in the removal of CO_2 from the muscle of these animals. In view of the other potential explanations for these observations, however, further study in this area is clearly warranted before any conclusions can be made on this issue.

EFFECTS OF CA INHIBITION OR INFUSION IN LAMPREY

Lamprey RBCs possess significant quantities of a CA isozyme with kinetic properties similar to those of the type I, slow turnover CA isozyme (Henry *et al.*, 1993). According to the predictive model proposed by Tufts and Boutilier (1989), lamprey CA should have an important role in CO_2 transport even in the absence of significant RBC anion exchange. An initial examination of the *in vivo* importance of RBC CA in lamprey, however, found that inhibition of RBC CA by acetazolamide resulted in only a very minor respiratory acidosis in the extracellular fluid (Henry *et al.*, 1995). Although these results are consistent with the presence of a relatively slow CA isozyme within the lamprey RBC (Henry *et al.*, 1993) and may also be indicative of a lower overall rate of CO_2 flux through the system (i.e., lower metabolic rate and CO_2 excretion rate), more convincing *in vivo* evidence of the central role of lamprey RBC CA in the CO_2 transport process was also warranted. Further study showed that when metabolic rates are elevated by exhaustive exercise in lamprey, inhibition of RBC CA results in a large and rapidly developing respiratory acidosis as compared to lamprey that were exercised without subsequent RBC CA inhibition (Tufts *et al.,* 1996). Thus, although inhibition of CA has a relatively minor impact on CO_2 removal from resting animals, the enzyme is critically important for CO_2 excretion under conditions when CO_2 production is elevated such as following exercise. These CA inhibition studies are therefore consistent with the predictive model for CO_2 transport in lamprey (Fig. 8) and support the view that RBC CA can still have a central role in the CO_2 transport process even in the absence of a functional Jacobs–Stewart cycle.

In view of the reported RBC anion exchange limitations in agnathan blood, it is somewhat surprising that CA infusion did not significantly enhance recovery of blood acid–base status or alter any of the blood CO_2 transport variables in exhaustively exercised lampreys (Tufts *et al.,* 1996). These results suggest that the CO_2 reactions in these animals may already be in equilibrium even though a previous investigation by Henry *et al.* (1993) found no evidence of plasma CA activity or of any CA on the endothelial membrane of the gill vasculature that could potentially catalyze the CO_2 reactions in the plasma. At present, there is no obvious explanation for these results, but several possibilities do exist. As mentioned earlier, the actual HCO_3^- permeability of lamprey RBCs could be higher than previously anticipated. This possibility seems unlikely based on current evidence, but has yet to be directly examined by extremely sensitive and well-established techniques. There has also been no attempt to determine whether plasma bicarbonate has access to any CA in the capillary beds of the locomotory muscle in agnathans. An experimental examination of this possibility may now be warranted since this could explain much of this apparent paradox. Finally, it should be noted that plasma CA activity was not monitored in the study by Tufts *et al.* (1996). Thus, if lamprey lack significant levels of endogenous CA inhibitor, any RBC lysis that was not visibly obvious would have gone undetected in these experiments and could also have influenced these results. Naturally occurring CA inhibitors from lower vertebrates do not show cross-reactivity with mammalian CA (Haswell *et al.,* 1983; Dimberg, 1994), and therefore it is unlikely that the lack of effect for infused CA in lamprey was due to inactivation by an endogenous plasma CA inhibitor.

CO_2 Transport in Swimming Hagfish

Although relatively little information is available regarding CO_2 transport and acid–base status under non-steady-state conditions in hagfish, Wells *et al.* (1986) did monitor some of these variables at rest and following forced swimming in both arterial and venous blood of *E. cirrhatus*. The measured P_{CO_2} values in both arterial and venous blood were low and resembled values in other fishes. Exercise caused P_{CO_2} to increase from 0.7

to 1.2 mm Hg in arterial blood and from 1.2 to 1.5 mm Hg in venous blood. Although these relatively small changes were not significant (probably because of the low n number), these variables returned to the levels in resting fish within 20 min of recovery. Swimming in this study appeared to be predominantly aerobic since the pH of arterial and venous blood, 7.92 and 7.77, respectively, and blood lactate levels were essentially unchanged after the exercise period.

POTENTIAL INSIGHTS FROM NON-STEADY-STATE
STUDIES OF AGNATHANS

It is interesting that the lack of rapid anion exchange in agnathan RBCs does not have any obviously negative consequences for the removal of CO_2 from blood or recovery of blood acid–base status in these animals following exercise. Recovery of muscle P_{CO_2} levels may be somewhat slower in agnathans than in more recent fishes, but the factors affecting removal of CO_2 from agnathan muscle require further study before any definite conclusions can be drawn from these observations. An important consequence of the lack of rapid RBC anion exchange in agnathans, however, may be the extremely low nonbicarbonate buffer value of the plasma since hemoglobin is not available to buffer metabolic acidoses in the plasma. During non-steady-state disturbances such as recovery from exercise, even moderate levels of metabolic protons will therefore produce relatively large pH disturbances in the plasma. Agnathan species may also be ideal models to examine the relative importance of RBC anion exchange limitations as a contributing factor toward the arterial P_{CO_2} increase that occurs in fish after exercise. Studies that incorporate direct measurements of arterial P_{CO_2} under these conditions will be required, however, in order to adequately address this issue. At present, the paucity of data currently available during non-steady-state conditions in hagfish makes it impossible to draw any definite conclusions regarding non-steady-state CO_2 transport in these animals.

IV. FUTURE DIRECTIONS

A number of significant advances that have improved our understanding of CO_2 transport and excretion in fish have taken place in recent years. As in most other fields, however, there are also several areas within this field where more information is required. For example, more studies incorporating real-time *in vivo* measurements of acid–base and CO_2 variables in both arterial and venous blood would be extremely valuable (although technically challenging), particularly under non-steady-state conditions.

Such studies should be carried out on both teleosts and nonteleosts. In groups such as the agnathans, such information may be crucial before a complete understanding of their unique strategies for CO_2 transport can be achieved. Although considerable effort has been directed toward understanding the mechanisms involved in CO_2 transport in some species of nonteleosts, additional *in vivo* and *in vitro* information from several phylogenetically significant nonteleost groups such as the elasmobranchs and the hagfish is still required. Additional *in vitro* measurements of processes such as RBC anion exchange under a variety of conditions, and within time frames similar to those occurring at the gill *in vivo* (i.e., <5 s), are also needed. Finally, incorporation of molecular approaches to conclusively resolve some issues such as the presence or absence of anion exchangerlike proteins in the membrane of agnathan RBCs would be an important contribution to this area.

REFERENCES

Albers, C., and Pleschka, K. (1967). Effects of temperature on CO_2 transport in elasmobranch blood. *Respir. Physiol.* **2,** 261–273.

Albers, C. (1970). Acid–base balance. *In* "Fish Physiology" (W. S. Hoar, and D. J. Randall, eds.), pp. 173–208. Academic Press, New York.

Alper, S. L. (1991). The band 3-related anion exchanger (AE) gene family. *Annu. Rev. Physiol.* **53,** 549–564.

Baumgarten-Schumann, D., and Piiper, J. (1968). Gas exchange in the gills of resting unanesthetized dogfish (*Scyliorhinus stellaris*). *Respir. Physiol.* **5,** 317–325.

Bidani, A., and Heming, T. A. (1991). Effects of perfusate buffer capacity on capillary CO_2-HCO^-_3-H^+ reactions: Theory. *J. Appl. Physiol.* **71,** 1460–1468.

Bindon, S. F., Gilmour, K. M., Fenwick, J. C., and Perry, S. F. (1994). The effect of branchial chloride cell proliferation on gas transfer in the rainbow trout, *Oncorhynchus mykiss*. *J. Exp. Biol.* **197,** 47–63.

Boutilier, R. G., Heming, T. A., and Iwama, G. K. (1984). Physiochemical parameters for use in fish respiratory physiology. *In* "Fish Physiology" (W. S. Hoar, and D. J. Randall, eds.), Vol. XA, pp. 403–430. Academic Press, New York.

Boutilier, R. G., Ferguson, R. A., Henry, R. P. and Tufts, B. L. (1993). Exhaustive exercise in the sea lamprey (*Petromyzon marinus*): Relationship between anaerobic metabolism and intracellular acid–base balance. *J. Exp. Biol.* **178,** 71–88.

Brauner, C. J., Gilmour, K. M., and Perry, S. F. (1996). Effect of haemoglobin oxygenation on Bohr proton release and CO_2 excretion in the rainbow trout. *Respir. Physiol.* **106,** 65–70.

Brauner, C. J., and Randall, D. J. (1996) The interaction between oxygen and carbon dioxide movements in fishes . *Comp. Biochem. Physiol. A* **113,** 83–90.

Brill, S. R., Musch, M. W., and Goldstein, L. (1992). Taurine efflux, band 3 and erythrocyte volume of the hagfish (*Myxine glutinosa*) and lamprey (*Petromyzon marinus*). *J. Exp. Zool.* **264,** 19–25.

Butler, P. J., and Metcalfe, J. D. (1988). Cardiovascular and respiratory systems. *In* "Physiology of Elasmobranch Fishes" (T. J. Shuttleworth, ed.), pp. 1–47, Springer-Verlag Berlin/Heidelberg/New York.

Cameron, B. A., and Tufts, B. L. (1994). *In vitro* investigation of the factors contributing to the unique CO_2 transport properties of blood in the sea lamprey (*Petromyzon marinus*). *J. Exp. Biol.* **197**, 337–348.

Cameron, B. A., Perry, S. F., Wu, C.-B., Ko, K., and Tufts, B. L. (1996). Bicarbonate permeability and immunological evidence for an anion exchanger-like protein in the red blood cells of the sea lamprey, *Petromyzon marinus*. *J. Comp. Physiol. B.* **166**, 197–204.

Cameron, J. N., and Davis, J. C. (1970). Gas exchange in rainbow trout (*Salmo gairdneri*) with varying blood oxygen capacity. *J. Fish Res. Bd. Can.* **27**, 1069–1085.

Cameron, J. N., and Polhemus, J. A. (1974). Theory of CO_2 exchange in trout gills. *J. Exp. Biol.* **60**, 183–194.

Cameron, J. N. (1978). Chloride shift in fish blood. *J. Exp. Zool.* **206**, 289–295.

Cameron, J. N. (1979). Excretion of CO_2 in water-breathing animals—A short review. *Mar. Biol. Lett.* **1**, 3–13.

Cameron, J. N., and Kormanik, G. A. (1982). Intracellular and extracellular acid–base status as a function of temperature in the freshwater channel catfish, *Ictalurus punctatus*. *J. Exp. Biol.* **99**, 127–142.

Carlsson, U., Kjellstrom, B., and Antonsson, B. (1980). Purification and properties of cyclostome carbonic anhydrase from erythrocytes of hagfish. *Biochim. Biophys. Acta.* **612**, 160–170.

Currie, S., and Tufts, B. L. (1993). An analysis of carbon dioxide transport in arterial and venous blood of the rainbow trout, *Oncorhynchus mykiss* following exhaustive exercise. *Fish Physiol. Biochem.* **12**, 183–192.

Currie, S., Kieffer, J. D., and Tufts, B. L. (1995). The effects of blood CO_2 reaction rates on CO_2 removal from muscle in exercised trout. *Respir. Physiol.* **100**, 261–269.

Dimberg, K., Hoglund, L. B., Knutsson, P. G., and Ridderstrale, Y. (1981). Histochemical localization of carbonic anhydrase in gill lamellae from young salmon (*Salmo salar* L.) adapted to fresh and salt water. *Acta. Physiol. Scand.* **112**, 218–220.

Dimberg, K. (1994). The carbonic anhydrase inhibitor in trout plasma: Purification and its effect on carbonic anhydrase activity and the Root effect. *Fish Physiol. Biochem.* **12**, 381–386.

Eddy, F. B. (1974). Blood gases of the tench (*Tinca tinca*) in well aerated and oxygen-deficient waters. *J. Exp. Biol.* **60**, 71–83.

Eddy, F. B. (1976). Acid–base balance in rainbow trout (*Salmo gairdneri*) subjected to acid stresses. *J. Exp. Biol.* **64**, 159–171.

Eddy, F. B., Lomholt, J. P., Weber, R. E., and Johansen, K. (1977). Blood respiratory properties of rainbow trout (*Salmo gairdneri*) kept in water of high CO_2 tension. *J. Exp. Biol.* **67**, 37–47.

Effros, R. M., Chang, R. S. Y., and Silverman, P. (1978). Acceleration of plasma bicarbonate conversion to carbon dioxide by pulmonary carbonic anhydrase. *Science* **199**, 427–429.

Ellory, J. C., Wolowyk, M. W., and Young, J. D. (1987). Hagfish (*Eptatretus stouti*) erythrocytes show minimal chloride transport activity. *J. Exp. Biol.* **129**, 377–383.

Enns, T. (1967). Facilitation by carbonic anhydrase of carbon dioxide transport. *Science* **155**, 44–47.

Evans, D. H. (1993). Osmotic and Ionic Regulation. *In* "The Physiology of Fishes" (D. H. Evans, ed.), pp. 315–341. CRC Press, Boca Raton, FL.

Ferguson, R. A., Sehdev, N., Bagatto, B., and Tufts, B. L. (1992). *In vitro* interactions between oxygen and carbon dioxide transport in the blood of the sea lamprey (*Petromyzon marinus*). *J. Exp. Biol.* **173**, 25–41.

Forster, R. E., Edsall, J. T., Otis, A. B., and Roughton, F. J. W. (1969). Carbon dioxide, carbonic acid, and bicarbonate ion: Physical properties and kinetics of interconversion. *In* "CO₂: Chemical, Biological and Physiological Aspects," pp. 15–27, NASA SP-188, Washington, D.C.

Fujiki, H., Braunitzer, G., and Rudloff, V. (1970). *N*-Formyl-proline as N-terminal amino acid of lamprey hemoglobin. *Hoppe-Seyler's Z. Physiol. Chem.* **351**, 901–902.

Gilmour, K. M., and Perry, S. F. (1994). The effects of hypoxia, hyperoxia or hypercapnia on the acid–base disequilibrium in the arterial blood of rainbow trout, *Oncorhynchus mykiss. J. Exp. Biol.* **192**, 269–284.

Gilmour, K. M., Randall, D. J., and Perry, S. F. (1994). Acid–base disequilibrium in the arterial blood of rainbow trout, *Oncorhynchus mykiss. Respir. Physiol.* **96**, 259–272.

Gilmour, K. M., Henry, R. P., Wood, C. M., and Perry, S. F. (1997). Extracellular carbonic anhydrase and an acid-base disequilibrium in the blood of the dogfish, *Squalus acanthias. J. Exp. Biol.* **200**, 173–183.

Gilmour, K. M., and Perry, S. F. (1996a). The effects of experimental anaemia on CO₂ excretion in rainbow trout, *Oncorhynchus mykiss. Fish Physiol. Biochem.* **15**, 83–94.

Gilmour, K. M., and Perry, S. F. (1996b). Effects of metabolic acid–base disturbances and elevated catecholamines on the acid–base disequilibrium in the arterial blood of rainbow trout. *J. Exp. Zool.* **274**, 281–290.

Gilmour, K. M. (1998). Causes and consequences of acid–base disequilibria. *In* "Fish Physiology: Haemoglobin and Respiration" (S. F. Perry, and B. L. Tufts, eds.), Vol. 17, pp. 321–348, Academic Press, New York.

Goss, G. G., Perry, S. F., Wood, C. M., and Laurent, P. (1992). Mechanisms of ion and acid–base regulation at the gills of freshwater fish. *J. Exp. Zool.* **263**, 143–159.

Goss, G. G., Perry, S. F., and Laurent, P. (1995). Gill morphology and acid–base regulation. *In* "Fish Physiology" (C. M. Wood, and T. J. Shuttleworth, eds.), Vol. 14. pp. 257–284, Academic Press, New York.

Goss, G. G., Perry, S. F., Fryer, J. N., and Laurent, P. (1998). Gill morphology and acid-base regulation in freshwater fishes. *Comp. Biochem. Physiol. A* **119**, 107–115.

Graham, M. S., Turner, J. D., and Wood, C. M. (1990). Control of ventilation in the hypercapnic skate *Raja ocellata. Respir. Physiol.* **80**, 259–277.

Greco, A. M., Gilmour, K. M., Fenwick, J. C., and Perry, S. F. (1995). The effects of soft-water acclimation on respiratory gas transfer in the rainbow trout, *Oncorhynchus mykiss. J. Exp. Biol.* **198**, 2557–2567.

Haswell, M. S., and Randall, D. J. (1978). The pattern of carbon dioxide excretion in the rainbow trout *Salmo gairdneri. J. Exp. Biol.* **72**, 17–24.

Haswell, M. S., Randall, D. J., and Perry, S. F. (1980). Fish gill carbonic anhydrase: Acid–base regulation or salt transport? *Am. J. Physiol.* **238**, 240–245.

Haswell, M. S., Raffin, J.-P., and LeRay, C. (1983). An investigation of the carbonic anhydrase inhibitor in eel plasma. *Comp. Biochem. Physiol A* **74**, 175–177.

Heisler, N. (1993) . Acid–base regulation. *In* "The Physiology of Fishes" (D. H. Evans, ed.), pp. 343–378. CRC Press, Boca Raton, FL.

Heming, T. A. (1984). The role of fish erythrocytes in transport and excretion of carbon dioxide. Ph.D. thesis, Univ. of British Columbia, Vancouver, BC.

Heming, T. A., Randall, D. J., Boutilier, R. G., Iwama, G. K., and Primmett, D. (1986). Ionic equilibria in red blood cells of rainbow trout (*Salmo gairdneri*): Cl⁻, HCO₃⁻ and H⁺. *Respir. Physiol.* **65**, 223–234.

Henry, R. P., Smatresk, N. J., and Cameron, J. N. (1988). The distribution of branchial carbonic anhydrase and the effects of gill and erythrocyte carbonic anhydrase inhibition in the channel catfish *Ictalurus punctatus. J. Exp. Biol.* **134**, 201–218.

Henry, R. P., Tufts, B. L., and Boutilier, R. G. (1993). The distribution of carbonic anhydrase type-I and type-II isozymes in lamprey and trout - possible co-evolution with erythrocyte chloride bicarbonate exchange. *J. Comp. Physiol. B* **163**, 380–388.

Henry, R. P., Boutilier, R. G., and Tufts, B. L. (1995). Effects of carbonic anhydrase inhibition on the acid–base status in lamprey and trout. *Respir. Physiol.* **99**, 241–248.

Henry, R. P., Wang, Y., and Wood, C. M. (1997a). Carbonic anhydrase facilitates CO_2 and NH_3 transport across the sarcolemma of trout white muscle. *Am. J. Physiol.* **272** (*41*), R1754–R1761.

Henry, R. P., Gilmour, K. M., Wood, C. M., and Perry, S. F. (1997b). Extracellular carbonic anhydrase activity and carbonic anhydrase inhibitors in the circulatory system of fish. *Physiol. Zool.* **700**, 650–659.

Henry, R. P. and Heming, T. A. (1998). Carbonic anhydrase and respiratory gas exchange. *In* "Fish Physiology: Haemoglobin and Respiration" (S. F. Perry, and B. L. Tufts, eds.), Vol. 17, pp. 75–111, Academic Press, New York.

Hildebrand, M. (1988). "Analysis of Vertebrate Structure," 3rd ed. Wiley, New York.

Holeton, G. F. (1970). Oxygen uptake and circulation by a hemoglobinless Antarctic fish (*Chaenocephalus aceratus* Lonnberg) compared with three red-blooded Antarctic fish. *Comp. Biochem. Physiol.* **34**, 457–471.

Hubner, S., Michel, F., Rudloff, V., and Appelhans, H. (1992). Amino acid sequence of band-3 protein from rainbow trout erythrocytes derived from cDNA. *Biochem. J.* **285**, 17–23.

Hughes, G. M. (1972). Morphometrics of fish gills. *Respir. Physiol.* **14**, 1–25.

Jensen, F. B. (1986). Pronounced influence of $Hb–O_2$ saturation on red cell pH in tench blood *in vivo* and *in vitro*. *J. Exp. Zool.* **238**, 119–224.

Jensen, F. B. (1989). Hydrogen ion equilibria in fish haemoglobins. *J. Exp. Biol.* **143**, 225–234.

Jensen, F. B. (1991). Multiple strategies in oxygen and carbon dioxide transport by haemoglobin. *In* "Physiological Strategies for Gas Exchange and Metabolism," Society for Experimental Biology Seminar Series (A. J. Woakes, M. K. Grieshaber, and C. R. Bridges, eds.), pp. 55–78. Cambridge Univ. Press, Cambridge.

Jensen, F. N., and Brahm, J. (1996). Kinetics of chloride transport across fish red blood cell membranes. *J. Exp. Biol.* **198**, 2237–2244.

Kay, M. M. B., Cover, C., Schluter, S. F., Bernstein, R. M., and Marchalonis, J. J. (1995). Band 3, the anion transporter, is conserved during evolution: implications for aging and vertebrate evolution. *Cell. Mol. Biol.* **41**(6), 833–842.

Klocke, R. A. (1978). Catalysis of CO_2 reactions by lung carbonic anhydrase. *J. Appl. Physiol.* **44**, 882–888.

Kopito, R. R. (1990). Molecular Biology of the anion exchanger gene family. *Int. Rev. Cytol.* **123**, 177–199.

Lenfant, C., and Johansen, K. (1966). Respiratory function in the elasmobranch *Squalus suckleyi* G. *Respir. Physiol.* **1**, 13–29.

Lin, H., and Randall, D. J. (1991). Evidence for the presence of an electrogenic proton pump on the trout gill epithelium. *J. Exp. Biol.* **161**, 119–134.

Lin, H., and Randall, D. J. (1993). H^+-ATPase activity in crude homogenates of fish gill tissue-inhibitor sensitivity and environmental and hormonal regulation. *J. Exp. Biol.* **180**, 163–174.

Lin, H., Pfeiffer, D. C., Vogl, A. W., Pan, J., and Randall, D. J. (1994). Immunolocalization of H^+-ATPase in the gill epithelia of rainbow trout. *J. Exp. Biol.* **195**, 169–183.

Lin, H., and Randall, D. J. (1995). Proton Pumps in Fish Gills. *In* "Fish Physiology: Cellular and Molecular Approaches to Fish Ionic Regulation" (C. M. Wood, and T. J. Shuttleworth, eds.), Vol. 14, pp. 229–255, Academic Press, New York.

Malte, H., and Weber, R. E. (1985). A mathematical model for gas exchange in the fish gill based on non-linear blood gas equilibrium curves. *Respir. Physiol.* **62**, 359–374.

Maren, T. H. (1967). Carbonic anhydrase: Chemistry, physiology and inhibition. *Physiol. Rev.* **47**, 598–781.

Maren, T. H., Friedland, B. R., and Rittmaster, R. S. (1980). Kinetic properties of primitive vertebrate carbonic anhydrases. *Comp. Biochem. Physiol.* **67**, 69–74.

Marshall, W. S. (1995). Transport processes in isolated teleost epithelia: opercular epithelium and urinary bladder. *In* "Fish Physiology: Cellular and Molecular Approaches to Fish Ionic Regulation" (C. M. Wood, and T. J. Shuttleworth, eds.). Vol. 14. pp. 1–23. Academic Press, New York.

Metcalfe, J. D., and Butler, P. J. (1988). The effects of alpha- and beta-adrenergic receptor blockade on gas exchange in the dogfish (*Scyliorhinus canicula* L.) during normoxia and hypoxia. *J. Comp. Physiol. B* **158**, 39–44.

Motais, R., Fievet, F., Garcia-Romeu, and S. Thomas. (1989). Na$^+$/H$^+$ exchange and pH regulation in red blood cells: Role of uncatalyzed H$_2$CO$_3$ dehydration. *Am. J. Physiol. C* **256**, 728–735.

Nikinmaa, M. (1986). Red cell pH of lamprey (*Lampetra fluviatilis*) is actively regulated. *J. Comp. Physiol. B* **156**, 747–750.

Nikinmaa, M., and Jensen, F. B. (1986). Blood oxygen transport and acid–base status of stressed trout (*Salmo gairdneri*): pre- and postbranchial values in winter fish. *Comp. Biochem. Physiol.* **84**, 391–396.

Nikinmaa, M., Kunnamo-Ojala, T., and Railo, E. (1986). Mechanisms of pH regulation in lamprey (*Lampetra fluviatilis*) red blood cells. *J. Exp. Biol.* **122**, 355–367.

Nikinmaa, M., and Railo, E. (1987). Anion movements across lamprey (*Lampetra fluviatilis*) red cell membrane. *Biochim. Biophys. Acta.* **899**, 134–136.

Nikinmaa. M. (1990). "Vertebrate Red Blood Cells." Springer-Verlag, Berlin Heidelberg.

Nikinmaa, M. (1992). Membrane transport and control of hemoglobin-oxygen affinity in nucleated erythrocytes. *Physiol. Rev.* **72**, 301–321.

Nikinmaa, M., and Mattsoff, L. (1992). Effects of oxygen saturation on the CO$_2$ transport properties of *Lampetra* red cells. *Respir. Physiol.* **87**, 219–230.

Nikinmaa, M., and Vihersaari, L. (1993). Pre- and postbranchial carbon dioxide content of rainbow trout (*Oncorhynchus mykiss*) blood after catecholamine injection. *J. Exp. Biol.* **180**, 315–321.

Nikinmaa, M., Tufts, B. L., and Boutilier, R. G. (1993). Volume and pH regulation in agnathan erythrocytes-comparisons between the hagfish, *Myxine glutinosa* and the lampreys, *Petromyzon marinus* and *Lampetra fluviatilis*. *J. Comp. Physiol. B.* **163**, 608–613.

Nikinmaa, M., Airaksinen, S., and Virkki, L. V. (1995). Haemoglobin function in intact lamprey erythrocytes: Interactions with membrane function in the regulation of gas transport and acid–base balance. *J. Exp. Biol.* **198**, 2423–2430.

Obaid, A. L., Critz, A. M., and Crandall, E. D. (1979). Kinetics of bicarbonate/chloride exchange in dogfish erythrocytes. *Am. J. Physiol.* **237**, 132–138.

Ohnishi, S. T., and Asai, H. (1985). Lamprey erythrocytes lack glycoproteins and anion transport. *Comp. Biochem. Physiol. B.* **81**, 405–407.

Perry, S. F., Haswell, M. S., Randall, D. J., and Farrell, A. P. (1981). Branchial ionic uptake and acid–base regulation in the rainbow trout, *Salmo gairdneri*. *J. Exp. Biol.* **92**, 289–303.

Perry, S. F., Davie, P. S., Daxboeck, C., and Randall, D. J. (1982). A comparison of CO$_2$ excretion in a spontaneously ventilating blood-perfused trout preparation and saline-perfused gill preparations: Contribution of the branchial epithelium and red blood cell. *J. Exp. Biol.* **101**, 47–60.

Perry, S. F., Payan, P., and Girard, J. P. (1984). The effects of perfusate HCO_3^- and P_{CO_2} on chloride uptake in perfused gills of rainbow trout (*Salmo gairdneri*). *Can. J. Fish. Aquat. Sci.* **41**, 1768–1773.

Perry, S. F., Daxboeck, C., Emmett, B., Hochachka, P. W., and Brill, R. W. (1985). Effects of exhausting exercise on acid–base regulation in skipjack tuna (*Katsuwonus pelamis*) blood. *Physiol. Zool.* **58**, 421–429.

Perry, S. F. (1986). Carbon dioxide excretion in fish. *Can. J. Zool.* **64**, 565–572.

Perry, S. F., and Wood, C. M. (1989). Control and coordination of gas transfer in fishes. *Can. J. Zool.* **67**, 2961–2970.

Perry, S. F., and Laurent, P. (1990). The role of carbonic anhydrase in carbon dioxide excretion, acid–base balance and ionic regulation in aquatic gill breathers. *In* "Transport, Respiration and Excretion: Comparative and Environmental Aspects" (J. P. Truchot, and B. Lahlou, eds.), pp. 39–57. Karger, Basel.

Perry, S. F., Wood, C. M., Thomas, S., and Walsh, P. J. (1991). Adrenergic inhibition of carbon dioxide excretion by trout red blood cells *in vitro* is mediated by activation of Na^+/H^+ exchange. *J. Exp. Biol.* **157**, 367–380.

Perry, S. F., and Gilmour, K. M. (1993). An evaluation of factors limiting carbon dioxide excretion by trout red blood cells *in vitro*. *J. Exp. Biol.* **180**, 39–54.

Perry, S. F., and McDonald, D. G. (1998). Gas exchange. *In* "The Physiology of Fishes" (D. H. Evans, ed.), pp. 251–278, CRC Press, Boca Raton, FL.

Perry, S. F. (1997a). Relationships between branchial chloride cells and gas transfer in freshwater fish. *Comp. Biochem. Physiol. A* **119**, 9–16.

Perry, S. F. (1997b). The chloride cell: Structure and function in the gill of freshwater fishes. *Annu. Rev. Physiol.* **59**, 325–347.

Perry, S. F., and Fryer, J. N. (1997). Proton pumps in the fish gill and kidney. *Fish Physiol. Biochem.* **17**, 363–369.

Perry, S. F., Reid, S. G., Wankiewicz, E., Iyer, V., and Gilmour, K. G. (1996a). Physiological responses of rainbow trout (*Oncorhynchus mykiss*) to prolonged exposure to softwater. *Physiol. Zool.* **69**, 1419–1441.

Perry, S. F., Wood, C. M., Walsh, P. J., and Thomas, S. (1996b). Fish red blood cell carbon dioxide excretion *in vitro*: A comparative study. *Comp. Biochem. Physiol. A* **113**, 121–130.

Perry, S. F., Brauner, C. J., Tufts, B. L., and Gilmour, K. M. (1997). Acid–base disequilibrium in the venous blood of rainbow trout (*Oncorhynchus mykiss*). *J. Exp. Biol. Online*, **2**, 1.

Peyreaud-Waitzenegger, M., and Soulier, P. (1989). Ventilatory and cardiovascular adjustments in the European eel (*Anguilla anguilla*) exposed to short term hypoxia. *Exp. Biol.* **48**, 107–122.

Piiper, J., Baumgarten, D., and Meyer, M. (1970). Effects of hypoxia upon respiration and circulation in the dogfish *Scyliorhinus stellaris*. *Comp. Biochem. Physiol. A* **36**, 513–520.

Piiper, J., Meyer, M., Worth, H., and Willmer, H. (1977). Respiration and circulation during swimming activity in the dogfish *Scyliorhinus stellaris*. *Respir. Physiol.* **30**, 221–239.

Piiper, J. (1989). Factors affecting gas transfer in respiratory organs of vertebrates. *Can. J. Zool.* **67**, 2956–2960.

Playle, R. C., Munger, R. S., and Wood, C. M. (1990). Effects of catecholamines on gas exchange and ventilation in rainbow trout (*Salmo gairdneri*). *J. Exp. Biol.* **152**, 353–367.

Powell, M. D., and Perry, S. F. (1996). Respiratory and acid–base disturbances in rainbow trout (*Oncorhynchus mykiss*) blood during exposure to chloramine-T, paratoluenesulphonamide and hypochlorite. *Can. J. Fish. Aquat. Sci.* **53**, 701–708.

Primmet, D. R. N., Randall, D. J., Mazeaud, M., and Boutilier, R. G. (1986). The role of catecholamines in erythrocyte pH regulation and oxygen transport in rainbow trout (*Salmo gairdneri*) during exercise. *J. Exp. Biol.* **122**, 139–148.

Rahim, S. M., Delaunoy, J. P., and Laurent, P. (1988). Identification and immunocytochemical localization of two different carbonic anhydrase isozymes in teleostean fish erythrocytes and gill epithelia. *Histochemistry* **89**, 451–459.

Randall, D. J., Burggren, W. W., Farrell, A. P., and Haswell, N. S. (1981) "The Evolution of Air Breathing in Vertebrates." Cambridge Univ. Press, Cambridge.

Randall, D. J., Perry, S. F., and Heming, T. A. (1982). Gas transfer and acid–base regulation in salmonids. *Comp. Biochem. Physiol. B* **73**, 93–103.

Randall, D. J., and Daxboeck, C. (1984). Oxygen and carbon dioxide transfer across fish gills. *In* "Fish Physiology" (W. S. Hoar, and D. J. Randall, eds.), Vol. XA, pp. 263–314, Academic Press, New York.

Randall, D. J., and Wright, P. A. (1989). The interaction between carbon dioxide and ammonia excretion and water pH in fish. *Can. J. Zool.* **67**, 2936–2942.

Randall, D. J. (1990). Control and co-ordination of gas exchange in water breathers. *In* "Advances in Comparative and Environmental Physiology" (R. G. Boutilier, ed.), pp. 253–278. Springer-Verlag, Berlin, Heidelberg.

Randall, D. J., and Brauner, C. (1991). Effects of environmental factors on exercise in fish. *J. Exp. Biol.* **160**, 113–126.

Randall, D. J., Lin, H., and Wright, P. A. (1991). Gill water flow and the chemistry of the boundary layer. *Physiol. Zool.* **64**, 26–38.

Riggs, A. (1970). Properties of fish hemoglobins. *In* "Fish Physiology" (W. S. Hoar, and D. J. Randall, eds.), Vol. IV, pp. 209–252. Academic Press, New York.

Romano, L., and Passow, H. (1984). Characterization of anion transport system in trout red blood cell. *Am. J. Physiol.* **246**, C330–C338.

Salhanny, J. M. (1990). "Erythrocyte Band-3 Protein" CRC Press, Boca Raton, FL.

Scarabello, M., Heigenhauser, G. J. F., and Wood, C. M. (1992). Gas exchange, metabolite status, and excess post-exercise oxygen consumption after repetitive bouts of exhaustive exercise in juvenile rainbow trout. *J. Exp. Biol.* **167**, 155–169.

Smith, K. L., and Hessler, R. R. (1974). Respiration of benthopelagic fishes: *in situ* measurements at 1230 metres. *Science* **184**, 72–73.

Steffensen, J. F., Tufts, B. L., and Randall, D. J. (1987). Effect of burst swimming and adrenaline infusion on O_2 consumption and CO_2 excretion in rainbow trout, *Salmo gairdneri. J. Exp. Biol.* **131**, 427–434.

Stevens, D. E., and Randall, D. J. (1967). Changes of gas concentrations in blood and water during moderate swimming activity in rainbow trout. *J. Exp. Biol.* **46**, 329–337.

Sullivan, G. V., Fryer, J. N., and Perry, S. F. (1995). Immunolocalization of proton pumps (H^+-ATPase) in pavement cells of rainbow trout gill. *J. Exp. Biol.* **198**, 2619–2629.

Sullivan, G. V., Fryer, J. N., and Perry, S. F. (1996). Localization of mRNA for proton pump (H^+-ATPase) and Cl^-/HCO_3^- exchanger in rainbow trout gill. *Can. J. Zool.* **74**, 2095–2103.

Swenson, E. R., and Maren, T. H. (1987). Roles of gill and red cell carbonic anhydrase in elasmobranch HCO_3^- and CO_2 excretion. *Am. J. Physiol.* **253**, R450–R458.

Swenson, E. R. (1990). Kinetics of oxygen and carbon dioxide exchange. *In* "Advances in Comparative and Environmental Physiology" (R. G. Boutilier, ed.). pp. 163–210. Springer-Verlag, Berlin Heidelberg.

Tang, Y., and Boutilier, R. G. (1991). White muscle intracellular acid-base and lactate status following exhaustive exercise: A comparison between freshwater- and seawater-adapted rainbow trout. *J. Exp. Biol.* **156**, 153–171.

Thomas, S., and Perry, S. F. (1992). Control and consequences of adrenergic activation of red blood cell Na^+/H^+ exchange on blood oxygen and carbon dioxide transport. *J. Exp. Zool.* **263**, 160–175.

Thomas, S., Fritsche, R., and Perry, S. F. (1994). Pre- and post-branchial blood respiratory status during acute hypercapnia or hypoxia in rainbow trout (*Oncorhynchus mykiss*). *J. Comp. Physiol* **164**, 451–458.

Truchot, J.-P., Toulmond, A., and Dejours, P. (1980). Blood acid–base balance as a function of water oxygenation: A study at two different ambient CO_2 levels in the dogfish, *Scyliorhinus canicula. Respir. Physiol.* **41**, 13–28.

Tufts, B. L., Ferguson, R. A., and Boutilier, R. G. (1988). *In vivo* and *in vitro* effects of adrenergic stimulation on chloride/bicarbonate exchange in rainbow trout erythrocytes. *J. Exp. Biol.* **140**, 301–312.

Tufts, B. L., and Boutilier, R. G. (1989). The absence of rapid chloride/bicarbonate exchange in lamprey erythrocytes: Implications for CO_2 transport and ion distributions between plasma and erythrocytes in the blood of *Petromyzon marinus. J. Exp. Biol.* **144**, 565–576.

Tufts, B. L., and Boutilier, R. G. (1990a). CO_2 transport properties of the blood of a primitive vertebrate, *Myxine glutinosa* (L.). *J. Exp. Biol.* **48**, 341–347.

Tufts, B. L., and Boutilier, R. G. (1990b). CO_2 transport in agnathan blood: evidence of erythrocyte Cl^-/HCO_3^- exchange limitations. *Respir. Physiol.* **80**, 335–348.

Tufts, B. L. (1991). Acid–base regulation and blood gas transport following exhaustive exercise in an agnathan, the sea lamprey, *Petromyzon marinus. J. Exp. Biol.* **159**, 371–385.

Tufts, B. L. (1992). *In vitro* evidence for sodium-dependent pH regulation in sea lamprey (*Petromyzon marinus*) red blood cells. *Can. J. Zool.* **70**, 411–416.

Tufts, B. L., Bagatto, B., and Cameron, B. (1992). *In vivo* analysis of gas transport in arterial and venous blood of the sea lamprey *Petromyzon marinus. J. Exp. Biol.* **169**, 105–119.

Tufts, B. L., Currie, S., and Kieffer, J. D. (1996). Relative effects of carbonic anhydrase infusion or inhibition on carbon dioxide transport and acid–base status in the sea lamprey, *Petromyzon marinus. J. Exp. Biol.* **199**, 933–940.

Tufts, B. L., Vincent, C. J., and Currie, S. (1998). Different red blood cell characteristics in a primitive agnathan (*M. glutinosa*) and a more recent teleost (*O. mykiss*) influence their strategies for blood CO_2 transport. *Comp. Biochem. Physiol.* **119A(2)**, 533–541.

Virkki, L. V., and Nikinmaa, M. (1994). Activation and physiological role of Na^+/H^+ exchange in lamprey (*Lampetra fluviatilis*) erythrocytes. *J. Exp. Biol.* **191**, 89–105.

Walsh, P. J., and Henry, R. P. (1991). Carbon dioxide and ammonia metabolism and exchange. *In* "Biochemistry and Molecular Biology of Fishes" (P. W. Hochachka, and T. P. Mommsen, eds.), Vol. 1, pp. 181–207. Elsevier, Amsterdam.

Weber, R. E., and Jensen, F. B. (1988). Functional adaptations in hemoglobins from ectothermic vertebrates. *Annu. Rev. Physiol.* **50**, 161–179.

Wells, R. M. G., Forster, M. E., Davison, W., Taylor, H. H., Davie, P. S., and Satchell, G. H. (1986). Blood oxygen transport in the free swimming hagfish, *Eptatretus cirrhatus. J. Exp. Biol.* **123**, 43–53.

Wilkie, M. P., J. Courturier and B. L. Tufts. (1998). Mechanisms of acid-base regulation in migrant sea lampreys (Petromyzon marinus) following exhaustive exercise. *J. Exp. Biol.* In Press.

Wood, C. M., McMahon, B. R., and McDonald, D. G. (1979). Respiratory gas exchange in the resting starry flounder, *Platichthys stellatus*: A comparison with other teleosts. *J. Exp. Biol.* **78**, 167–179.

Wood, C. M., and Shelton, G. (1980). The reflex control of heart rate and cardiac output in the rainbow trout: Interactive influences of hypoxia, haemorrhage, and systemic vasomotor tone. *J. Exp. Biol.* **87**, 271–284.

Wood, C. M., McDonald, D. G., and McMahon, B. R. (1982). The influence of experimental anaemia on blood acid–base regulation *in vivo* and *in vitro* in the starry flounder (*Platichthys stellatus*) and the rainbow trout (*Salmo gairdneri*). *J. Exp. Biol.* **96**, 221–237.

Wood, C. M., and Perry, S. F. (1985). Respiratory, circulatory, and metabolic adjustments to exercise in fish. *In* "Circulation, Respiration, Metabolism" (R. Gilles, ed.), pp. 2–22. Springer-Verlag, Berlin.

Wood, C. M., Turner, J. D., Munger, S., and Graham, M. S. (1990). Control of ventilation in the hypercapnic skate *Raja ocellata* II. Cerebrospinal fluid and intracellular pH in the brain and other tissues. *Respir. Physiol.* **80,** 279–297.

Wood, C. M. (1991). Acid–base and ion balance, metabolism and their interactions, after exhaustive exercise in fish. *J. Exp. Biol.* **160,** 285–308.

Wood, C. M., and Perry, S. F. (1991). A new in vitro assay for CO_2 excretion by trout red blood cells: Effects of catecholamines. *J. Exp. Biol.* **157,** 349–366.

Wood, C. M. (1994). HCO_3^- dehydration by the blood of rainbow trout following exhaustive exercise. *Respir. Physiol.* **98,** 305–318.

Wood, C. M., and Munger, R. S. (1994). Carbonic anhydrase injection provides evidence for the role of blood acid–base status in stimulating ventilation after exhaustive exercise in rainbow trout. *J. Exp. Biol.* **194,** 225–253.

Wood, C. M., Perry, S. F., Walsh, P. J., and Thomas, S. (1994). HCO_3^- dehydration by the blood of an elasmobranch in the absence of a Haldane effect. *Respir. Physiol.* **98,** 319–337.

Wood, C. M., and Simmons, H. (1994). The conversion of plasma bicarbonate to CO_2 by rainbow trout red blood cells in vitro: Adrenergic inhibition and the influence of oxygenation status. *Fish Physiol. Biochem.* **12,** 445–454.

Wright, P. A., Heming, T., and Randall, D. J. (1986). Downstream pH changes in water flowing over the gills of rainbow trout. *J. Exp. Biol.* **126,** 499–512.

Wright, P. A., Randall, D. J., and Perry, S. F. (1989). Fish gill water boundary layer: A site of linkage between carbon dioxide and ammonia excretion. *J. Comp. Physiol. B* **158,** 627–635.

Wyman, J. (1964). Linked functions and reciprocal effects in hemoglobin: A second look. *Adv. Protein Chem.* **19,** 223–286.

8

THE LINKAGE BETWEEN OXYGEN AND CARBON DIOXIDE TRANSPORT

C. J. BRAUNER
D. J. RANDALL

I. Introduction
II. Fundamental Basis for the Linkage
    A. Bohr Effect
    B. Haldane Effect
III. Hb Characteristics That Influence the Magnitude of the Linkage
    A. Magnitude of Bohr–Haldane Coefficients in Different Fish Species
    B. Root Effect
    C. Hb Buffer Values
IV. Physiological Basis for the Linkage
    A. Theoretical Optimal Bohr–Haldane Coefficient
    B. Gas Exchange and the Linkage between O_2 and CO_2 *in Vivo*
    C. Effect of Physiological States and Environmental Conditions on the Linkage
V. Conclusion
    References

I. INTRODUCTION

Animals consume oxygen and produce carbon dioxide, usually in about the same quantities. In the blood, oxygen is transported to the tissues and carbon dioxide is carried away from the tissues. The interaction between oxygen (O_2) and carbon dioxide (CO_2) exchange in the blood of vertebrates has been a subject of intense investigation for most of this century. In 1904, the pioneering work of Bohr, Hasselbalch, and Krogh demonstrated that CO_2 reduced the affinity of hemoglobin (Hb) for O_2 (Bohr effect) in the blood of humans. Shortly thereafter, it was discovered that deoxygenated blood contained a greater CO_2 content than oxygenated blood (Haldane effect; Christiansen *et al.,* 1914), and it was realized that Hb plays a central role in both O_2 and CO_2 transfer. The magnitude of the interaction between

Fish Physiology, Volume 17:
FISH RESPIRATION

O_2 and CO_2 exchange has been examined in all vertebrate classes, but no single class possesses quite the diversity of that seen in Pisces. In some fish, there is little interaction either due to the absence of Hb in the blood entirely (i.e., ice fish) or due to the presence of Hb with very small Bohr–Haldane effects (i.e., dogfish). In other fish (i.e., carp and tench), the magnitude of the Bohr–Haldane effects is among the highest observed in vertebrates. Most fish species possess a number of different Hb isomorphs (multiple Hbs), all of which are present within each red cell (Brunori *et al.*, 1974). The magnitude of the interaction *in vivo* is a result of the sum of the individual components and the magnitude of their Bohr–Haldane coefficients. In contrast with the blood of most vertebrates, the magnitude of the Bohr–Haldane effect varies with O_2 saturation in the blood of some teleost fishes, which has implications for the interaction between O_2 and CO_2 exchange and acid–base homeostasis. The great diversity in Hb characteristics in the class Pisces results in large variations in the extent of the interaction between O_2 and CO_2 exchange in the blood of fish.

Another factor increasing the complexity of the interaction between O_2 and CO_2 is that these gases are not always exchanged across the same respiratory surface. In many air-breathing fish, most of the carbon dioxide is excreted across the gills, whereas oxygen is taken up across the air-breathing organ. In addition, tissues do not always produce similar quantities of the two gases. The swim bladder gas gland produces carbon dioxide through the pentose phosphate shunt but consumes little oxygen; this system is involved in oxygen secretion into the swim bladder. There is a complex series of interactions between oxygen and carbon dioxide that facilitate oxygen secretion into the swim bladder against large oxygen differences between blood and swim bladder (see Pelster and Randall, Chapter 3 this volume). The delivery of oxygen to the eye of some fish shows similarity to oxygen secretion in the swim bladder. Both the eye and the swim bladder involve vascular countercurrent exchangers that result in the development of localized high oxygen partial pressures in the blood.

Thus, because of their wide range of hemoglobin types, diversity of breathing modes, and respiratory exchange sites, as well as their capacity to develop high oxygen levels in some tissues, fish constitute an interesting group in which to investigate the interactions between oxygen and carbon dioxide transport.

II. FUNDAMENTAL BASIS FOR THE LINKAGE

A. Bohr Effect

The affinity of Hb for oxygen is decreased when ligands such as Cl^-, organic phosphates, H^+, and CO_2 bind to specific sites on the Hb molecule (see Jensen, Fago, and Weber, Chapter 1 this volume). The Bohr effect (Riggs, 1988) describes the change (Δ) in Hb–O_2 affinity that results from a change in H^+ concentration and is often expressed numerically as $\Delta \log P_{50}/\Delta pH$ (Bohr coefficient), where P_{50} is the P_{O_2} at which 50% of the Hb molecule is saturated, and pH refers to that of the Hb solution. When the Bohr coefficient is measured in whole blood, ΔpH of the plasma is often used rather than that of the red cell. The use of ΔpH of the plasma underestimates the true pH dependence of Hb–O_2 affinity because blood consists of two compartments where $\Delta pH_{red\ cell}/\Delta pH_{plasma}$ is less than unity (Albers et al., 1983). Standardization of the Bohr coefficient to red cell pH solves this problem (Nikinmaa, 1990), but requires an understanding of the relationship between red cell and plasma pH, which varies interspecifically and with changes in intracellular NTP (Albers et al., 1983; Jensen, 1988). Further ambiguity in the use of P_{50} in the Bohr effect calculation arises due to the presence of the Root effect (reduction in maximal Hb–O_2 saturation at low pH; see Pelster and Randall, Chapter 3 this volume) in the blood of many fishes. The Root effect refers to the fact that at low pH, Hb cannot be completely saturated even in pure oxygen at several atmospheres (Root, 1931; Scholander and Van Dam, 1954). Thus, in Root effect Hbs at low pH, P_{50} is often calculated as the P_{O_2} at which 50% of the Hb can be saturated under the experimental conditions, which may be quite different from the P_{O_2} required to saturate 50% of the heme groups (Nikinmaa, 1990). The presence of the Root effect only complicates the calculation of the Bohr coefficient significantly at acid pH values. Provided that the Bohr coefficient is calculated at pH values near and above in vivo pH values, or that true P_{50} values are calculated, this problem can be avoided.

The magnitude of the Bohr effect can be measured in different ways. The fixed-acid Bohr effect is calculated by measuring the change in Hb–O_2 affinity when pH is changed by adding a strong acid or base at fixed P_{CO_2}. The CO_2 Bohr effect is measured by changing the P_{CO_2} of the gas incubating the sample and results from a combination of H^+ and CO_2 binding (Hlastala, 1984). In the Hbs of many vertebrates, CO_2 binds directly to the terminal amino groups of the Hb chains (oxylabile carbamate formation) inducing a conformational change in the Hb molecule, which reduces Hb–O_2 affinity. The Hbs of most fish, however, do not form much oxylabile carbamate, and there is only a small difference between the fixed acid and CO_2 Bohr effects in Hb solutions in the presence of organic phosphates (see Weber and Jensen, 1988). Oxylabile carbamate formation is minimal in fish because the terminal amines of the α-Hb subunits are acetylated and unavailable, whereas those on the β-chains preferentially bind with organic phosphates present within the red cell (Gillen and Riggs, 1973; Farmer, 1979; Weber and Lykkeboe, 1978). Finally, P_{CO_2} levels in fish are very low, further reducing the likelihood of carbamate formation *in vivo* (Heming *et al.,* 1986).

The magnitude of the fixed-acid Bohr effect varies markedly with pH and has two parts in the pH range between 5 and 10 in the blood of mammals. Above pH 6.5, it is called the "alkaline Bohr effect," which is generally characterized by a decrease in $Hb-O_2$ affinity with a reduction in pH. The presence of organic phosphates (usually ATP and/or GTP in fish, collectively referred to as NTP) generally increases the magnitude of the alkaline Bohr effect. In Hb components of some fish, such as the eel (*Anguilla anguilla*), the alkaline Bohr effect may be opposite (referred to as a "reverse alkaline Bohr effect") whereby $Hb-O_2$ affinity increases with a reduction in pH. The reverse alkaline Bohr effect is virtually eliminated in the presence of organic phosphates at concentrations found *in vivo* (Fago *et al.,* 1995; Feuerlein and Weber, 1996). The acid Bohr effect occurs at pH values less than 6.5 and, in most vertebrates, is characterized by a reverse Bohr effect. This phenomenon is virtually absent in fish Hbs, many of which have a marked Root effect where $Hb-O_2$ affinity continues to decrease with a reduction in pH. This review is centered on the interaction between O_2 and CO_2 exchange *in vivo,* and thus deals predominantly with the alkaline Bohr effect, which will be referred to simply as the Bohr effect from this point forward.

B. Haldane Effect

The Haldane effect refers to the fact that deoxygenated blood contains more CO_2 than oxygenated blood (Christiansen *et al.,* 1914). Deoxygenation of Hb results in an increase in the pK of specific ionizable groups resulting in the binding of protons and the formation of carbamate (see Klocke, 1987, for a review). As described, oxylabile carbamate formation is minimal in the Hb of fish, and the Haldane effect results predominantly from oxylabile proton binding (Weber and Jensen, 1988). The Haldane effect can be quantified directly by measuring the difference in CO_2 content between completely oxygenated and deoxygenated whole blood at constant P_{CO_2} (normalized for Hb concentration), or by measuring the distance between fixed acid titration curves for oxygenated and deoxygenated blood at constant pH and P_{CO_2} (mole H^+/mole of Hb). Figure 1 illustrates the Haldane effect in the most common (anodic) Hb component of the eel. GTP increases the magnitude of the Haldane effect and brings the pH at which the maximum effect is seen into the physiological pH range of the red cell. This action of GTP has also been demonstrated in the blood of tench (Jensen and Weber, 1985) and carp (Chien and Mayo, 1980).

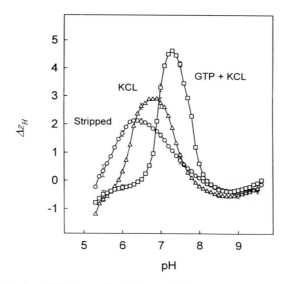

Fig. 1. The fixed acid Haldane effect (ΔZ_H, mol H[+] mol Hb$_4$ oxygenated) as a function of pH for the anodic Hb component of *Anguilla anguilla* in distilled water (stripped, circles), in 0.1 M KCl (triangles) and in 0.1 M KCl with GTP at a molar ratio of $3:1$ GTP:Hb$_4$ (squares). ΔZ_H was calculated from the vertical distance between the oxygenated and deoxygenated titration curves. Each data point indicates the mean from three curves with the standard error of the mean indicated by the vertical bars every 0.5 pH unit. Brauner and Weber, unpublished data.

Factors that alter the Haldane effect will also similarly influence the Bohr effect because at the molecular level, they are mirror images of the same phenomena. While the Bohr effect describes the change in Hb–O_2 affinity that results from a change in H^+ concentration (and therefore Hb–H^+ binding), the Haldane effect describes the change in Hb–H^+ affinity that results from a change in O_2 partial pressure (and therefore Hb–O_2 saturation). The linkage between the Bohr and Haldane effects has long been recognized and is illustrated by the classic Wyman linkage equation (Wyman, 1973)

$$(\log P_{O_2}/\text{pH})_Y = (H^+/Y)_{\text{pH}},$$

where Y is Hb–O_2 saturation, and H^+ is the number of protons (referred to as Bohr protons from this point forward) bound per heme. Assuming that the shape of the oxygen equilibrium curve (OEC) is symmetrical and pH independent and that proton release is linearly related to O_2 binding (Wyman, 1973) the preceding linkage equation is commonly reduced to

$$-\Delta\log P_{50}/\Delta pH = \Delta H^+,$$

where ΔH^+ is the number of protons released per mole of O_2 bound to Hb (Haldane coefficient). Thus, the Bohr and Haldane coefficients are thermodynamically equivalent in Hb solutions. Whole blood is a much more complicated system than a Hb solution because it consists of two phases. Sigaard-Anderson (1974) has demonstrated, however, that the Bohr and Haldane coefficients of whole blood are proportional to those of Hb solutions.

Typically, a single Bohr and Haldane coefficient is calculated for the entire range of Hb–oxygen saturations (i.e., the Haldane coefficient is calculated from two Hb saturations, oxygenated and deoxygenated). Experimental evidence for the equivalence of the Bohr and Haldane coefficients over a wide range of oxygenation levels has been obtained in human Hb (Tyuma *et al.*, 1973; Tyuma and Ueda, 1975). Few similar data are available for the Hbs of fish. The only study in which Bohr and Haldane coefficients have been independently, but simultaneously, measured in fish indicates that the negative Bohr coefficient is similar to, but slightly larger than, the Haldane coefficient between pH 6 and 8 (Jensen and Weber, 1985).

The relationship between Bohr and Haldane coefficients over discrete ranges of Hb oxygenation has not been investigated in fish Hbs. Owing to the presence of the Root effect, the OEC is very pH-dependent and proton release during oxygen binding is far from linear. At least some fish Hbs, therefore, do not meet the assumptions of the linkage equation, and the equivalence of the Bohr and Haldane coefficients over discrete ranges of Hb oxygenation remains to be investigated. To simplify the discussion, it will be assumed that the Bohr and Haldane coefficients are identical over discrete ranges of the OEC, and they will collectively be referred to as the Bohr–Haldane coefficient and be reported positive from this point forward.

III. HB CHARACTERISTICS THAT INFLUENCE THE MAGNITUDE OF THE LINKAGE

The binding and release of protons as hemoglobin is deoxygenated and oxygenated will influence the extent of carbon dioxide hydration/dehydration within the red blood cell and influence conditions for O_2 uptake and delivery. Thus, the magnitude of the interaction between O_2 and CO_2 exchange *in vivo* will be determined to a large extent by the magnitude of the Bohr–Haldane effect, which in some fish species is nonlinear over the OEC. In addition, the magnitude of the interaction will be influenced by the Root effect and the buffer value of the Hb at fixed oxygenation status.

A. Magnitude of Bohr–Haldane Coefficients in Different Fish Species

The magnitude of the Bohr–Haldane coefficient in fish varies dramatically between species. Most of the Bohr coefficients reported in the literature have been calculated as the change in log P_{50} relative to plasma pH, which underestimates the true Bohr–Haldane coefficient as described earlier. Where possible, true Bohr–Haldane coefficients in the following section have been calculated by expressing the reported values relative to red cell pH. The true values are indicated in parentheses following the reported values. In some elasmobranchs (Lai *et al.,* 1990; Wood *et al.,* 1994), the Bohr–Haldane effect is virtually absent. In teleost fishes, however, it can range from values of around 0.35 (Lapennas, 1983), which are typical of other vertebrates, to moderate values, such as 0.52 (1.10 assuming delta pH red cell/delta pH plasma = 0.47; Heming *et al.,* 1986) in trout (*Oncorhynchus mykiss;* Weber *et al.,* 1976a) and 0.59 in the tuna (*Euthynnus affinis;* Jones *et al.,* 1986), and to extremely high values, such as 0.73 (1.2, assuming $\Delta pH_{red\ cell}/\Delta pH_{plasma} = 0.61$) in tench (*Tinca tinca;* Jensen and Weber, 1982), 0.98 (1.4, assuming $\Delta pH_{red\ cell}/\Delta pH_{plasma} = 0.69$; Albers *et al.,* 1983) in carp (*Cyprinus carpio;* Weber and Lykkeboe, 1978), and 1.10 in the intertidal rockpool blennid (*Blennius pholis,* Bridges *et al.,* 1984). In a survey of 34 species of fishes from the Amazon, spanning 32 genera, the Bohr–Haldane coefficients were less than 0.2 in 6 species (predominantly catfish), between 0.2 and 0.3 in 8 species, between 0.3 and 0.4 in 8 species, between 0.4 and 0.5 in 4 species, between 0.5 and 0.6 in 8 species, and greater than 0.6 in 3 species (Powers *et al.,* 1979). Thus, there was a rather even distribution of fish possessing Bohr–Haldane coefficients (relative to plasma pH) of between 0.2 and 0.6. Assuming a constant $\Delta pH_{red\ cell}/\Delta pH_{plasma}$ in these Amazonian species of 0.61, the value found in tench (Jensen and Weber, 1982), this corresponds to a roughly even distribution of Bohr–Haldane coefficients between a value of 0.33 and 0.98.

The blood system in most fishes comprises multiple Hb types, and the values reported here are a result of this mixture. Furthermore, it has traditionally been assumed that the Bohr–Haldane coefficient calculated over the entire range of the OEC is constant over discrete ranges of Hb oxygenation. In the blood of many teleost fishes, however, this is clearly not the case. The nonlinear Bohr–Haldane effect has important consequences in terms of the linkage between O_2 and CO_2 exchange *in vivo.*

Nonlinear Bohr–Haldane Effect in Teleost Fishes

Under physiological conditions, the Haldane effect may occur predominantly in the upper region of the OEC. That is, in both trout and tench, the majority of Bohr protons are released between about 50 and 100% saturation (Jensen, 1986; Fig. 2a). The nonlinear Haldane effect may arise from the presence of multiple Hb components with different O_2 affinities and Bohr–Haldane coefficients, or it could be due to a truly nonlinear release of Bohr protons within a single Hb component.

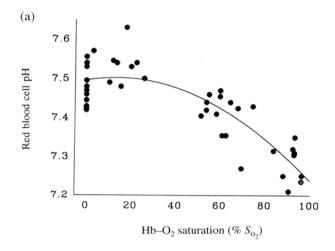

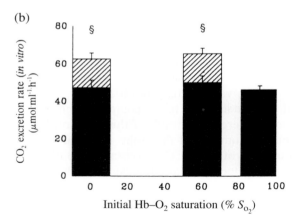

Fig. 2. (a) Relationship between red cell pH (pH_i) and Hb–O_2 saturation ($\%S_{O_2}$) in the blood of rainbow trout *in vitro* at constant P_{CO_2}. (b) Whole blood CO_2 excretion rate measured *in vitro* with (cross-hatch) and without (solid bars) oxygenation following incubation at respective Hb–O_2 saturation ($\%S_{O_2}$). [§]statistically significant difference between solid and cross-hatch bars at respective incubation saturation. Modified from Brauner *et al.,* (1996).

Multiple Hb Components. The blood of most fishes comprises a number of different Hb isomorphs. In a survey of teleost fishes from the Amazon, of 77 genera examined, only 8% possessed a single Hb component, whereas the mean value was four components per species (Fyhn *et al.,* 1979). Within an individual, Hbs can be separated into two general categories depending upon electrophoretic characteristics. The cathodic components generally exhibit high oxygen affinities and small Bohr–Haldane coefficients, whereas the anodic components are characterized by a low oxygen affinity, large Bohr–Haldane coefficients, and often the presence of a Root effect (Gillen and Riggs, 1973; Weber *et al.,* 1976a, 1976b; Giles and Randall, 1980). The large difference in pH sensitivity between the two components indicates that there may be a division of labor with respect to oxygen transport, with the cathodic component securing oxygen uptake at the gills following the development of a metabolic acidosis (Weber, 1990).

In whole blood containing multiple Hb components, the Bohr–Haldane coefficient at different $Hb-O_2$ saturations is determined by the fractional oxygen saturation and the magnitude of the Bohr–Haldane coefficient of the respective components at that saturation. That is, at low P_{O_2}, the cathodic component(s) may be completely saturated, whereas the anodic component(s) is only partially saturated resulting in a small Haldane coefficient at that saturation. As P_{O_2} (and therefore saturation of the anodic component) increases, so will the magnitude of the Haldane coefficient, resulting in a nonlinear Bohr–Haldane effect over the OEC. The nonlinearity will be significant only in fish that possess considerable quantities of both components such as in adult eel (*Anguilla anguilla*) and trout (*O. mykiss*), where about two-thirds of the components are anodic (Pelster and Weber, 1990).

Nonlinearity within a Component. The nonlinear Bohr–Haldane effect can also occur in the blood when a different number of Bohr protons are released at each of the four steps of tetramer oxygenation. Bohr proton release at each step of tetramer oxygenation can be estimated by the successive oxygenation theory of Adair (1925). The Hbs of most vertebrates exhibit some degree of disparity in Bohr proton release at the successive steps of oxygenation, but this will be observed in whole blood (as a nonlinear Bohr–Haldane effect) only if the cooperativity of oxygen binding is low (indicated by the Hill number, n), where populations of Hb can exist at various intermediate stages of oxygenation. If n is high, the Hb tetramer will be either completely oxygenated or deoxygenated, and Bohr proton release in a Hb solution or whole blood will appear linear with oxygenation regardless of H^+ release at individual steps (Imai and Yonetani, 1975). For example, in human stripped Hb, there are almost no protons released in the fourth step of oxygenation and approximately equal numbers in steps 1, 2, and 3 (Chu *et al.*, 1984). In a Hb solution, however, Bohr proton release is virtually linear between 20 and 95% S_{O_2} with slightly more protons being released between 0 and 20% S_{O_2}, and slightly less between 95 and 100% S_{O_2} (Tyuma *et al.*, 1973; Tyuma and Ueda, 1975). The measured nonlinear Haldane effect observed in humans and most other vertebrates is minor in comparison with that observed in the blood of teleost fishes.

In the stripped Hb of tuna *(Thunnus thynnus)* at pH 8, when cooperativity is quite high (nH = 2.8), Bohr proton release appears linear with Hb oxygenation despite the greatest release of Bohr protons at step 3 of oxygenation (Ikeda-Saito *et al.*, 1983). With a reduction in pH, n decreases, which is accompanied by an increase in Bohr proton release at steps 3 and 4 of oxygenation, thereby resulting in the majority of Bohr protons being released in the upper region of the OEC in a Hb solution. At pH 7, when n is about 1.5, all proton release occurs between 40 and 100% S_{O_2}. A similar observation has been made in stripped Hb of carp (Chien and Mayo, 1980).

In stripped hemolysates of menhaden (*Brevoortia tyrannus*) at physiological pH, about 70% of the Bohr protons are released between 50 and 100% saturation (Saffran and Gibson, 1981). In tench hemolysates at physiological pH, more than 50% of Bohr protons are released at the 3rd step of Hb oxygenation alone, and this increases in the presence of ATP or GTP (Weber *et al.*, 1987). In tench red cells *in vitro,* almost all Bohr protons are released between 50 and 100% S_{O_2}, which is not a result of multiple Hbs because 90% of tench Hb consists of one component (Jensen and Weber, 1982). The nonlinear release of Bohr protons within a given Hb component in many teleost fishes results from a relatively large release of Bohr protons in the latter steps of Hb oxygenation coupled with a low n.

The low n (2 or less) is a general characteristic of fish Hbs (Brittain, 1991), and an n of 1 (or less) at low pH is a defining characteristic of the Root effect (Brittain, 1987). In addition, Root effect Hbs are characterized by a larger release of Bohr protons at step 3 (and step 4 at low pH; Ikeda-saito *et al.*, 1983; Weber *et al.*, 1987) than those of non-Root effect Hbs (Imai and Yonetani, 1975). Thus, there may be a general relationship between the presence of a Root shift and the large nonlinear release of Bohr protons in whole blood; however, further experiments are required to validate this idea.

B. Root Effect

The physiological definition of the Root effect is that a reduction in pH of the blood reduces Hb saturation at atmospheric oxygen tension (Root, 1931; Brittain, 1987). The Root effect is exclusive to the blood of fishes and is best correlated with the presence of a rete where a countercurrent system results in a localized acidosis in the blood that drives oxygen from Hb to the eye or swim bladder (see Pelster and Randall, Chapter 3, this volume). Although large reductions in pH are observed in the rete, the Root effect operates at *in vivo* pH values observed in the general circulation in many fishes. At the molecular level, the Root effect is generally considered to be an exaggerated alkaline Bohr effect (Brittain, 1987) where a reduction in pH shifts the OEC so far to the right that complete saturation of Hb cannot be achieved even in the presence of 100 atmospheres of pure O_2 (Scholander and Van Dam, 1954). The presence of organic phosphates increases the magnitude of the Root effect and increases the pH at which the Root effect is observed (Pelster and Weber, 1990).

C. Hb Buffer Values

The most significant buffers in the blood of most vertebrates are bicarbonate and Hb. In fish, P_{CO_2} and HCO_3^- levels are much lower than those in air-breathing vertebrates, and consequently the bicarbonate buffering capacity of the blood is much lower. The buffer value of Hb at constant oxygenation status varies between fish species. In dogfish Hb, the buffer value is quite high and is greater than that of pig Hb (Jensen, 1989) and human Hb (Breepoel *et al.*, 1980) at physiological pH values. In teleost fishes such as tench (Jensen and Weber, 1985), carp, trout (Jensen, 1989), eel (Breepoel *et al.*, 1980), and tuna (Brunori, 1966), Hb buffer values are considerably lower. At physiological pH values, the buffer value of Hb is determined by imidazole groups of histidine residues and the terminal amines. Low buffer values in teleost Hbs correlate with acetylation of the α-amino groups of the α chains and a lower number of histidines than in other vertebrates (Jensen, 1989). Only some of the histidines are titratable. For example, in carp, only 7 of a total of 18 histidine residues present are titratable (Jensen, 1989), whereas in the hemolyzates of eel, the number of titratable histidines may be as low as 4 (calculated from Breepoel *et al.*, 1980). This is considerably lower than the 20 titratable histidines of a total of 38 in human Hb and 22 of 38 in horse Hbs (Janssen *et al.*, 1972).

All teleost species mentioned earlier are characterized by large Haldane effects, and there appears to be an inverse relationship between buffer values and the magnitude of the Haldane coefficient (Jensen, 1989). That is, carp and eel, which have the lowest buffer value, have the largest Haldane effects; trout is intermediate in both parameters; and dogfish, which have a high buffer value, have virtually no Haldane effect (Jensen, 1989). A combination of a low Hb buffer value and a large Haldane effect results in a large influence of Hb oxygenation status on red cell pH. For example, in trout, red cell pH in oxygenated blood at constant P_{CO_2} is 0.21 pH units lower than that in deoxygenated blood (Brauner et al., 1996). In carp, the difference in red cell pH between oxygenated and deoxygenated blood is 0.27 pH unit (Albers et al., 1983), and in tench, the difference is 0.35 pH unit (Jensen, 1986). For comparison, in human red cells incubated in the absence of CO_2, there is a decrease in red cell pH of only 0.07 pH unit at pH_e 7.4 and 0.09 pH unit at pH_e 7.9 (Duhm and Gerlach, 1971) due to the relatively greater Hb buffer value and lower Haldane effect.

IV. PHYSIOLOGICAL BASIS FOR THE LINKAGE

A. Theoretical Optimal Bohr–Haldane Coefficient

The Bohr–Haldane effect will influence respiratory gas transport, as well as blood acid–base homeostasis. At the tissue level, metabolically produced CO_2 acidifies the blood, which induces a right shift of the OEC (Bohr shift) and facilitates oxygen delivery to the tissues. The benefit to oxygen delivery is determined by the product of the Bohr coefficient and the magnitude of the pH change in the blood. Conventionally, a large Bohr coefficient has been assumed to be beneficial for oxygen delivery; however, the associated large Haldane effect will limit the magnitude of the pH change as oxygen is delivered to the tissues. Thus, because the Bohr and Haldane effects are linked functions, a large Bohr coefficient does not occur in conjunction with a large arterial–venous difference in pH in normally metabolizing tissues (i.e., tissues that possess a respiratory quotient (RQ) between 0.7 and 1.0). Consequently, a number of analyses have been conducted on the Hbs of mammals to determine whether the Bohr–Haldane effect functions predominantly for O_2 delivery or for buffering pH and P_{CO_2} changes in the blood. The results of several studies have led to the conclusion that the Bohr effect plays a much greater role in buffering pH and P_{CO_2} changes than in promoting oxygen delivery to the tissues (Hill et al., 1973) because of the minimal change in blood pH during capillary blood transit in the absence of an acute acid–base disturbance (Grant, 1982). In contrast, Bartels (1972) estimated a large benefit of the Bohr effect for oxygen delivery to the tissues, especially during heavy exercise when the measured arterial–venous pH difference in humans approaches 0.14 to 0.2 pH unit. Lapennas (1983) has argued that the Bohr effect is optimized for oxygen delivery in most animals based upon the equivalence between measured Bohr coefficients and that which was theoretically determined to be optimal for oxygen delivery.

The optimal Bohr coefficient for tissue oxygen delivery was determined to be $0.5 \times RQ$ (Lapennas, 1983). Because RQ is generally between 0.7 and 1.0, the optimal Bohr coefficient for O_2 delivery is between 0.35 and 0.5 (relative to plasma pH), respectively. Lapennas's analysis applies to blood that has a constant Bohr–Haldane effect over the entire OEC and does not form oxylabile carbamate. Although the latter is true in the blood of many fishes (Gillen and Riggs, 1973; Farmer, 1979; Weber and Lykkeboe, 1978; Weber and Jensen, 1988), the application of this optimal value in fish blood *in vivo* is complicated by the existence of the nonlinear Bohr–Haldane effect. Furthermore, there is no doubt that the Root effect (which is usually correlated with a Bohr–Haldane coefficient much greater than 0.5), which exists in the blood of many fishes, is a characteristic that optimizes O_2 delivery in the swim bladder and the eye (see Pelster and Randall, Chapter 3, this volume). These structures have acid-generating mechanisms and at a given level of blood acidification in the rete, the greater the Root effect (and, therefore, the Bohr–Haldane coefficient), the greater the amount of oxygen driven to the respective structure. Lapennas's analysis, however, does provide a useful framework for inferring how different Bohr–Haldane coefficients, between species and over different regions of the OEC, influence conditions for gas exchange and acid–base homeostasis at the gills and the tissues *in vivo,* as discussed next.

B. Gas Exchange and the Linkage between O_2 and CO_2 *in Vivo*

O_2 and CO_2 exchange in blood is a complex combination of processes including diffusion, convective mixing, and chemical reactions (see Tufts and Perry, Chapter 7 this volume). Some of these processes occur in parallel, many occur in series, and none can be viewed in isolation because a change at one point may have effects throughout the entire system. Most observations are made under steady-state conditions, and, in most cases, we can only infer a probable consequence of interactions between the various reactions. For example, we do not know the exact time course of carbon dioxide removal as blood flows through the gills and how it relates to the time course of oxygen uptake. Thus, the following must be viewed as possible consequences of interactions between oxygen and carbon dioxide transfer.

INFLUENCE OF HB OXYGENATION ON CO_2 EXCRETION

When blood enters the gills, physically dissolved CO_2 that exists in prebranchial blood (Pb_{O_2}) rapidly diffuses across the respiratory epithelium into the ventilatory water (Fig. 3). The remaining CO_2 excreted is dependent upon HCO_3^- dehydration. In trout, and some other teleost fishes, carbonic anhydrase (CA), which catalyzes HCO_3^- dehydration, is not accessible to the plasma in the gills (Perry *et al.*, 1982; Henry *et al.*, 1988; Perry and Laurent, 1990; Perry *et al.*, 1997), but exists in very high concentrations within the red blood cell. At the low temperatures and rapid transit times seen in fish gills, plasma bicarbonate dehydration is negligible, and virtually all bicarbonate dehydration occurs within the red blood cell. Dogfish, on the other hand, have carbonic anhydrase activity in the plasma, (Wood *et al.*, 1994) and somewhat lower red blood cell carbonic anhydrase activity than that observed in trout erythrocytes. Presumably, bicarbonate dehydration in plasma is much more important during CO_2 excretion in dogfish than in trout (Perry *et al.*, 1996).

As the concentration of intracellular HCO_3^- is reduced, HCO_3^- enters the red cell in exchange for Cl^- via the band 3 protein, the slowest single step in CO_2 excretion (Wieth *et al.*, 1982; see Tufts and Perry, Chapter 7 this volume). Continued CO_2 excretion is dependent upon the rate at which HCO_3^- and protons can be replenished within the red cell (Perry and Gilmour, 1993). The rate of HCO_3^- entry into the red cell is proportional to the HCO_3^- gradient across the red cell membrane, which in turn is determined by the pH gradient across the red cell. The removal of protons

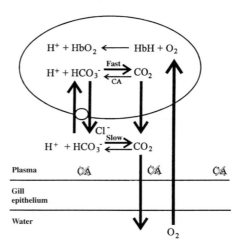

Fig. 3. A diagrammatic representation of gas exchange at the gills. Oxygen diffuses into the red cell and binds to Hb, releasing Bohr protons (H^+). The Bohr protons are consumed during HCO_3^- dehydration, and CO_2 subsequently diffuses into the ventilatory water. The reverse scenario occurs in the tissues. CA, carbonic anhydrase; Hb, hemoglobin. From Brauner, (1995).

during HCO_3^- dehydration will raise red cell pH and reduce the pH gradient, and therefore the HCO_3^- gradient, across the red cell unless protons can be replaced. Proton flux across the red cell is slow relative to blood transit time through the gills, and therefore protons are titrated from Hb, augmented by Hb oxygenation. The Hbs of teleost fishes characteristically possess a low buffer value and a large Haldane effect, and therefore protons required during CO_2 excretion are supplied predominantly by the Haldane effect. In contrast, the high buffering value of dogfish plasma (Graham *et al.*, 1990) will reduce carbon dioxide flux through the red cell and therefore the coupling of oxygen and carbon dioxide transport in the blood. Thus, in the dogfish, protons consumed during HCO_3^- dehydration are supplied predominantly from the high buffer value of both Hb and plasma, the contribution of the Haldane effect being small.

The Haldane effect has long been implicated in CO_2 excretion in vertebrates (Klocke, 1987). In 1914, Christiansen, Douglas, and Haldane stated: "The oxygenation of blood in the lungs helps to drive out CO_2 and increases by 50% or slightly more the amount of CO_2 given off at each round of the circulation." The relatively larger Haldane effect and lower Hb buffer value in many teleost fishes (Jensen, 1989, 1991) implies an even greater dependence of CO_2 excretion on $Hb–O_2$ binding than that in mammals, which in principle could compensate for a relatively slow rate of Cl^-/HCO_3^- exchange (Jensen, 1989; Jensen and Brahm, 1995).

The influence of Hb oxygenation on red cell HCO_3^- entry and subsequent CO_2 excretion from blood can be qualitatively demonstrated using the *in vitro* CO_2 excretion assay of Wood and Perry (1991), which measures the rate of appearance of radiolabeled $^{14}CO_2$ in a trap following the addition of $H^{14}CO_3^-$ to the plasma under controlled gas tensions. Although the assay does not exactly simulate gas exchange at the gills *in vivo* (see Perry and Gilmour, 1993, for a critique of the assay), it does provide a relative index of HCO_3^- flux across the red cell under the conditions of the assay. In four species of fishes, there is a linear relationship between the magnitude of the Haldane effect and the magnitude of the oxygenation-induced elevation in CO_2 excretion rate (Fig. 4). Thus, under the conditions of the assay, Hb oxygenation clearly elevates CO_2 excretion in fish that possess a large Haldane effect and low buffer value. Interestingly, the *in vitro* CO_2 excretion rates obtained using this assay in the presence or absence of oxygenation were the lowest in dogfish that possessed the smallest Haldane effect and largest plasma and Hb buffer values (Graham *et al.*, 1990; Jensen, 1991). Given that the rate of Cl^-/HCO_3^- exchange in the red cells of dogfish is similar to that of human red cells (Obaid *et al.*, 1979) and slightly faster than that in teleost fishes (Jensen and Brahm, 1995), a large Haldane effect

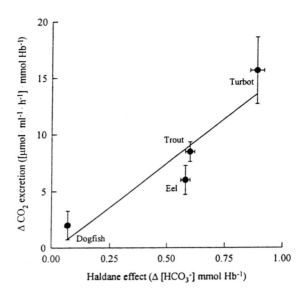

Fig. 4. The relationship between the magnitude of the Haldane effect and the elevation in CO_2 excretion rate associated with oxygenation of the blood (ΔCO_2) during the *in vitro* assay ($r^2 = 0.90$). The Haldane effect was measured as the difference in true plasma HCO_3^- between oxygenated and deoxygenated blood, normalized for Hb concentration. ΔCO_2 was measured as the difference in rate of CO_2 excretion between deoxygenated blood and deoxygenated blood that was rapidly oxygenated. From Perry *et al.* (1996), reprinted by permission of Elsevier Science Inc.

and low Hb buffer value may be associated with an increased capacity for CO_2 excretion in the blood of fishes.

In rainbow trout blood, oxygenation of deoxygenated blood elevates CO_2 excretion rate by between 30 and 40% over that measured in the absence of oxygenation (Perry and Gilmour, 1993; Brauner *et al.*, 1996). This oxygenation-induced elevation in CO_2 excretion rate occurs between 60 and 100% S_{O_2}, consistent with the region of the OEC where Bohr protons are released in trout (Fig. 2). Thus, in resting fish, where venous S_{O_2} is around 50%, the entire Haldane effect may be exploited during blood transit through the gills.

In many vertebrates, oxylabile carbamate (reversibly bound CO_2) may constitute between 13 and 20% (Klocke, 1987) of total CO_2 excreted; however, this pathway is virtually absent in most teleost fishes (Gillen and Riggs, 1973; Farmer, 1979; Weber and Lykkeboe, 1978; Weber and Jensen, 1988). Furthermore, there is no plasma accessible CA in the gills of teleost fishes (Perry *et al.*, 1982; Henry *et al.*, 1988; Perry and Laurent, 1990; Perry *et al.*, 1997), whereas that present in the lungs of vertebrates permits HCO_3^- dehydration in the plasma compartment, which may account for as much as 10% of the total CO_2 excreted (Crandall and Bidani, 1981). The low dependence on oxylabile carbamate and lack of plasma-accessible CA in the gills increase the dependence on red cell Cl^-/HCO_3^- exchange during CO_2 excretion relative to that in air-breathing vertebrates and fish such as dogfish. The increased dependence on Cl^-/HCO_3^- exchange during CO_2 excretion, in combination with a low Hb buffer value and large Haldane effect, results in a tight interaction between O_2 and CO_2 exchange in the blood of many fishes.

IMPORTANCE OF CO_2 Removal to O_2 Uptake at
the Gills

Oxygen uptake at the gills is dependent upon CO_2 removal due to the presence of the Root effect that exists in the blood of many fishes. In trout blood maintained at constant P_{CO_2}, the red cell pH of oxygenated blood is 0.21 pH units lower than deoxygenated blood (Brauner *et al.*, 1996). This pH difference illustrates the magnitude of the acidosis associated with Bohr proton release during oxygenation of the blood in the absence of CO_2 removal. A reduction in pH of this magnitude could, in theory, significantly reduce blood oxygen-carrying capacity in trout due to the presence of a large Root effect (Brauner and Randall, 1996). A limitation in oxygen uptake to this degree will never be realized *in vivo* because the Bohr protons are consumed during HCO_3^- dehydration and subsequent CO_2 removal (with the exception of air-breathing fishes), but this example serves to illustrate the sensitivity of O_2 uptake to CO_2 removal. Thus, the combination of a large Root and Haldane effect and low buffer value, observed in the blood of many teleost fishes, results in a tight interaction between O_2 and CO_2 exchange in the red cell at the gills.

LINKAGE AT THE TISSUES

At the tissues, the interaction between O_2 and CO_2 exchange is important in facilitating oxygen delivery and enhancing CO_2 removal by driving CO_2 hydration and buffering the resulting protons. Lapennas's (1983) steady-state analysis indicates that a Bohr–Haldane coefficient of between 0.35 and 0.5 ($0.5 \times RQ$) results in optimal O_2 delivery to the tissues, whereas a larger Bohr–Haldane coefficient acts to minimize pH and P_{CO_2} changes during blood transit through the muscle. The presence of a nonlinear Bohr–Haldane effect changes the magnitude of the Bohr–Haldane coefficient according to the region of the OEC used for gas exchange *in vivo*. This influences the role of the Bohr–Haldane effect in gas exchange and acid–base homeostasis at both the gills and the tissues. In general, if the Bohr–Haldane effect is small, then the interaction between oxygen and carbon dioxide transfer will be reduced. This is the case in dogfish. In addition, environmental conditions that influence the overall magnitude of the Bohr–Haldane effect will also influence the magnitude of the interaction between O_2 and CO_2 transfer.

C. Effect of Physiological States and Environmental Conditions on the Linkage

AEROBIC EXERCISE

During aerobic exercise, a number of physiological adjustments permit the greatly elevated flux of O_2 from the environmental water to the tissues. At the gills, gas exchange is optimized by an increase in ventilation volume (Stevens and Randall, 1967), cardiac output (Kiceniuk and Jones, 1977; Randall and Daxboeck, 1984; Thorarensen *et al.*, 1996), and perfused surface area of the respiratory epithelium, and a reduction in epithelial thickness (Farrell *et al.*, 1979; Jones and Randall, 1978). At the tissues, the increase in cardiac output and perfusion of the red muscle (Randall and Daxboeck, 1982; Wilson and Egginton, 1994) are not sufficient to meet the demand for O_2 delivery and consequently arterial–venous O_2 saturation difference ($Sa-v_{O_2}$) increases with swimming velocity (Kiceniuk and Jones, 1977). In teleost fish that exhibit a nonlinear Bohr–Haldane effect, the increase in $Sa-v_{O_2}$ with exercise alters the magnitude of the interaction between O_2 and CO_2 and influences conditions for gas exchange and acid–base homeostasis at both the gills and the tissues.

In resting or slowly swimming trout, venous oxygen saturation (Sv_{O_2}) is slightly greater than 50% (Kiceniuk and Jones, 1977; Table 1), and almost the entire Haldane effect exhibited by trout Hb may be exploited during gas exchange. The magnitude of the Bohr–Haldane coefficient (moles of H^+ released per mole of O_2 bound to Hb) in the region of the OEC used for gas exchange *in vivo* can be calculated from measured steady-state arterial and venous blood-gas parameters (Brauner, 1995) and is 0.99 in slowly swimming trout (Table 1). At this swimming velocity, the Bohr–Haldane coefficient is greater than the respiratory exchange ratio (0.76, Brauner, 1995). Thus, during blood passage through the gills, more protons will be released during oxygenation than consumed during HCO_3^- dehydration, resulting in a slight acidosis in arterial relative to venous blood (Milligan and Wood, 1986; Brauner, 1995). The slight acidosis in arterial blood is not sufficient to reduce oxygen uptake at the gills, and stoichiometrically all CO_2 excreted is related to O_2 uptake through the release of Bohr protons (Table 1).

Table 1

O_2 and CO_2 Transport Parameters in Rainbow Trout at Different Levels of Sustained Exercise

% U_{crit}	S_aO_2	S_vO_2	P_aCO_2 (mmHg)	Hald. coeff.	%$\dot{M}CO_2$: $\dot{M}O_2$	%Pb_{CO_2}
15.8	0.90	0.55§	2.49	0.99	116.0	0.6
(1.7)	(0.05)	(0.06)	(0.16)	(0.16)	(17.2)	(0.9)
9	9	9	9	7	7	7
55.4	0.85	0.42§	2.73	0.67	95.8	1.6
(1.5)	(0.05)	(0.05)	(0.15)	(0.08)	(9.4)	(0.6)
9	9	9	9	7	7	8
90.9	0.87	0.29*§	3.21*	0.52	54.5*	5.2*
(1.2)	(0.06)	(0.06)	(0.22)	(0.20)	(19.2)	(1.4)
8	8	8	8	6	6	7
98.8	0.83	0.20*§	3.89*	0.43*	45.7*	9.0*
(0.9)	(0.05)	(0.03)	(0.21)	(0.13)	(9.4)	(1.6)
9	8	8	9	5	5	7

Note. Where U_{crit} refers to critical swimming velocity (or maximum sustained swimming velocity); Hald, coeff. refers to the Haldane coefficient (moles of protons released per mole of O_2 that binds to Hb) calculated over the region of the Hb–O_2 equilibrium curve used for gas exchange *in vivo;* %$\dot{M}CO_2$: $\dot{M}O_2$ refers to the maximum proportion of CO_2 excretion ($\dot{M}CO_2$) linked with O_2 uptake ($\dot{M}O_2$) via the release of Bohr protons during Hb oxygenation [i.e., $\dot{M}CO_2/(\dot{M}O_2 \times$ Hald. coeff.) $\times$ 100]; %Pb_{CO_2} refers to the proportion of total CO_2 excreted that was solely due to physically dissolved CO_2 that existed in prebranchial blood (i.e., did not involve HCO_3^- dehydration). Values represent mean values with SEM in brackets and number of fish (n) beneath.
* Statistically different from the lowest swimming velocity.
§ Statistically different from respective arterial value. From Brauner (1995).

In the tissues, the reverse processes occur during capillary blood transit where the large Haldane effect promotes CO_2 hydration in the red cell, reducing changes in P_{CO_2} during tissue capillary blood transit. The slight increase in pH due to a larger Haldane effect than RE may actually reduce O_2 delivery under these conditions, albeit to a minor extent. The situation is complex because if carbon dioxide enters the blood more rapidly than oxygen leaves, the initial effect could be an acidosis, promoting a Root off effect, that could raise blood P_{O_2} and enhance oxygen transfer (Brauner and Randall, 1996). The subsequent desaturation of the hemoglobin would raise pH and tend to ameliorate oxygen transfer. Thus, changes in pH in blood during transit through the capillaries are likely to be complex. At present, however, there are no direct measurements of these changes.

As exercise intensity progressively increases, Sa_{O_2} remains constant (Kiceniuk and Jones, 1977; Table 1), but there is a decrease in Sv_{O_2} and a decrease in the overall magnitude of the Bohr–Haldane coefficient in the region of the OEC used for gas exchange *in vivo* (Table 1). At maximal swimming velocity, the mean Bohr–Haldane coefficient is 0.43, which approaches the "optimal value" calculated by Lapennas ($0.5 \times RQ$) for oxygen delivery. Thus, CO_2 movement from the tissues to the blood enhances oxygen delivery to the tissues, relative to that in resting fish, at the expense of CO_2 hydration and acid–base homeostasis. Owing to the reduced "effective buffer capacity" of the blood associated with the reduced Bohr–Haldane coefficient, a given CO_2 excretion rate relative to blood flow will elevate Pv_{CO_2} relative to that observed in slowly swimming fish. A progressive increase in Pv_{CO_2} and arterial–venous P_{CO_2} (Pa–v_{CO_2}) difference was observed at greater swimming velocities consistent with this theory (Brauner, 1995).

At the gills, the reduction in Bohr–Haldane coefficient results in only 46% of the CO_2 excreted being stoichiometrically related to O_2 uptake causing an elevation in Pa_{CO_2} and total arterial CO_2 levels (Fig. 5). The respiratory acidosis was compensated for by HCO_3^- uptake over the 3 h duration of the experiment (fish were acclimated to and swum in seawater),

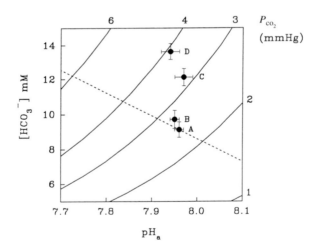

Fig. 5. A pH/HCO_3^- plot of changes in blood acid–base status of rainbow trout during different levels of sustained exercise. The data points represent mean values (error bars represent sem) for arterial plasma pH (pH_a) and [HCO_3^-] from fish swimming at 16 (A), 55 (B), 91 (C), and 99 (D) % of critical swimming velocity (U_{crit}). The whole blood buffer line (dotted line) was calculated from Wood *et al.* (1982). From Brauner, (1995). Data from the same group of fish are also displayed in Table 1.

however no significant changes in RE were observed. In order to maintain CO_2 excretion, the diffusion of physically dissolved CO_2 from prebranchial blood (Pb_{CO_2}) plays a greater role in total CO_2 excretion (i.e., the arterial–venous P_{CO_2}, difference increases), reaching a maximum of about 9% of total CO_2 excreted near U_{crit}, relative to 0.6% at the lowest swimming velocity (Table 1). Thus, the reduction in the magnitude of the overall Bohr–Haldane coefficient during exercise (due to an increased $Sa-v_{O_2}$ difference and nonlinear Bohr–Haldane effect) alters the pattern of CO_2 excretion at the gills and results in an elevation of Pa_{CO_2} and HCO_3^- levels. The increased bicarbonate levels elevate HCO_3^- buffering capacity of the tissues, and the elevation in Pa_{CO_2} may play a role in the elevation in ventilation associated with and following violent exercise in fish (Wood and Munger, 1994).

In fish that possess a low Bohr–Haldane effect but a large buffer value, such as dogfish, there is minimal interaction between O_2 and CO_2 exchange during exercise, and acid–base homeostasis at the tissues will be determined predominantly by the buffer value of the blood and changes in P_{CO_2}. Few measurements of arterial and venous CO_2 levels have been made in dogfish during exercise, but in the leopard shark (*Triakis semifasciata),* which also possesses a small Bohr–Haldane coefficient, there is only a small increase in Pa_{CO_2} during exercise and a minor increase in $Pa-v_{CO_2}$ difference relative to that observed in trout (Lai *et al.,* 1990).

ENVIRONMENTAL HYPOXIA

Environmental hypoxia results in an elevation in ventilation volume with almost no change in cardiac output despite a marked bradycardia in many fish species (Holeton and Randall, 1967; Smith and Jones, 1982; Randall, 1990). The increase in the ventilation:perfusion ratio enhances convective conditions for CO_2 removal by increasing the P_{CO_2} gradient between the water and the blood. In fish that possess a nonlinear Bohr–Haldane effect, a reduction in Sa_{O_2} will reduce Bohr proton release during oxygenation in the gills. Consequently, HCO_3^- dehydration during CO_2 excretion will titrate protons from the weakly buffered Hb, elevating red cell pH and reducing CO_2 flux through the red blood cell relative to that which would occur if the entire Haldane effect could be exploited.

In rainbow trout subjected to an environmental P_{O_2} of 30–35 torr for 6 h, arterial saturation is reduced to 50%, virtually eliminating the Haldane effect during gas exchange *in vivo.* Despite the elimination of the Haldane effect and, therefore, reduced bicarbonate flux through the red blood cell, the associated hyperventilation reduces arterial P_{CO_2} from 2.6 to 2.1 torr and total CO_2 content from 8.5 to 6.0 mM (Lessard *et al.,* 1995). Anaerobic metabolism may supply some protons for bicarbonate dehydration during hypoxia, but because the plasma reaction is uncatalyzed and protons are transferred into the red blood cell slowly, the effect of a metabolic acidosis on bicarbonate stores is limited. Although not present in trout, plasma-accessible CA would circumvent the influence of the nonlinear Haldane effect on CO_2 levels in the blood and eliminate the rate limitation of the Cl^-/HCO_3^- exchanger provided sufficient protons are present in the plasma for HCO_3^- dehydration. Consequently, continuous CA infusion into trout during hypoxia resulted in a further reduction in arterial P_{CO_2} (which decreased to 1.7 torr) and total CO_2 (which decreased to 4.2 mM; Lessard *et al.*, 1995). Thus, CA infusion during hypoxia illustrates the protective effect on blood and tissue CO_2 levels during hypoxia of confining HCO_3^- dehydration to the red cell and uncoupling it (due to the nonlinear Bohr–Haldane effect and low arterial saturation) from O_2 uptake.

In dogfish, which have a relatively high Hb buffer value (but low Haldane effect) and plasma-accessible CA, hypoxia would be expected to result in a greater "washout" of arterial CO_2 than that seen in trout given a similar ventilation : perfusion volume ratio. This remains to be examined in detail.

As the degree of hypoxia becomes greater, gas exchange in the blood of fish will occur at progressively lower levels on the OEC. Thus, during oxygen delivery to the tissues, the number of Bohr protons bound per unit oxygen consumed will decrease. Thus, for a given CO_2 tissue excretion rate at constant blood flow, the greater the degree of hypoxia, the greater the venous P_{CO_2} will be due to the reduced "effective buffering capacity" of the blood. This has been observed in trout, where hypoxia sufficient to reduce Sa_{O_2} to 40% results in an elevation in Pv_{CO_2} despite a reduction in arterial P_{CO_2} (Thomas *et al.*, 1994). At the gills, an increase in $Pa-v_{CO_2}$ implies a greater dependence on the diffusion of physically dissolved CO_2 from prebranchial blood during CO_2 excretion, indicating that the pattern of CO_2 excretion has been altered, as is observed during aerobic exercise. Hypoxia is not always associated with an increase in Pv_{CO_2}, but the nonlinear Haldane effect will elevate Pv_{CO_2} over that which would otherwise occur, resulting in a conservation of blood and tissue CO_2 and HCO_3^- stores that may be important in maintaining osmotic balance during exposure to hypoxia.

At some degree of hypoxia (probably when Sa_{O_2} is between 60 and 80%), the Bohr–Haldane coefficient in the region of the OEC used for gas exchange will approach Lapennas's theoretical optimum for oxygen delivery to the tissues. Further reductions in Sa_{O_2} (i.e., below 50%), however, will reduce the Bohr coefficient below this theoretical optimum and will be of little benefit to tissue oxygen delivery.

During acclimation to hypoxia in fish, there is a reduction in erythrocytic ATP (Tetens and Lykkeboe, 1981) or GTP (Wood and Johansen, 1972; Weber and Lykkeboe, 1978; Jensen and Weber, 1982) levels that increase Hb-O_2 affinity, maximizing O_2 uptake at the gills in the face of reduced water P_{O_2}. Modulation of the NTP:Hb ratio *in vivo* also reduces the magnitude of the Bohr–Haldane effect is some fish species (see Val and Val, 1995). In rainbow trout acclimated to hypoxia for two weeks, however, there was no significant change in the Bohr coefficient (from -0.49 in normoxia to -0.45 in hypoxia) despite almost a 50% reduction in red cell ATP concentration (Bushnell *et al.*, 1984). The degree to which a change in NTP:Hb ratio during hypoxia exposure influences the interaction between O_2 and CO_2 exchange *in vivo* remains to be investigated but is likely species-specific.

ANEMIA

During experimental anemia induced by phenylhydrazine hydrochloride exposure or replacement of red cells with saline (Cameron and Wohlschlag, 1969; Cameron and Davis, 1970; Wood *et al.*, 1979; Smith and Jones, 1982), there is a progressive reduction in venous P_{O_2} and O_2 content, resulting in an increase in Sa–v_{O_2} with minimal changes in metabolic rate, ventilation volume, and cardiac output, provided that Hct is above about 5–10% (Wood *et al.*, 1979). The increased Sa–v_{O_2} influences the interaction between O_2 and CO_2 exchange in a similar manner as that described for aerobic exercise. Generally, during anemia in fish, there is a graded elevation in NTP:Hb ratio (Lane, 1984; Vorger and Ristori, 1985), which can more than double 96 h following the development of anemia (Val *et al.*, 1994). The increase in NTP:Hb ratio reduces Hb-O_2 affinity, facilitating tissue oxygen delivery, but will slightly increase the magnitude of the Bohr–Haldane coefficient. The degree to which the Bohr–Haldane coefficient increases *in vivo* has not been measured.

Further reductions in Hct elicit an increase in cardiac output, and often an increase in ventilation volume, to maintain oxygen consumption rate, but there is a dramatic reduction in the proportion of O_2 transported in the blood bound to Hb. Wood *et al.* (1979), calculated that 9% of O_2 is carried in physical solution in control fish, and this increased to more than 70% in severely anemic fish (Hct = 1%). The reduced dependence on Hb for O_2 transport results in a drastic reduction in the interaction between O_2 and CO_2 exchange. In addition, the reduction in Hct reduces the number of Cl^-/HCO_3^- exchanger sites available for CO_2 exchange and consequently arterial and venous P_{CO_2} and total CO_2 content increase, as does the arterial–venous pH difference.

HYPEROXIA

Hyperoxic exposure also reduces the magnitude of the interaction between O_2 and CO_2 exchange by increasing the proportion of O_2 transported by the blood as dissolved O_2. Assuming a resting $Ca-v_{O_2}$ difference of 1.47 mM (Kiceniuk and Jones, 1977), and a plasma O_2 solubility (and erythrocyte O_2 solubility to simplify the analysis) of 1.98 M/torr (Boutilier et al., 1984), a P_{O_2} of just under 400 torr would be required for 50% of the O_2 consumed to be transported as dissolved O_2, provided that cardiac output and metabolic rate remained constant. A reduction in the magnitude of the interaction during hyperoxia exposure would be expected to lead to an elevation in arterial P_{CO_2} and total CO_2 content, which has been observed in trout, but this has been attributed largely to a reduction in ventilation volume and lamellar vasoconstriction (Wood and Jackson, 1980).

TEMPERATURE CHANGES

Temperature can have a marked effect on the magnitude of the Bohr–Haldane coefficient. In blood taken from tidepool blennids (*Blennius pholis* L.) acclimated to 13°C, the Bohr–Haldane effect of the blood was dramatically reduced *in vitro* at 25°C relative to 12.5°C (Bridges et al., 1984). In fish acclimated to different temperatures, the NTP:Hb ratio is often modulated, which will also affect the magnitude of the Bohr–Haldane coefficient. In the armored catfish, *Hoplosternum littorale,* the 2,3-DPG:Hb ratio (where 2,3-DPG, 2,3-diphosphoglyerate, is the primary allosteric effector of Hb in this fish species) increases with an increase in temperature from 25 to 35°C resulting in an increase in Bohr–Haldane coefficient from 0.11 to 0.17 (Affonso et al., unpublished, in Val and Val, 1995). Thus, in this species of fish, the magnitude of Bohr–Haldane effect increases with temperature *in vivo*. In eel (*anguilla anguilla*) the highest GTP:Hb ratio was observed in fish acclimated to an intermediate temperature of 17°C, with lower levels observed at 2 and 29°C (Laursen et al., 1985). In general, changes in NTP:Hb level with temperature vary intraspecifically and are confounded by other environmental conditions such as oxygen availability (see Weber, 1996, for a review) but will undoubtedly have an effect on the magnitude of the interaction between O_2 and CO_2 exchange.

CATECHOLAMINES

Catecholamines are released in many fish during exposure to hypoxia (Boutilier et al., 1988), hypercapnia (Perry et al., 1989), anemia (Iwama et al., 1987), exhaustive exercise (Primmett et al., 1986), and following air exposure or physical disturbance such as tail grabbing. The release of catecholamines during exposure to these environmental or physical disturbances elicit numerous physiological effects that maintain or elevate energy turnover and oxygen uptake at the gills (see Randall and Perry, 1992, for a review). Some of the effects of catecholamine release, however, will influence the interaction between O_2 and CO_2 exchange.

Red Cell Na^+/H^+ Exchange. Circulating catecholamines bind to the β-adrenergic receptors of red cells initiating Na^+/H^+ exchange across the red cell membrane, which ultimately elevates red cell pH (Nikinmaa, 1990). In the presence of a large Root and Bohr effect, the increase in red cell pH secures oxygen uptake at the gills in the face of an extracellular metabolic acidosis and/or when oxygen tensions are reduced.

The adrenergically induced increase in red cell pH (and reduction in the pH_e–pH_i difference), reduces the pH and HCO_3^- gradient across the red cell, potentially influencing the rate of Cl^-/HCO_3^- exchange during blood passage through the gills. The increase in venous red cell pH, however, elevates red cell HCO_3^- concentration (Currie and Tufts, 1993) reducing the dependence on Cl^-/HCO_3^- exchange during CO_2 excretion. As described earlier, CO_2 excretion is also limited by proton availability (Perry and Gilmour, 1993), and during gill blood passage there will be a competition between the proton-consuming processes of the red cell: adrenergically activated Na^+/H^+ exchange and HCO_3^- dehydration. During Hb oxygenation, some of the released Bohr protons will undoubtedly be removed from the red cell by Na^+/H^+ exchange, reducing the physiological interaction between O_2 and CO_2. However, localization of the Na^+/H^+ exchange to the red cell membrane permits most of the Bohr protons released during Hb oxygenation to be consumed during HCO_3^- dehydration at the gills (Randall and Brauner, 1991). Although the presence of catecholamines in whole blood has often been demonstrated to reduce the conversion rate of plasma bicarbonate to carbon dioxide *in vitro* (Wood and Perry, 1991; Wood, 1994; Wood and Simmons, 1994), catecholamine infusion *in vivo* has little or no effect on carbon dioxide excretion in resting trout (Steffensen *et al.,* 1987; Playle *et al.,* 1990; Nikinmaa and Vihersaari, 1993).

During conditions when arterial saturation is moderately reduced by a stressor, the subsequent release of catecholamines raises arterial saturation, and a higher region of the OEC is used for gas exchange. Thus, in fish that possess a nonlinear Bohr–Haldane effect, catecholamines will increase the interaction between O_2 and CO_2. Catecholamines also increase the hematocrit and tend to reduce red cell NTP levels, both of which potentially could alter the interaction between oxygen and carbon dioxide transfer.

NTP and Hb Concentration Changes. Catecholamines result in the release of erythrocytes from the spleen (Nilsson and Grove, 1974), increasing the total number of circulating red cells. The increase in hematocrit could alter the portion of the OEC being used for gas exchange. In addition, red cells released from the spleen are immature and characterized by reduced levels of NTP and Hb (Speckner *et al.,* 1989). Furthermore, adrenergic-induced swelling of the red cell results in a decrease in Hb and NTP concentrations, with a further reduction in NTP content depending upon the degree to which blood is deoxygenated (Ferguson and Boutilier, 1989; Nikinmaa, 1990). At constant NTP:Hb ratio but reduced Hb and NTP concentrations, complexing of NTP with Hb is reduced, which will reduce the magnitude of the Bohr–Haldane effect. An overall reduction in NTP:Hb ratio will also reduce the magnitude of the Bohr–Haldane effect and therefore the interaction between O_2 and CO_2 exchange.

The degree to which catecholamines will influence the interaction between O_2 and CO_2 transport *in vivo* due to NTP changes in the red cell and activation of the Na^+/H^+ exchanger is not known and clearly requires further study.

Indirect Interaction via Hb and Band 3. The band 3 protein of the red cell composes about 25% of the total membrane protein and exists in some 1 million copies per red cell. The 90- to 100-kDa protein consists of two structural and functional domains. The membrane domain spans the plasma membrane several times and contains the Cl^-/HCO_3^- exchange site. The cytosolic domain contains the binding sites for cytoskeletal proteins, glycolytic enzymes, and hemoglobin (see Salhany, 1990, and Wang, 1994, for reviews, see also Jensen, Fago, and Weber, Chapter 1, this volume). In humans, deoxygenated Hb binds more strongly to the cytosolic domain of band 3 *in vitro* than does oxygenated Hb (Walder *et al.,* 1984; Tsuneshige *et al.,* 1987). It has been hypothesized that Hb binding to the cytosolic domain could induce a conformational change in band 3 that increases the rate of Cl^-/HCO_3^- exchange (see Salhany, 1990; Jensen, Fago, and Weber, Chapter 1 this volume), the slowest single step in CO_2 excretion. Thus, there could be an indirect interaction between O_2 and CO_2 exchange mediated through the oxygenation status of Hb and its interaction with band 3.

There is indirect evidence that Hb oxygenation status may influence the rate of Cl^-/HCO_3^- exchange in fish red cells. It has been observed that at constant oxygenation status, there is an increased CO_2 evolution rate in deoxygenated blood of trout relative to oxygenated blood *in vitro* (Perry and Gilmour, 1993; Wood and Simmons, 1994). Although the *in vitro* CO_2 excretion assay of Wood and Perry (1991) incorporates a number of steps between the initial addition of $H^{14}CO_3$ to the plasma and the final trapping of $^{14}CO_2$, the assay qualitatively demonstrates rates of Cl^-/HCO_3^- exchange. Although Perry *et al.* (1996) demonstrated a greater rate of CO_2 evolution *in vitro* in deoxygenated relative to oxygenated blood in three teleost fishes, they were unable to demonstrate an effect in dogfish. Neither Brauner *et al.* (1996), using the same *in vitro* CO_2 excretion assay on trout blood, nor Jensen and Brahm (1995), directly measuring band 3 mediated Cl^- fluxes in four species of teleost fishes, were able to demonstrate an influence of Hb oxygenation status on Cl^- or HCO_3^- fluxes. Further research is required to address the functional interaction of Hb and band 3. Interestingly, if deoxygenated Hb elevates the rate of Cl^-/HCO_3^- exchange *in vivo,* this may counteract the effect of the nonlinear Haldane effect on Cl^-/HCO_3^- exchange *in vivo,* when Hb is partially deoxygenated.

Ontogenetic Changes. In many fish species, there is a transition in the relative proportions of different Hb isomorphs during development (see Giles and Rystephanuk, 1989), which will influence the magnitude of the whole blood Bohr–Haldane coefficients at different life stages. In juvenile coho salmon (*Oncorhynchus kisutch*), almost all the Hbs present are anodic components. In adults, only 55% of the Hb components are anodic, with the most abrupt loss of anodic components associated with parr–smolt transition (Giles and Vanstone, 1976). As described, the anodic components are characterized by large Bohr–Haldane and Root effects, which are virtually absent in the cathodic components (Sauer and Harrington, 1988). Thus, in fry, the whole blood Bohr–Haldane effect is quite large (the Bohr coefficient was determined to be 1.729 in the stripped anodic components of fry) and decreases progressively during development. The functional significance of such a large ontogenetic change in Bohr–Haldane coefficient remains unclear. It has been proposed that the large Bohr–Haldane effect in fry would facilitate oxygen unloading to the tissues, which is an attractive hypothesis because oxygen consumption rate (expressed per unit body weight) is much greater in fry than adults due to the inverse scaling of metabolic rate with body size. Given such a large Haldane effect (i.e., more than double the RE), however, it is unlikely that blood would be acidified during tissue blood transport because of the large uptake of protons by Hb during deoxygenation. Thus, the large Bohr–Haldane coefficient may actually result in an impairment of oxygen delivery depending upon the relative rate of CO_2 and O_2 movement described earlier (section IV,C). The large Bohr–Haldane effect occurs in conjunction with a large Root effect, and it is possible that the greater abundance of the anodic components in fry reflects the need for greater oxygen delivery to the eye during development; however, this is purely speculative.

Air-Breathing Fishes. In some fishes, air-breathing is an important adaptive response to hypoxia, water turbidity, and drought (see Johansen, 1970), as well as acidic and hydrogen sulfide–rich waters (Brauner *et al.,* 1995).

Structural adaptations for air-breathing are remarkably diverse and may consist of a vascularized buccal cavity, a modified stomach and intestine, a vascularized swim bladder, or a true lung (see Johansen, 1970, and Dehadrai and Tripathi, 1976, for reviews). The dependence upon the air-breathing organ (ABO) during oxygen uptake varies interspecifically and between environmental conditions.

In facultative air-breathers, gas exchange normally occurs across the gills, and aerial respiration is used to augment oxygen uptake under less desirable conditions. During hypoxia, for instance, there is a greater reliance upon aerial O_2 uptake; however, almost all CO_2 is still excreted across the gills. In obligate air-breathers, the ABO is used to some degree for O_2 uptake under all conditions but also secures an increasing proportion of the total O_2 uptake during hypoxia. For example, in *Arapaima gigas,* an obligate air-breathing teleost from the Amazon, about 80% of O_2 uptake occurs across the air-breathing organ in normoxia (Stevens and Holeton, 1978; Brauner and Val, 1996), which increases to 100% in hypoxia (Stevens and Holeton, 1978). In normoxia, about 80% of CO_2 excreted diffuses across the gills (Randall *et al.,* 1978; Brauner and Val, 1996). A similar scenario exists in other obligate air-breathing fishes (see Singh, 1976, for a review), and in some cases, as in *Electrophorus electricus,* a large proportion of CO_2 elimination may also occur across the skin (Val and Val, 1995). Thus, during aerial respiration in both facultative and obligate air-breathing fishes, gas exchange is spatially uncoupled to some degree. That is, the majority of O_2 uptake occurs across the ABO, whereas the majority of CO_2 excretion occurs across the gills or skin.

In water-breathing fish that possess a low Hb buffer value and large Haldane and Root effects, uncoupling of gas exchange has been hypothesized to impair both O_2 uptake and CO_2 excretion. In the absence of CO_2 removal at the gills of trout, the oxygenation-induced acidosis of the red cell (due to Bohr proton release) could, in theory, reduce oxygen binding to Hb via the Root effect by up to 48% (Brauner and Randall, 1996). Conversely, the absence of Hb oxygenation at the gills would be expected to impair CO_2 excretion given that all CO_2 excreted in resting rainbow trout is stoichiometrically related to O_2 uptake (see Table 1) and Hb oxygenation significantly elevates the rate of CO_2 excretion from blood *in vitro* (Figs. 2 and 4). Thus in air-breathing fishes, gas exchange during aerial respiration could be compromised due to the spatial uncoupling of O_2 and CO_2 exchange.

In the face of spatial uncoupling of O_2 and CO_2 exchange, in order for O_2 uptake not to be limited by the absence of CO_2 removal, the Hbs of air-breathing fishes should exhibit small Root and Haldane effects. In fact, most obligate air-breathing fishes do not exhibit a pronounced Root effect (see Johansen, 1970), and most obligate and facultative air-breathers possess moderate whole blood Haldane coefficients (Johansen *et al.,* 1978). One exception among obligate air-breathers is *A. gigas,* where a reduction in pH may reduce blood oxygen-carrying capacity by up to 45% (Val and Val, 1995; Brauner and Val, 1996). In addition, the Root effect is quite common in facultative air-breathers (see Val and Val, 1995, and Pelster and Weber, 1991, for reviews). Although the Hb buffer value at fixed oxygenation status is low in the few air-breathing fishes examined to date (Brauner and Val, 1996), total CO_2 content and therefore HCO_3^- buffering capacity of the blood is quite high during aerial respiration (Wood *et al.,* 1979; Brauner and Val, 1996). Thus, pH changes in the red cell associated with oxygenation in the absence of CO_2 removal will be relatively minor, and therefore, it is unlikely that spatial uncoupling of O_2 and CO_2 exchange will result in a significant impairment of oxygen uptake in the ABO.

Owing to the moderate size of the Haldane effect in most air-breathing fishes, the influence of spatial uncoupling of O_2 and CO_2 exchange will not have a pronounced effect on CO_2 excretion. Though the Haldane effect undoubtedly plays a role during CO_2 excretion when O_2 and CO_2 exchange are spatially coupled (37% of gas exchange in *A. gigas;* Fig. 6), it is possible that Bohr protons released in the ABO may be used for CO_2 excretion in the gills. This is possible only if the blood transit time from the ABO is rapid relative to the rate of proton flux across the red cell, which is not known.

In general, there is an inverse relationship between the magnitude of the Haldane effect and Hb buffer value at fixed oxygenation status in vertebrates (Jensen, 1989). In *A. gigas* and *Lipposarcus pardalis* (a facultative air-breathing catfish that uses its stomach and intestine as an ABO), the Haldane effect and Hb buffer value are both low, seemingly maladaptive

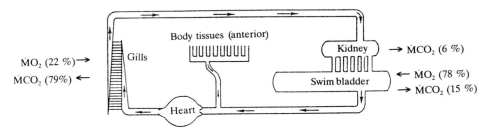

Fig. 6. Schematic representation of the blood vascular system in *Arapaima gigas* (after Greenwood and Liem, 1984) with the relative proportion of O_2 and CO_2 exchange that occurs across the respective surfaces. From Brauner and Val, (1996).

for CO_2 transport and excretion (Brauner and Val, 1996). In *A. gigas* and other obligate air-breathing fishes, both P_{CO_2} and HCO_3^- levels in the blood are high. Thus, CO_2 excretion may be more dependent on physically dissolved CO_2 in prebranchial blood than is observed in water-breathing teleosts.

During aerial respiration in facultative air-breathers, P_{CO_2} and HCO_3^- levels are also high (Wood *et al.*, 1979); however, when these fish are breathing exclusively water, blood CO_2 and HCO_3^- levels are low, typical of other water-breathing fishes. Thus, in *L. pardalis* (which has a low buffer value and low Bohr–Haldane effect), the blood has a limited capacity for acid–base homeostasis in the tissues, and a large difference in pH is observed between arterial and venous blood, even at rest (0.17 pH unit; Brauner and Val, 1996). *L. pardalis* tolerates large reductions in blood pH during hypercapnia exposure (Randall *et al.*, 1996), indicating that the tissues may be resistant to large fluctuations in pH of the blood.

Finally, in addition to the general uncoupling associated with air-breathing, the magnitude of the Bohr coefficient may change during air exposure. In *Synbranchus marmaratus*, air exposure for between 21 and 44 h resulted in a decrease in the magnitude of the Bohr–Haldane coefficient from 0.45 to 0.23 (Johansen *et al.*, 1978). Interestingly, this was associated with an increase in red cell ATP and GTP content, which normally increases the magnitude of the Bohr–Haldane coefficient.

V. CONCLUSION

Within the class Pisces, there is tremendous diversity in the magnitude of the linkage between O_2 and CO_2 exchange. This diversity arises from absolute differences in Hb characteristics as well as the mode of breathing (i.e., water- versus air-breathing). Intraspecific variation in the degree of linkage may result during acclimation to different environmental conditions or from developmental changes. The greatest degree of interaction is associated with a large Bohr–Haldane and Root effect and low Hb buffer value, which are common Hb characteristics in many teleost fishes. In many instances, the Bohr–Haldane effect is nonlinear with the majority of the Bohr–Haldane effect occurring in the upper region of the OEC (between 50 and 100% S_{O_2}). Thus, the linkage between O_2 and CO_2 exchange also depends upon the region of the OEC used for gas exchange *in vivo,* which is influenced by activity level and other environmental conditions. Although, the linkage between O_2 and CO_2 exchange quantitatively elevates transfer of the two gases at both the gills and the tissues, it undoubtedly slows the overall equilibration of the reactions (Hlastala, 1984) and may contribute to the acid–base disequilibria in the arterial and venous blood of fish (see Gilmour, Chapter 9, this volume).

REFERENCES

Adair, G. S. (1925). The hemoglobin system. VI. The oxygen dissociation curve of hemoglobin. *J. Biol. Chem.* **63,** 529–545.

Albers, C., Goetz, K.-H., and Hughes, G. M. (1983). Effect of acclimation temperature on intraerythrocytic acid–base balance and nucleoside triphosphates in the carp, *Cyprinus carpio. Respir. Physiol.* **34,** 145–159.

Bartels, H. (1972). The biological significance of the Bohr effect. *In* "Oxygen Affinity of Hemoglobin and Red Cell Acid–Base Status" (M. Rorth and P. Astrup, eds.), pp. 717–735. Munksgaard, Copenhagen.

Bohr, C., Hasselbalch, K., and Krogh, A. (1904). Ueber einen in biologischer beziehung wichtigen einfluss, den die Kohlensauerspannung des blutes auf dessen sauerstoffbindung uebt. *Skand. Arch. Physiol.* **16,** 402–412.

Boutilier, R. G., Heming, T. A., and Iwama, G. K. (1984). Appendix: Physicochemical parameters for use in fish respiratory physiology. *In* "Fish Physiology" (W. S. Hoar and D. J. Randall, eds.), Vol. 10A, pp. 403–430. Academic Press, New York.

Boutilier, R. G., Dobson, G., Hoeger, U., and Randall, D. J. (1988). Acute exposure to graded levels of hypoxia in rainbow trout *Salmo gairdneri*: Metabolic and respiratory adaptations. *Respir. Physiol.* **71,** 69–82.

Brauner, C. J. (1995). An analysis of the transport and interaction of oxygen and carbon dioxide in fish. Ph.D. thesis. Univ. of British Columbia, Vancouver, B.C.

Brauner, C. J., and Randall, D. J. (1996). The interaction between oxygen and carbon dioxide movements in fishes. *Comp. Biochem. Physiol. A* **113,** 83–90.

Brauner, C. J., and Val, A. L. (1996). The interaction between O_2 and CO_2 exchange in the obligate air breather, *Arapaima gigas,* and the facultative air breather, *Lipossarcus pardalis. In* "Physiology and Biochemistry of the fishes of the Amazon" (A. L. Val, V. M. F. Almeida-Val, and D. J. Randall, eds.), pp. 101–110. INPA, Manaus, Brazil.

Brauner, C. J., Ballantyne, C. L., Randall, D. J., and Val, A. L. (1995). Air breathing in the armoured catfish (*Hoplosternum littorale*) as an adaptation to hypoxic, acidic, and hydrogen sulphide rich waters. *Can. J. Zool.* **73,** 739–744.

Brauner, C. J., Gilmour, K. M., and Perry, S. F. (1996). Effect of haemoglobin oxygenation on Bohr proton release and CO_2 excretion in the rainbow trout. *Respir. Physiol.* **106,** 65–70.

Breepoel, P. M., Kreuzer, F., and Hazevoet, M. (1980). Studies of the hemoglobins of the eel (*Anguilla anguilla* L.). I. Proton binding of stripped hemolysate: Separation and properties of two major components. *Comp. Biochem. Physiol. A* **65,** 69–75.

Bridges, C. R., Taylor, A. C., Morris, S. J., and Grieshaber, M. K. (1984). Ecophysiological adaptations in *Blennius pholis* (L.) blood to intertidal rockpool environments. *J. Exp. Mar. Biol. Ecol.* **77,** 151–167.

Brittain, T. (1987). The Root effect. *Comp. Biochem. Physiol. B* **86,** 473–481.

Brittain, T. (1991). Cooperativity and allosteric regulation in non-mammalian vertebrate haemoglobins. *Comp. Biochem. Physiol. B* **99,** 731–740.

Brunori, M. (1966). The carbon monoxide Bohr effect in hemoglobin from *Thunnus thynnus. Arch. Biochem. Biophys.* **114,** 195–199.

Brunori, M., Giardina, B., Antonini, E., Benedetti, P. A., and Bianchini, G. (1974). Distribution of the haemoglobin components of trout blood among the erythrocytes: Observations by single-cell spectroscopy. *J. Mol. Biol.* **86,** 165–169.

Bushnell, P. G., Steffensen, J. F., and Johansen, K. (1984). Oxygen consumption and swimming performance in hypoxia acclimated rainbow trout *Salmo gairdneri. J. Exp. Biol.* **113,** 225–235.

Cameron, J. N., and Wohlschlag, D. E. (1969). Respiratory response to experimentally induced anaemia in the pinfish (*Lagodon rhomboides*). *J. Exp. Biol.* **50,** 307–317.

Cameron, J. N., and Davis, J. C. (1970). Gas exchange in rainbow trout (*Salmo gairdneri*) with varying blood oxygen capacity. *J. Fish. Res. Board Can.* **27,** 1069–1085.

Chien, J. C. W., and Mayo, K. H. (1980). Carp hemoglobin. II. The alkaline Bohr effect. *J. Biol. Chem.* **255,** 9800–9806.

Christiansen, J., Douglas, C. G., and Haldane, J. S. (1914). The absorption and dissociation of carbon dioxide by human blood. *J. Physiol. Lond.* **48,** 244–277.

Chu, A. H., Turner, B. W., and Ackers, G. K. (1984). Effects of protons on the oxygenation-linked subunit assembly in human hemoglobin. *Biochemistry* **23,** 604–617.

Crandall, E. D., and Bidani, A. (1981). Effects of red blood cell HCO_3^-/Cl^- exchange kinetics on lung CO_2 transfer: Theory. *J. Appl. Physiol.* **50,** 265–271.

Currie, S., and Tufts, B. L. (1993). An analysis of carbon dioxide transport in arterial and venous blood of the rainbow trout, *Oncorhynchus mykiss,* following exhaustive exercise. *Fish Physiol. Biochem.* **12** (3),183–192.

Dehadrai, P. V., and Tripathi, S. D. (1976). Environment and ecology of freshwater air-breathing teleosts. *In* "Respiration of Amphibious Vertebrates" (G. M. Hughes, ed.), pp. 39–72. Academic Press, New York.

Duhm, J., and Gerlach, E. (1971). On the mechanisms of the hypoxia-induced increase of 2,3-diphosphoglycerate in erythrocytes. Studies on rat erythrocytes *in vivo* and on human erythrocytes *in vitro*. *Pflügers Arch.* **326,** 254–269.

Fago, A., Carratore, V., DiPrisco, G., Feuerlein, R. J., Sottrup-Jensen, L., and Weber, R. E. (1995). The cathodic hemoglobin of *Anguilla anguilla:* Amino acid sequence and oxygen equilibria of a reverse Bohr effect hemoglobin with high oxygen affinity and high phosphate sensitivity. *J. Biol. Chem.* **270,** 18897–18902.

Farmer, M. (1979). The transition from water to air breathing: Effects of CO_2 on hemoglobin function. *Comp. Biochem. Physiol. A* **62,** 109–114.

Farrell, A. P., Daxboeck, C., and Randall, D. J. (1979). The effect of input pressure and flow on the pattern and resistance to flow in the isolated perfused gill of a teleost fish. *J. Comp. Physiol.* **133,** 233–240.

Ferguson, R. A., and Boutilier, R. G. (1989). Metabolic-membrane coupling in red blood cells of trout: Effects of anoxia and adrenergic stimulation. *J. Exp. Biol.* **143,** 149–164.

Feuerlein, R. J., and Weber, R. E. (1996). Oxygen equilibria of cathodic eel hemoglobin analysed in terms of the MWC model and Adair's successive oxygenation theory. *J. Comp. Physiol.* **165,** 597–606.

Fyhn, E. H., Fyhn, H. J., Davis, B. J., Powers, D. A., Fink, W. L., and Garlick, R. L. (1979). Hemoglobin heterogeneity in Amazonian fishes. *Comp. Biochem. Physiol. A* **62,** 39–66.

Giles, M. A., and Vanstone, W. E. (1976). Ontogenetic variation in the multiple hemoglobins of coho salmon (*Oncorhynchus kisutch*) and effect of environmental factors on their expression. *J. Fish. Res. Board Can.* **33,** 1144–1149.

Giles, M. A., and Randall, D. J. (1980). Oxygenation characteristics of the polymorphic hemoglobins of coho salmon (*Oncorhynchus kisutch*) at different developmental stages. *Comp. Biochem. Physiol. A* **65,** 265–271.

Giles, M. A., and Rystephanuk, D. M. (1989). Ontogenetic variation in the multiple hemoglobins of Arctic charr, *Salvelinus alpinus. Can. J. Fish. Aquat. Sci.* **46,** 804–809.

Gillen, R. G., and Riggs, A. (1973). Structure and function of the isolated hemoglobins of the American eel, (*Anguilla rostrata*). *J. Biol. Chem.* **248,** 1961–1969.

Graham, M. S., Turner, J. D., and Wood, C. M. (1990). Control of ventilation in the hypercapnic skate *Raja ocellata.* I. Blood extradural fluid. *Respir. Physiol.* **80,** 259–277.

Grant, B. J. B. (1982). Influence of Bohr–Haldane effect on steady-state gas exchange. *J. Appl. Physiol.* **52** (5), 1330–1337.

Greenwood, P. H., and Liem, K. F. (1984). Aspiratory respiration in *Arapaima gigas* (Teleostei, Osteoglossomorpha): A reappraisal. *J. Zool. (London)* **203,** 411–425.

Heming, T. A., Randall, D. J., Boutilier, R. G., Iwama, G. K., and Primmett, D. (1986). Ionic equilibria in red blood cells of rainbow trout (*Salmo gairdneri*): Cl$^-$, HCO$_3^-$ and H$^+$. *Respir. Physiol.* **65,** 223–234.

Henry, R. P., Smatresk, N. J., and Cameron, J. N. (1988). The distribution of branchial carbonic anhydrase and the effects of gill and erythrocyte carbonic anhydrase inhibition in the channel catfish *Ictalurus punctatus. J. Exp. Biol.* **134,** 201–218.

Hill, E. P., Power, G. G., and Longo, L. D. (1973). Mathematical simulation of pulmonary O$_2$ and CO$_2$ exchange. *Am. J. Physiol.* **224,** 904–917.

Hlastala, M. P. (1984). Interactions between O$_2$ and CO$_2$ in blood. *In* "Advances in the Biosciences: Oxygen Transport in Red Blood Cells" (C. Nicolau, ed.), Vol. 54, pp. 95–103. Pergamon Press, Oxford.

Holeton, G. F., and Randall, D. J. (1967). The effect of hypoxia upon the partial pressure of gases in the blood and water afferent and efferent to the gills of rainbow trout. *J. Exp. Biol.* **46,** 317–327.

Ikeda-Saito, M., Yonetani, T., and Gibson, Q. H. (1983). Oxygen equilibrium studies on hemoglobin from the bluefin tuna (*Thunnus thynnus*). *J. Mol. Biol.* **168,** 673–686.

Imai, K., and Yonetani, T. (1975). pH dependence of the Adair constants of human hemoglobin: Nonuniform contribution of successive oxygen binding to the alkaline Bohr effect. *J. Biol. Chem.* **250,** 2227–2231.

Iwama G. K., Boutilier R. G., Heming T. A., Randall D. J., and Mazeaud M. (1987) The effects of altering gill water flow on gas transfer in rainbow trout. *Can. J. Zool.* **65,** 2466–2470.

Janssen, L. H. M., DeBruin, S. H., and Van Os, G. A. J. (1972). Titration behavior of histidines in human, horse, and bovine hemoglobins. *J. Biol. Chem.* **247,** 1743–1749.

Jensen, F. B. (1986). Pronounced influence of Hb–O$_2$ saturation on red cell pH in tench blood *in vivo* and *in vitro. J. Exp. Zool.* **238,** 119–124.

Jensen, F. B. (1988). Red-cell pH in tench. Interacting effects of cellular nucleoside triphosphates, Hb-oxygenation and extracellular pH. *Acta. Physiol. Scand.* **132,** 431–437.

Jensen, F. B. (1989). Hydrogen ion equilibria in fish haemoglobins. *J. Exp. Biol.* **143,** 225–234.

Jensen, F. B. (1991). Multiple strategies in oxygen and carbon dioxide transport by haemoglobin. *In* "Physiological Strategies for Gas Exchange and Metabolism." (A. J. Woakes, M. K. Greishaber, and C. R. Bridges, eds.), pp. 55–78. Cambridge Univ. Press, Cambridge.

Jensen, F. B., and Weber, R. E. (1982). Respiratory properties of tench blood and hemoglobin. Adaptation to hypoxic–hypercapnic water. *Mol. Physiol.* **2,** 235–250.

Jensen, F. B., and Weber, R. E. (1985). Proton and oxygen equilibria, their anion sensitivities and interrelationships in tench hemoglobin. *Mol. Physiol.* **7,** 41–50.

Jensen, F. B., and Brahm, J. (1995). Kinetics of chloride transport across fish red blood cell membranes. *J. Exp. Biol.* **198,** 2237–2244.

Johansen, K. (1970). Air breathing in fishes. *In* "Fish Physiology" (W. S. Hoar, and D. J. Randall, eds.), Vol. 4, pp. 361–411. Academic Press, New York.

Johansen, K., Mangum, C. P., and Lykkeboe, G. (1978). Respiratory properties of the blood of Amazon fishes. *Can. J. Zool.* **56,** 891–897.

Jones, D. R., and Randall, D. J. (1978). The respiratory and circulatory systems during exercise. *In* "Fish Physiology" (W. S. Hoar and D. J. Randall, eds.), Vol. VII, pp. 425–501. Academic Press, New York.

Jones, D. R., Brill, R. W., and Mense, D. C. (1986). The influence of blood gas properties on gas tensions and pH of ventral and dorsal aortic blood in free-swimming tuna, *Euthynnus affinis. J. Exp. Biol.* **120,** 201–213.

Kiceniuk, J. W., and Jones, D. R. (1977). The oxygen transport system in trout (*Salmo gairdneri*) during sustained exercise. *J. Exp. Biol.* **69,** 247–260.

Klocke, R. A. (1987). Carbon dioxide transport. *In* "Handbook of Physiology: The Respiratory System" (A. P. Fishman, L. E. Farhi, S. M. Tenney, and S. R. Geiger, eds.), Vol. IV, pp. 173–197. American Physiological Society, Bethesda, MD.

Lai, N. C., Graham, J. B., and Burnett, L. (1990). Blood respiratory properties and the effect of swimming on blood gas transport in the leopard shark *Triakis semifasciata. J. Exp. Biol.* **151,** 161–173.

Lane, H. C. (1984). Nucleoside triphosphate changes during the peripheral life-span of erythrocytes of adult rainbow trout (*Salmo gairdneri*). *J. Exp. Zool.* **231,** 57–62.

Lapennas, G. N. (1983). The magnitude of the Bohr coefficient: Optimal for oxygen delivery. *Respir. Physiol.* **54,** 161–172.

Laursen, J. S., Andersen, N. A., and Lykkeboe, G. (1985). Temperature acclimation and oxygen binding properties of blood of the European eel, *anguilla anguilla. Comp. Biochem. Physiol. A* **81,** 79–86.

Lessard, J., Val, A. L., Aota, A., and Randall, D. J. (1995). Why is there no carbonic anhydrase activity available to fish plasma? *J. Exp. Biol.* **198,** 31–38.

Milligan, C. L., and Wood, C. M. (1986). Intracellular and extracellular acid–base status and H$^+$ exchange with the environment after exhaustive exercise in the rainbow trout. *J. Exp. Biol.* **123,** 93–121.

Nikinmaa, M. (1990). "Vertebrate Red Cells: Adaptations of Function to Respiratory Requirements." Springer-Verlag, Berlin.

Nikinmaa, M., and Vihersaari, L. (1993). Pre- and postbranchial carbon dioxide content of rainbow trout (*Oncorhynchus mykiss*) blood after catecholamine injection. *J. Exp. Biol.* **180,** 315–321.

Nilsson, S., and Grove, D. J. (1974). Adrenergic and cholinergic innervation of the spleen of the cod: *Gadus morhua. Eur. J. Pharmac.* **28,** 135–143.

Obaid, A. L., Critz, A. M., and Crandall, E. D. (1979). Kinetics of bicarbonate/chloride exchange in dogfish erythrocytes. *Am. J. Physiol.* **237** (3), R132–R138.

Pelster, B., and Weber, R. E. (1990). Influence of organic phosphates on the Root effect of multiple fish haemoglobins. *J. Exp. Biol.* **149,** 425–437.

Pelster, B., and Weber, R. E. (1991). The physiology of the Root effect. *In* "Advances in Comparative and Environmental Physiology," Vol. 8, pp. 51–77. Springer-Verlag, Berlin.

Perry, S. F., and Laurent, P. (1990). The role of carbonic anhydrase in carbon dioxide excretion, acid–base balance and ionic regulation in aquatic gill breathers. *In* "Animal Nutrition and Transport Processes. 2. Transport, Respiration and Excretion: Comparative and Environmental Aspects" (J. P. Truchot and B. Lahlou, eds.), Vol. 6, pp. 39–57. Karger, Basel.

Perry, S. F., and Reid, S. D. (1992). Relationship between blood O$_2$-content and catecholamine levels during hypoxia in rainbow trout and American eel. *Am. J. Physiol.* **263** (32), R240–R249.

Perry, S. F., and Gilmour, K. (1993). An evaluation of factors limiting carbon dioxide excretion by trout red blood cells *in vitro*. *J. Exp. Biol.* **180,** 39–54.

Perry, S. F., Davie, P. S., Daxboeck, C., and Randall, D. J. (1982). A comparison of CO_2 excretion in a spontaneously ventilating blood-perfused trout preparation and saline-perfused gill preparations: Contribution of the branchial epithelium and red blood cell. *J. Exp. Biol.* **101,** 47–60.

Perry, S. F., Kinkead, R., Gallaugher, P., and Randall, D. J. (1989). Evidence that hypoxemia promotes catecholamine release during hypercapnic acidosis in rainbow trout (*Salmo gairdneri*). *Respir. Physiol.* **77,** 351–364.

Perry, S. F., Wood, C. M., Walsh, P. J., and Thomas, S. (1996). Fish red blood cell carbon dioxide excretion *in vitro*. A comparative study. *Comp. Biochem. Physiol. A* **113,** 121–130.

Perry, S. F., Brauner, C. J., Tufts, B., and Gilmour, K. M. (1997). Acid–base disequilibrium in the venous blood of rainbow trout (*Oncorhynchus mykiss*). *Exp. Bio. Online,* **2,** 1.

Playle, R. C., Munger, S., and Wood, C. M. (1990). Effects of catecholamines on gas exchange and ventilation in rainbow trout (*Salmo gairdneri*). *J. Exp. Biol.* **152,** 353–367.

Powers, D. A., Fhyn, H. J., Fyhn, U. E. H., Martin, J. P., Garlick, R. L., and Wood, S. C. (1979). A comparative study of the oxygen equilibria of blood from 40 genera of Amazonian fishes. *Comp. Biochem. Physiol. A* **62,** 67–85.

Primmett, D. R. N., Randall, D. J., Mazeaud, M., and Boutilier, R. G. (1986). The role of catecholamines in erythrocyte pH regulation and oxygen transport in rainbow trout (*Salmo gairdneri*) during exercise. *J. Exp. Biol.* **122,** 139–148.

Randall, D. (1990). Control and co-ordination of gas exchange in water breathers. *In* "Advances in Comparative and Environmental Physiology" (R. G. Boutilier, ed.), Vol. 6, pp. 253–278. Springer-Verlag, Berlin, Heidelberg.

Randall, D. J., and Daxboeck, C. (1982). Cardiovascular changes in the rainbow trout (*Salmo gairdneri* Richardson) during exercise. *Can. J. Zool.* **60,** 1135–1140.

Randall, D. J., and Daxboeck, C. (1984). Oxygen and carbon dioxide transfer across fish gills. *In* "Fish Physiology" (W. S. Hoar and D. J. Randall, eds.), Vol. 10A, pp. 263–307. Academic Press, New York.

Randall, D., and Brauner, C. (1991). Effects of environmental factors on exercise in fish. *J. Exp. Biol.* **160,** 113–126.

Randall, D. J., and Perry, S. F. (1992). Catecholamines. *In* "Fish Physiology" (W. S. Hoar, D. J. Randall, and A. P. Farrell, eds.), Vol. 12, pp. 255–300. Academic Press, New York.

Randall, D. J., Farrell, A. P., and Haswell, M. S. (1978). Carbon dioxide excretion in the pirarucu (*Arapaima gigas*), an obligate air-breathing fish. *Can. J. Zool.* **56,** 977–982.

Randall, D., Brauner, C., and Wilson, J. (1996). Acid excretion in Amazonian fish. *In* "Physiology and Biochemistry of the Fishes of the Amazon" (A. L. Val, V. M. F. Almeida-Val, and D. J. Randall, eds.), pp. 91–99. INPA, Manaus, Brazil.

Riggs, A. F. (1988). The Bohr effect. *Annu. Rev. Physiol.* **50,** 181–204.

Root, R. W. (1931). The respiratory function of the blood of marine fishes. *Biol. Bull. Mar. Biol. Lab. Woods Hole* **61,** 427–456.

Saffran, W. A., and Gibson, Q. H. (1981). Asynchronous ligand binding and proton release in a Root effect hemoglobin. *J. Biol. Chem.* **256,** 4551–4556.

Salhany, J. M. (1990). "Erythrocyte Band 3 Protein." CRC Press, Boca Raton, 213.

Sauer, J., and Harrington, J. P. (1988). Hemoglobins of the sockeye salmon, *Oncorhynchus nerca. Comp. Biochem. Physiol. A* **91,** 109–114.

Scholander, P. F., and Van Dam, L. (1954). Secretion of gases against high pressures in the swimbladder of deep sea fishes. I. Oxygen dissociation in blood. *Biol. Bull. Mar. Biol. Lab. Woods Hole* **107,** 247–259.

Sigaard-Anderson, O. (1974). "The Acid-Base Status of the Blood," 4th ed. Munksgaard, Copenhagen; Williams and Wilkins, New York.

Singh, B. N. (1976). Balance between aquatic and aerial respiration. *In* "Respiration of Amphibious Vertebrates" (G. M. Hughes, ed.), pp. 125–164. Academic Press, New York.

Smith, F., and Jones, D. (1982). The effect of changes in blood oxygen-carrying capacity on ventilation volume in the rainbow trout (*Salmo gairdneri*). *J. Exp. Biol.* **97**, 325–334.

Speckner, W., Schindler, J. F., and Albers, C. (1989). Age-dependent changes in volume and haemoglobin content of erythrocytes in the carp (*Cyprinus carpio*, L.). *J. Exp. Biol.* **141**, 133–149.

Steffensen, J. F., Tufts, B. L., and Randall, D. J. (1987). Effect of burst swimming and adrenaline infusion on CO_2 excretion in rainbow trout, *Salmo gairdneri. J. Exp. Biol.* **131**, 427–434.

Stevens, E. D., and Randall, D. J. (1967). Changes of gas concentrations in blood and water during moderate swimming activity in rainbow trout. *J. Exp. Biol.* **46**, 329–337.

Stevens, E. D., and Holeton, G. F. (1978). The partitioning of oxygen uptake from air and from water by the large obligate air-breathing teleost pirarucu (*Arapaima gigas*). *Can. J. Zool.* **56**, 974–976.

Tetens, V., and Lykkeboe, G. (1981). Blood respiratory properties of rainbow trout, *Salmo gairdneri*: Responses to hypoxia acclimation and anoxic incubation of blood *in vitro. J. Comp. Physiol.* **134**, 117–125.

Thomas, S., Fritsche, R., and Perry, S. F. (1994). Pre- and post-branchial blood respiratory status during acute hypercapnia or hypoxia in rainbow trout (*Oncorhynchus mykiss*). *J. Comp. Physiol.* **164**, 451–458.

Thorarensen, H., Gallaugher, P., and Farrell, A. P. (1996). Cardiac output in swimming rainbow trout, *Oncorhynchus mykiss,* acclimated to seawater. *Physiol. Zool.* **69**, 139–153.

Tsuneshige, A., Imai, K., and Tyuma, I. (1987). The binding of hemoglobin to red cell membrane lowers its oxygen affinity. *J. Biochem.* **101**, 695–704.

Tyuma, I., and Ueda, Y. (1975). Non-linear relationship between oxygen saturation and proton release, and equivalence of the Bohr and Haldane coefficients in human hemoglobin. *Biochim. Biophys. Acta* **65**, 1278–1283.

Tyuma, I., Kamigawara, Y., and Imai, K. (1973). pH dependence of the shape of the hemoglobin–oxygen equilibrium curve. *Biochim. Biophys. Acta* **310**, 317–320.

Val, A. L., and Almeida-Val, V. M. F. (1995). "Fishes of the Amazon and Their Environment: Physiological and Biochemical Aspects." Springer-Verlag, Berlin.

Val, A. L., Mazur, C. F., De Salvo-Souza, R. H., and Iwama, G. K. (1994). Effects of experimental anaemia on intra-erythrocytic phosphate levels in rainbow trout, *Oncorhynchus mykiss. J. Fish Biol.* **45**, 269–277.

Vorger, P., and Ristori, M. T. (1985). Effects of experimental anemia on the ATP content and the oxygen affinity of the blood in the rainbow trout (*Salmo gairdneri*). *Comp. Biochem. Physiol. A* **82** (1), 221–224.

Walder, J. A., Chatterjee, R., Steck, T. L., Low, P. S., Musso, G. F., Kaiser, E. T., Rogers, P. H., and Arnone, A. (1984). The interaction of hemoglobin with the cytoplasmic domain of band 3 of the human erythrocyte membrane. *J. Biol. Chem.* **259**, 10238–10246.

Wang, D. N. (1994). Band 3 protein: Structure, flexibility and function. *FEBS Letters* **346**, 26–31.

Weber, R. E. (1990). Functional significance and structural basis of multiple hemoglobins with special reference to ectothermic vertebrates. *In* "Comparative Physiology; Animal Nutrition and Transport Processes. 2. Transport, Respiration and Excretion: Comparative and Environmental Aspects; II. Blood Oxygen Transport: Adjustment to Physiological and Environmental Conditions" (J. P. Truchot and B. Lahlou, eds.), Vol. 6, pp. 58–75. Karger, Basel.

Weber, R. E. (1996). Hemoglobin adaptations in Amazonian and temperate fish with special reference to hypoxia, allosteric effectors and functional heterogeneity. *In* "Physiology and Biochemistry of the Fishes of the Amazon" (A. L. Val, V. M. F. Almeida-Val, and D. J. Randall, eds.), pp. 75–90. INPA, Manaus, Amazonas, Brazil.

Weber, R. E., and Jensen, F. B. (1988). Functional adaptations in hemoglobins from ectothermic vertebrates. *Annu. Rev. Physiol.* **50,** 161–179.

Weber, R. E., and Lykkeboe, G. (1978). Respiratory adaptations in carp blood: Influences of hypoxia, red cell organic phosphates, divalent cations and CO_2 on hemoglobin–oxygen affinity. *J. Comp. Physiol.* **128,** 127–137.

Weber, R. E., Wood, S.C., and Lomholt. J. P. (1976). Temperature acclimation and oxygen-binding properties of blood and multiple haemoglobins of rainbow trout. *J. Exp. Biol.* **65,** 333–345.

Weber, R. E., Lykkeboe, G., and Johansen, K. (1976b). Physiological properties of eel haemoglobin: Hypoxic acclimation, phosphate effects and multiplicity. *J. Exp. Biol.* **64,** 75–88.

Weber, R. E., Jensen, F. B., and Cox, R. P. (1987). Analysis of teleost hemoglobin by Adair and Monod–Wyman–Changeux models: Effects of nucleotide triphosphates and pH on oxygenation of tench hemoglobin. *J. Comp. Physiol.* **157,** 145–152.

Wieth, J. O., Andersen, O. S., Brahm, J., Bjerrum, P. J., and Borders, Jr., C. L. (1982). Chloride-bicarbonate exchange in red blood cells: Physiology of transport and chemical modification of binding sites. *Phil. Trans. R. Soc. Lond. B* **299,** 383–399.

Wilson, R. W., and Egginton, S. (1994). Assessment of maximum sustainable swimming performance in rainbow trout (*Oncorhynchus mykiss*). *J. Exp. Biol.* **192,** 299–305.

Wood, C. M. (1994). HCO_3^- dehydration by the blood of rainbow trout following exhaustive exercise. *Respir. Physiol.* **98,** 305–318.

Wood, C. M., and Jackson, E. B. (1980). Blood acid–base regulation during environmental hyperoxia in the rainbow trout (*Salmo gairdneri*). *Respir. Physiol.* **42,** 351–372.

Wood, C. M., and Perry, S. F. (1991). A new *in vitro* assay for carbon dioxide excretion by trout red blood cells: Effects of catecholamines. *J. Exp. Biol.* **157,** 349–366.

Wood, C. M., and Munger, R. S. (1994). Carbonic anhydrase injection provides evidence for the role of blood acid–base status in stimulating ventilation after exhaustive exercise in rainbow trout. *J. Exp. Biol.* **194,** 225–253.

Wood, C. M., and Simmons, H. (1994). The conversion of plasma HCO_3^- to CO_2 by rainbow trout red blood cells *in vitro*: Adrenergic inhibition and the influence of oxygenation status. *Fish Physiol. Biochem.* **12,** 445–454.

Wood, C. M., McMahon, B. R., and McDonald, D. G. (1979). Respiratory, ventilatory, and cardiovascular responses to experimental anaemia in the starry flounder, *Platichthys stellatus*. *J. Exp. Biol.* **82,** 139–162.

Wood, C. M., McDonald, D. G., and McMahon, B. R. (1982). The influence of experimental anaemia on blood acid–base regulation *in vivo* and *in vitro* in the starry flounder (*Platichthys stellatus*) and the rainbow trout (*Salmo gairdneri*). *J. Exp. Biol.* **96,** 221–237.

Wood, C. M., Perry, S. F., Walsh, P. J., and Thomas, S. (1994). HCO_3^- dehydration by the blood of an elasmobranch in the absence of a Haldane effect. *Respir. Physiol.* **98,** 319–337.

Wood, S. C., and Johansen, K. (1972). Adaptation to hypoxia by increased HbO_2 affinity and decreased red cell ATP concentration. *Nature* **237,** 278–279.

Wood, S. C., Weber, R. E., and Davis, B. J. (1979). Effects of air-breathing on acid–base balance in the catfish, *Hypostomus sp. Comp. Biochem. Physiol. A* **62,** 185–187.

Wyman, J. (1973). Linked functions and reciprocal effects in haemoglobin: A second look. *Adv. Protein Chem.* **19,** 223–286.

9

CAUSES AND CONSEQUENCES OF ACID–BASE DISEQUILIBRIA

KATHLEEN M. GILMOUR

I. Introduction
II. Postbranchial Disequilibria
    A. Causes
    B. Variations on the Basic Theme
III. Prebranchial Disequilibria
IV. Consequences of Acid–Base Disequilibria
    References

I. INTRODUCTION

The question of whether CO_2–HCO_3^-–H^+ reactions in the blood are in equilibrium following CO_2 loss at the gas exchange surface has only recently come under scrutiny in fish, although it has been a subject of debate in mammalian systems for some time. Furthermore, the physiological significance of postbranchial or postcapillary changes in blood pH remains uncertain, despite the intense research effort directed toward this question. A related issue is whether CO_2–HCO_3^-–H^+ reactions are in equilibrium following the addition of CO_2 to the blood in the tissues. This chapter will review the work that has been carried out to date on CO_2–HCO_3^-–H^+ disequilibria in fish blood, with an emphasis on the mechanisms behind disequilibrium events. Detailed reviews of the situation in mammalian systems have been provided by Bidani and Crandall (1988) and Klocke (1988). For an in-depth examination of CO_2 excretion in fish, the reader is referred to Chapter 7 by Tufts and Perry (this volume), while Chapter 3 by Henry and Heming (this volume) provides detailed information on the distribution and function of carbonic anhydrase in fish.

Fish Physiology, Volume 17:
FISH RESPIRATION

II. POSTBRANCHIAL DISEQUILIBRIA

A. Causes

It was Roughton (1935) who first recognized that changes in plasma pH should lag behind alterations of P_{CO_2} when carbonic anhydrase (CA) activity is not accessible to plasma reactions. Subsequently, computer models for mammalian systems predicted that, in the absence of CA activity available to plasma reactions, plasma pH should continue to increase in the postcapillary blood following CO_2 excretion in the lungs (Hill et al., 1973; Forster and Crandall, 1975; Bidani et al., 1978; Zock et al., 1981). Owing to the presence of lung capillary endothelial CA activity that is available to plasma reactions (Crandall and O'Brasky, 1978; Effros et al., 1978; Klocke, 1978; Hanson et al., 1981), however, these models can only be tested experimentally in the presence of CA inhibitors. Crandall et al. (1977) used an in vivo preparation in which the withdrawal of arterial blood through an external chamber containing a pH electrode was suddenly stopped; by theory, pH should remain constant when the flow of blood is stopped if the blood is in an equilibrium state. In the presence of the CA inhibitor benzolamide, a postcapillary increase in plasma pH of 0.02–0.04 unit was observed when the flow of blood was stopped, a result that agreed well with the predictions of the computer models. A similar disequilibrium was manifest when an appropriate dose of the CA inhibitor acetazolamide was employed (Bidani and Crandall, 1978). In vitro preparations, consisting of isolated lungs perfused with saline solutions in which the effluent perfusate is passed through a stopflow pH electrode apparatus, have also been used to test the computer models. The results of these experiments consistently demonstrated an increase in outflowing perfusate pH under stopflow conditions when CA inhibitors such as acetazolamide were included in the inflowing perfusate (Crandall and O'Brasky, 1978; Bidani et al., 1983; Heming et al., 1986, 1994; Heming and Bidani, 1990, 1992).

The results of in vitro experiments on mammalian lung preparations were also consistent in demonstrating an equilibrium condition in outflowing perfusate in the absence of CA inhibitors (Table 1), a state that was ascribed to the existence of lung capillary endothelial CA activity accessible to the perfusate (Crandall and O'Brasky, 1978; Bidani et al., 1983; Heming et al., 1986, 1994; Heming and Bidani, 1990, 1992). Corresponding in vivo experiments, however, produced more variable results, including a smaller than expected increase in plasma pH (Hill et al., 1977; Bidani and Crandall, 1978; Parrott et al., 1979; Ponte and Purves, 1980; Takahashi and Phillipson, 1991), no change (Crandall et al., 1977), or a decrease in plasma pH (Rispens et al., 1980; Chakrabarti et al., 1983) in the postcapillary blood (Table 1).

Table 1
Acid–Base Disequilibria in Blood Leaving the Gas Exchange Surface

Preparation	Animal	ΔpH during stopflow	Ref.
In vivo	Cat or dog	0.000 (24)	Crandall et al., 1977
In vivo	Dog	0.008 ± 0.001 (62)	Hill et al., 1977
In vivo	Cat or dog	0.010 ± 0.004 (51)	Bidani and Crandall, 1978
In vivo	Dog	−0.015 to −0.035 (4)[a]	Rispens et al., 1980
In vivo	Dog	−0.0036 ± 0.0012 (22)	Chakrabarti et al., 1983
Saline-perfused isolated lung	Rat	0.001 ± 0.0004 (11)	Crandall and O'Brasky, 1978
Saline-perfused isolated lung	Rabbit	0.000 ± 0.000 (4)	O'Brasky and Crandall, 1980
Saline-perfused isolated lung	Guinea pig	0.000 ± 0.000 (4)	O'Brasky and Crandall, 1980
Isolated lung perfused with RBC suspension (hct = 20%)	Rat	0.000 ± 0.000 (19)	Crandall et al., 1981
Saline-perfused lungs	Turtle	0.00 (8)	Stabenau et al., 1996
Lung perfused with RBC suspension (hct = 15%)	Turtle	0.0 (16)	Stabenau, 1994
In vivo	Rainbow trout	0.032 ± 0.003 (84)	Gilmour et al., 1994, 1996; Gilmour and Perry, 1994, 1996
In vivo	Spiny dogfish	−0.028 ± 0.003 (27)	Gilmour et al., 1996
In vivo	Spiny dogfish	−0.028 ± 0.009 (6)	Wilson, 1995

[a] Difference in pH between two measuring sites rather than ΔpH during stopflow period.

These inconsistencies have yet to be fully resolved, underlining the difficulties inherent in measuring small pH differences *in vivo* (Swenson, 1984, 1990), as well as the complex origins of disequilibria *in vivo*. The attainment of an equilibrium condition in a multicompartment solution, such as blood, depends on equilibrium being reached within each compartment, that is, within both the red blood cell (RBC) and the plasma, and also between the compartments, by the exchange of acid–base equivalents across the RBC membrane (Bidani and Crandall, 1988). Because vertebrate erythrocytes contain high concentrations of CA (Klocke, 1988; Henry and Heming,

1998), ensuring very rapid equilibration of RBC $CO_2-HCO_3^--H^+$ reactions, and because the band 3 Cl^-/HCO_3^- exchanger present in most vertebrate RBC membranes (Bidani and Crandall, 1988; Tufts and Perry, 1998) provides for rapid movement of acid–base equivalents across the RBC membrane (but see Section II,B), attention has been focused largely on the role of plasma $CO_2-HCO_3^--H^+$ reactions in the generation of an acid–base disequilibrium. Computer models of postcapillary plasma pH changes in the presence of endothelial CA activity variously predict an increase or decrease in plasma pH, with the direction of the disequilibrium probably dependent upon the extent of catalysis of plasma reactions by endothelial CA (Crandall and Bidani, 1981; Zock et al., 1981; Bidani and Crandall, 1982). The occurrence of plasma CA inhibitors in some species, including dogs, rats, and rabbits (Hill, 1986; Roush and Fierke, 1992), all of which have been used in disequilibrium studies, provides an additional complicating factor, making it difficult to predict to what extent lung capillary endothelial CA activity is available to catalyze plasma $CO_2-HCO_3^--H^+$ reactions.

Fish provide a useful alternative to mammalian systems in which to test predictions about plasma pH changes in blood leaving the gas exchange surface because, in contrast to other vertebrates, which all appear to possess CA activity that is available to plasma reactions at the gas exchange surface (mammals: Crandall and O'Brasky, 1978; Effros et al., 1978; Klocke, 1978; reptiles: Fain and Rosen, 1973; Stabenau, 1994; Stabenau et al., 1996; amphibians: Fain and Rosen, 1973), at least some groups of fish do not (Henry et al., 1988, 1993; Perry and Laurent, 1990; Perry et al., 1997). This apparent difference in the distribution of CA activity between air- and water-breathers raises questions about the physiological significance of CA activity at the gas exchange surface and its impact on the status of $CO_2-HCO_3^--H^+$ reactions. Interestingly, recent evidence suggests that plasma reactions in the gills of certain elasmobranch fish may have access to CA activity (Wood et al., 1994; Swenson et al., 1995; Wilson, 1995; Gilmour et al., 1997).

RAINBOW TROUT

Plasma reactions in the gills of teleost fish, such as rainbow trout, do not have access to CA activity. Branchial tissue has been shown to contain CA, but the activity is restricted to the cytoplasm of the epithelial cells and is not associated with the pillar cell membranes (Henry et al., 1988, 1993; Rahim et al., 1988). The blood plasma of rainbow trout does not exhibit CA activity (Wood et al., 1994) and actually contains a CA inhibitor (Dimberg, 1994). According to the models developed for mammalian systems, then, a slow increase in the plasma pH might be expected to occur downstream of gas exchange in the gills. Such a disequilibrium has in fact been documented in the postbranchial blood of rainbow trout using an extracorporeal preparation in combination with a stopflow technique (Gilmour and Perry, 1994, 1996; Gilmour *et al.*, 1994, 1997; Perry *et al.*, 1997). Blood was pumped from the coeliac artery of the fish through a sealed external circuit containing pH, P_{O_2}, and P_{CO_2} electrodes. When the flow of blood through the extracorporeal circulation was stopped by turning off the pump, pH rose, indicating that plasma CO_2–HCO_3^-–H^+ reactions were not in equilibrium (Fig. 1).

The slow pH rise in the postbranchial blood is the result of an imbalance in CO_2–HCO_3^-–H^+ reactions (Fig. 2). As blood enters the branchial vasculature, CO_2 is lost from both the plasma and the RBC to the ventilatory water, driving CO_2–HCO_3^-–H^+ reactions toward HCO_3^- dehydration and resulting in the transfer of plasma HCO_3^- ions into the RBC via the chloride shift (Perry, 1986). Because the branchial CA activity is sequestered within

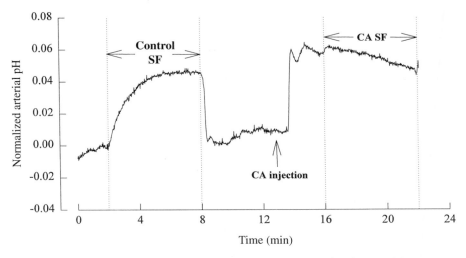

Fig. 1. A representative, continuous recording of arterial pH (pH_a) in a rainbow trout is presented to illustrate the increase in pH_a that occurs when the flow of blood through the extracorporeal circulation is stopped under control conditions (control SF). After the flow of blood through the external circuit was restarted, bovine CA (10 mg in 1 ml teleost saline) was injected into the dorsal aorta. Injection of CA established an equilibrium condition in the blood, causing arterial pH in flowing blood to increase to its equilibrium value and eliminating the disequilibrium when the flow of blood was stopped (CA SF). Stopflow (SF) periods are marked by the dotted lines. Data were acquired at a sampling frequency of 1 s^{-1} using Biopac Systems Inc. analog-to-digital conversion system and AcqKnowledge III software and were normalized by the subtraction from each data point in the recording of pH_a the value at the beginning of the control stopflow period. The extracorporeal circulation used was as described by Gilmour *et al.* (1994).

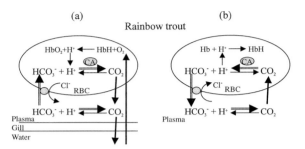

Fig. 2. A schematic model of processes involved in (a) CO_2 excretion at the gill and (b) the generation of an acid–base disequilibrium in the postbranchial blood of rainbow trout. Because plasma reactions do not have access to CA activity (a), the plasma HCO_3^- dehydration reaction continues in the postbranchial blood (b), effectively transferring protons from the plasma to the RBC and producing a positive acid–base disequilibrium.

the gill cells and is not available to plasma reactions, the plasma dehydration reaction takes place at the uncatalyzed rate, which has a half-time of 25–90 s (Perry, 1986) under these conditions, yet the residence time for blood in the gill is probably only 0.5–2.5 s (Cameron and Polhemus, 1974; Hughes *et al.*, 1981; Randall, 1982). Thus, as the blood leaves the branchial vasculature, the plasma CO_2–HCO_3^-–H^+ reactions will be out of equilibrium. Imbalances are also likely to exist between the plasma and RBC with respect to H^+ ion concentrations due to the rapid consumption of protons by the catalyzed HCO_3^- dehydration reaction within the RBC versus the negligible depletion of H^+ ions by the uncatalyzed plasma reaction. The correction of these imbalances in the postbranchial blood by means of the Jacobs–Stewart cycle, which has as its slowest step the uncatalyzed dehydration of HCO_3^- in the plasma (Bidani *et al.*, 1978), gives rise to the disequilibrium; protons are consumed, causing plasma pH to increase, and CO_2 is formed. The formation of CO_2 in the closed system of the postbranchial vasculature should be detectable as an increase in P_{CO_2} in the extracorporeal loop under stopflow conditions, but owing to the slow response time of P_{CO_2} electrodes at low temperature and the small changes anticipated in blood P_{CO_2}, the results of P_{CO_2} measurements under stopflow conditions have been inconsistent (Gilmour and Perry, 1994; Gilmour *et al.*, 1994).

An equilibrium condition was produced by the intraarterial injection of bovine CA. The addition of extracellular CA activity to the plasma enables plasma CO_2–HCO_3^-–H^+ reactions to proceed at the catalyzed rate and come to equilibrium before the blood reaches the pH electrode; i.e., CO_2–HCO_3^-–H^+ reactions are in equilibrium in the postbranchial blood

(Gilmour *et al.*, 1994). In addition to eliminating the disequilibrium, injection of bovine CA generally resulted in a very rapid increase in arterial pH under conditions of blood flow through the external circuit (Fig. 1). The magnitude of this pH increase corresponded closely to that of the pH rise under stopflow conditions, suggesting that the increase in pH during blood flow reflected the establishment of the equilibrium condition in the circulatory system. That is, administration of bovine CA caused the arterial pH to increase to the equilibrium value, the pH measured under control conditions at the end of the stopflow period. The increase in pH was generally not accompanied by a decrease in arterial P_{CO_2}.

DOGFISH

In contrast to the situation in rainbow trout, plasma reactions in dogfish do appear to have access to CA activity. Wood *et al.* (1994) used a radioisotopic HCO_3^- dehydration assay (Wood and Perry, 1991) to demonstrate that the rate of HCO_3^- dehydration of separated plasma from the dogfish *Scyliorhinus canicula* was significantly reduced by the CA inhibitor acetazolamide. Similar results were obtained by Gilmour *et al.* (1997) for *Squalus acanthias,* using both an electrometric ΔpH assay (Henry and Kormanik, 1985; Henry, 1991) and the radioisotopic dehydration assay. It should be noted that the blood plasma of the spiny dogfish does not contain a CA inhibitor and that the extracellular CA activity may arise from natural, endogenous RBC lysis (Henry *et al.,* 1997a). Calculation of the factor by which the rate of the uncatalyzed CO_2 hydration reaction was enhanced in the presence of plasma versus RBC lysate indicated that the extracellular CA activity in dogfish plasma (enhancement factor of approximately 12-fold) is very low in comparison to the CA activity in the RBC (enhancement factor of nearly 20,000-fold) (Henry *et al.,* 1997a). Indeed, the extracellular CA activity on its own was found to be insufficient to avoid the generation of a disequilibrium *in vitro* when two dogfish blood samples of differing CO_2 tensions were mixed (Gilmour *et al.,* 1997). However, plasma CO_2–HCO_3^-–H^+ reactions in the branchial vasculature may have access to gill membrane-bound CA activity in addition to the extracellular CA activity. Wilson (1995) utilized a saline-perfused gill preparation, similar in concept to the saline-perfused isolated lung preparations employed by mammalian researchers (Crandall and O'Brasky, 1978; Klocke, 1978), to obtain evidence for pillar cell membrane-associated CA activity in the gills of *Squalus acanthias.* Further evidence was provided by *in vivo* experiments in which CA inhibitors of different tissue permeabilities were used to distinguish between the roles of gill membrane-bound CA and cytoplasmic CA during recovery from a metabolic acidosis in *S. acanthias* (Swenson *et al.,* 1995, 1996). Although the degree of gill membrane-bound CA activity has not yet been quantified (nor has the CA isozyme been identified), it is evident that plasma CO_2–HCO_3^-–H^+ reactions in the branchial vasculature of dogfish will be catalyzed to a greater extent than those in the postbranchial blood. The impact of these CA activities on the acid–base status of blood leaving the gills of dogfish was examined by Wilson (1995) and Gilmour *et al.* (1997) using *in vivo* preparations consisting of an extracorporeal blood circulation in combination with a stopflow technique.

In both studies, pH *decreased* when the flow of blood was stopped under control conditions (Fig. 3). Gilmour *et al.* (1997) clearly demonstrated that the *absence* of a pH *rise* in the postbranchial blood of the dogfish was directly attributable to the extracellular and membrane-bound CA activities, since pH did increase under stopflow conditions when these CA activities were

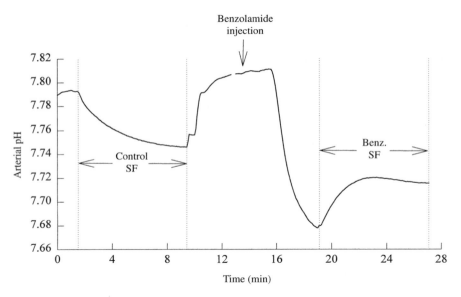

Fig. 3. A representative, continuous recording of arterial pH (pH$_a$) in a spiny dogfish illustrating the decrease in pH$_a$ that occurs when the flow of blood through the extracorporeal circulation is stopped under control conditions (control SF). The fish was treated with benzolamide (1.3 mg kg^{-1}) when the flow of blood through the external circuit was restarted after the control stopflow period. Benzolamide selectively inhibited the extracellular and gill membrane-bound CA activities, leading to the production of a positive disequilibrium in the postbranchial blood (benz. SF). The decrease in pH$_a$ in flowing blood following the injection of benzolamide was probably a consequence of the establishment of the new disequilibrium condition, since the equilibrium pH$_a$ value (pH$_a$ at the end of a stopflow period) was similar under both control and benzolamide conditions (Gilmour *et al.*, 1997). Stopflow (SF) periods are marked by the dotted lines. The extracorporeal circulation and data acquisition systems were described by Gilmour *et al.* (1997).

selectively inhibited with benzolamide (Fig. 3). The basis of the *decrease* in plasma pH under stopflow conditions, however, was less certain. It seems likely that the slow downstream pH decrease arises from a reequilibration of protons and CO_2 across the RBC membrane in the postbranchial blood by the Jacobs–Stewart cycle (Fig. 4) (Crandall and Bidani, 1981; Wilson, 1995; Gilmour *et al.*, 1997). This reequilibration of protons and CO_2 is necessary because a significant portion of HCO_3^- dehydration is likely to occur in the plasma in the presence of CA activity available to plasma reactions. Owing to the low buffering capacity of the plasma in comparison to that of the RBC (Wood *et al.*, 1994), the plasma HCO_3^- dehydration reaction will result in a relatively greater depletion of protons from the plasma than from the RBC during the transit of the blood through the gills, necessitating the transfer of protons from RBC to plasma in the postbranchial blood. The movement of protons by means of the Jacobs–Stewart cycle would be perceptible as a disequilibrium embodying a fall in plasma pH in the postbranchial blood. Although the rate of the Jacobs–Stewart cycle is generally thought to be limited by the speed of the uncatalyzed plasma hydration–dehydration reaction (Bidani *et al.*, 1978), given the presence of extracellular CA activity in the dogfish, albeit limited activity, it is also possible that the Jacobs–Stewart cycle in dogfish could be constrained by the rate of Cl^-/HCO_3^- exchange (see Section II,B).

Interestingly, the model developed by Crandall and Bidani (1981) for mammalian systems predicts that slow decreases in plasma pH will be detectable in the postpulmonary blood even in the presence of large

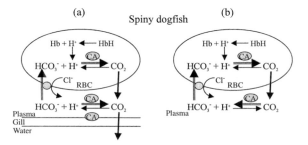

Fig. 4. A diagrammatic representation of processes involved in (a) CO_2 excretion at the gill and (b) the production of an acid–base disequilibrium in the postbranchial blood of the spiny dogfish. In (a), the availability of extracellular and gill membrane-bound CA activities to plasma reactions in the branchial vasculature, coupled with the low buffer capacity of the plasma relative to the RBC, results in a relatively greater depletion of protons from the plasma than from the RBC. The transfer of protons from the RBC to the plasma in the postbranchial blood (b) is probably the basis of the negative disequilibrium. Modified from Gilmour *et al.* (1997).

amounts of lung capillary endothelial CA because the plasma is relatively less well buffered than the RBCs. As with the rainbow trout, infusion of bovine CA eliminated the disequilibrium in dogfish, in apparent disagreement with the mammalian model. The discrepancy is resolved by the fact that the mammalian model stipulates that CA activity is available to plasma reactions only in the lung capillaries, whereas intraarterial injection of bovine CA results in CA activity being available to plasma reactions throughout the circulatory system.

B. Variations on the Basic Theme

Despite the obvious difference between rainbow trout and dogfish with respect to the direction of the postbranchial blood acid–base disequilibria, the underlying basis of the disequilibrium appears to be similar in both: the reequilibration of H^+ ions across the RBC membrane by the Jacobs–Stewart cycle as well as the reestablishment of an equilibrium condition in plasma CO_2–HCO_3^-–H^+ reactions (Figs. 2 and 4). The difference lies in the direction in which the Jacobs–Stewart cycle proceeds, which, in turn, is a function of the availability of CA activity to plasma reactions. Alterations in the disequilibrium in response to various factors can be interpreted in light of these similarities and differences.

ACETAZOLAMIDE AND BENZOLAMIDE

Treatment of dogfish with acetazolamide resulted in either an increase in the magnitude of the pH *fall* under stopflow conditions (Gilmour *et al.*, 1997) or the establishment of a disequilibrium in which pH *increased* (Wilson, 1995), whereas in trout treated with acetazolamide, the direction of the disequilibrium was reversed so that plasma pH *decreased* when the flow of blood past the pH electrode was stopped (Gilmour *et al.*, 1994). The explanation for these apparently contradictory results (Table 2) lies in the dose of acetazolamide employed and the time after administration of the drug at which the disequilibrium was measured. At high concentrations and given sufficient time to equilibrate across the RBC membrane, acetazolamide, methazolamide, and benzolamide all inhibit RBC CA activity significantly, resulting in the development of a respiratory acidosis (Swenson and Maren, 1987; Henry *et al.*, 1988, 1995). In dogfish (Gilmour *et al.*, 1997) or trout (Gilmour *et al.*, 1994) treated with 30 mg kg^{-1} acetazolamide, RBC CA activity is significantly inhibited, and CO_2 production continues after the blood leaves the branchial vasculature. Because the postbranchial blood is in a closed system, the CO_2 formed in the RBC diffuses into the plasma, driving plasma CO_2–HCO_3^-–H^+ reactions toward CO_2 hydration and the generation of protons (at the uncatalyzed rate) and yielding a large disequi-

Table 2

Acid–Base Disequilibria in the Postbranchial Blood of Fish Following CA Inhibition

Fish	CA inhibitor	Dose (mg kg^{-1})	CA presumably inhibited	Control ΔpH	Treatment ΔpH
Trout[a]	Acetazolamide	30	RBC	0.036 ± 0.009 (6)	−0.041 ± 0.005 (6)*
Trout[b]	Benzolamide	1.3	—	0.038 ± 0.004 (6)	0.038 ± 0.009 (6)
Dogfish[c]	Acetazolamide	30	RBC, extracellular, gill membrane-bound	−0.019 ± 0.006 (6)	−0.045 ± 0.007 (6)*
Dogfish[c]	Benzolamide	13	RBC, extracellular, gill membrane-bound	−0.047 (1)	−0.136 (1)
Dogfish[c]	Benzolamide	1.3	Extracellular, gill membrane-bound	−0.053 ± 0.012 (6)	0.059 ± 0.016 (6)*
Dogfish[d]	Acetazolamide	1	Extracellular, gill membrane-bound	−0.028 ± 0.009 (6)	0.042 ± 0.011 (6)*

[a] Gilmour et al., 1994.
[b] Perry et al., 1997.
[c] Gilmour et al., 1996.
[d] Wilson, 1995.
*Significant difference between ΔpH during the control and treatment stopflow periods.

librium characterized by a fall in plasma pH. If, however, the dose of acetazolamide is kept low enough to avoid significant inhibition of RBC CA activity, then the resultant selective inhibition of the extracellular and gill membrane-bound CA activities transforms the dogfish into a trout with respect to CO_2 excretion and leads to a slow plasma pH rise in blood downstream of the gills (Wilson, 1995). Similarly, a pH increase under stopflow conditions was measured in dogfish treated with a low dose (1.3 mg kg^{-1}) of benzolamide (Gilmour et al., 1997). The absence of extracellular and gill membrane-bound CA activities in rainbow trout implies that treatment with a low dose of benzolamide should have no effect on the acid–base disequilibrium in the postbranchial blood, and this was indeed shown to be the case (Perry et al., 1997).

A Possible Role for Cl^-/HCO_3^- Exchange

The entry of plasma HCO_3^- into the RBC via the band 3 Cl^-/HCO_3^- exchanger is generally thought to be the rate-limiting step in CO_2 excretion in fish (Perry, 1986; Perry and Gilmour, 1993; Wood and Munger, 1994). This step could also contribute to disequilibrium events in the postbranchial blood owing to the similarity between the reaction time for Cl^-/HCO_3^- exchange (120–400 ms for 63% equilibration of ion flux: Cameron, 1978; Obaid et al., 1979; Jensen and Brahm, 1995) and the residence time of blood in the gills (0.5–2.5 s: Cameron and Polhemus, 1974; Hughes et al., 1981; Randall, 1982; Butler and Metcalfe, 1988); it has been suggested (Jensen and Brahm, 1995) that to avoid a disequilibrium condition, the gill transit time should be about a factor of 50 greater than the anion exchange 63% reaction time. Theoretical (Crandall and Bidani, 1981) and experimental (Crandall et al., 1981) work on mammalian and reptilian (Stabenau, 1994) systems indicated that a large disequilibrium in which plasma pH decreases should exist in the postpulmonary blood when RBC Cl^-/HCO_3^- exchange rates are slowed down. In fish, the magnitude of the pH fall under stopflow conditions in the postbranchial blood of dogfish was significantly increased following treatment with the anion exchange inhibitor 4,4'-diisothiocyano-2,2'-disulfonic stilbene (DIDS) (Gilmour et al., 1997). The reequilibration of CO_2 and protons across the RBC membrane, which is probably the basis of the disequilibrium in dogfish postbranchial blood under control conditions, proceeds at a reduced rate in DIDS-treated dogfish because the speed of the Jacobs–Stewart cycle is limited by the anion exchange step in addition to the limitation, which is also present under control conditions, imposed by the speed of the plasma CO_2 hydration reaction. Four of the DIDS-treated dogfish examined by Gilmour et al. (1997) were subsequently injected with bovine CA. Although bovine CA infusion following DIDS treatment did not eliminate the disequilibrium as had CA treatment alone, it did significantly reduce the magnitude of the pH decrease under stopflow conditions (Fig. 5) (K. M. Gilmour, R. P. Henry, C. M. Wood, and S. F. Perry, unpublished data). Infusion of bovine CA should relieve all but the anion exchange limitation on the Jacobs–Stewart cycle, such that the disequilibrium observed in DIDS + CA-treated fish is due to the slow rate of anion exchange alone. By inference, then, the absence of a persistent disequilibrium in dogfish treated only with bovine CA suggests that the pH decrease observed under control (stopflow) conditions is a result of the slow rate of plasma CO_2–HCO_3^-–H^+

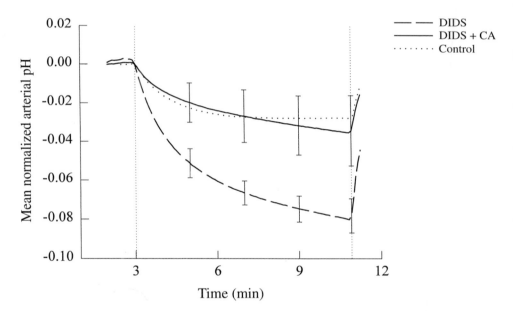

Fig. 5. Mean normalized values for arterial pH (pH_a; $N = 4$) in the postbranchial blood of spiny dogfish during sequential stopflow periods (marked by the dotted lines) under control conditions, following treatment with DIDS (final concentration in the blood of 2×10^{-5} mol L^{-1}) and following the subsequent administration of bovine CA (2 mg kg^{-1}). The arterial pH at 11 min in the DIDS response was significantly different from the control and DIDS + CA values (one-way ANOVA followed by Fisher's LSD multiple comparisons test, $P < 0.05$). Because the magnitude of pH changes in the stopflow period was small relative to individual variability in pH_a, data for individual fish were normalized by subtracting from each point in a response the value at the beginning of the stopflow period. Error bars represent SEMs and are shown only every 2 min on the DIDS and DIDS + CA responses for clarity. The extracorporeal circulation and data acquisition system were those detailed by Gilmour *et al.* (1997).

reactions, i.e., that the Jacobs–Stewart cycle under control conditions is limited by the rate of the plasma hydration–dehydration reaction, despite the presence of extracellular CA activity, and not by the rate of Cl^-/HCO_3^- exchange. In the context of this argument, it is noteworthy that Bhargava *et al.* (1992) determined a gill blood transit time of 6.5 s in the leopard shark, *Triakis semifasciata,* a value considerably longer than that estimated for teleost fish and about 40 times greater than the reaction time to 63% equilibration for anion exchange across dogfish RBCs (160 ms, estimated from data presented in Obaid *et al.,* 1979; Jensen and Brahm, 1995).

A reduction in the rate of Cl^-/HCO_3^- exchange might be predicted to reverse the direction of the postbranchial disequilibrium in rainbow trout. The inhibition of anion exchange in rainbow trout will result in post-branchial imbalances not only in the plasma CO_2–HCO_3^-–H^+ reactions, the basis of the disequilibrium observed under control conditions (see Section II,A), but also in the RBC CO_2–HCO_3^-–H^+ reactions. Loss of CO_2 from the RBC by diffusion coupled with the release of Bohr protons from hemoglobin during oxygenation will create conditions favoring HCO_3^- dehydration, but the slow entry of HCO_3^- into the RBC via the band 3 exchanger will limit the rate of this normally rapid equilibration process. Under these conditions, the RBC HCO_3^- dehydration reaction will continue in the postbranchial blood. Further, an imbalance in H^+ across the RBC membrane opposite to that normally expected may exist owing to the release of Bohr protons not consumed in HCO_3^- dehydration, and the equilibration of these Bohr protons across the RBC membrane would also lead to slow decreases in plasma pH in the postbranchial blood. The effect of a reduction in the rate of Cl^-/HCO_3^- exchange on the disequilibrium in rainbow trout postbranchial blood has not yet been tested experimentally, but somewhat unexpected results have been obtained in experiments carried out on turtle lungs perfused with a RBC suspension (Stabenau, 1994). Turtle lungs possess pulmonary capillary endothelial CA activity, and a pH decrease under stopflow conditions was apparent in the effluent perfusate following addition of the anion exchange inhibitor DIDS to the perfusate, a situation corresponding to that for DIDS-treated dogfish (Gilmour *et al.,* 1997). However, inclusion of the membrane-impermeant CA inhibitor quaternary ammonium sulfanilamide (QAS) in the perfusate to inhibit the lung capillary endothelial CA activity, thereby producing circumstances parallel to those that might be expected in a DIDS-treated trout, resulted in the measurement of a disequilibrium in which pH *increased* rather than the pH *decrease* predicted by theory (Stabenau, 1994). Further *in vivo* disequilibrium experiments on rainbow trout may help to resolve this apparent discrepancy.

It is interesting to speculate whether the lamprey can be considered an extreme case with respect to the inhibition of anion exchange since this agnathan may lack a functional RBC membrane Cl^-/HCO_3^- exchanger (Nikinmaa and Railo, 1987; Tufts and Boutilier, 1989, 1990). Two additional differences between lamprey and other fish suggest that the postbranchial blood of lamprey may exhibit an acid–base status other than the large pH decrease that might have been predicted to occur based solely on a consideration of the apparent absence of a band 3 anion exchanger. In lamprey, unlike in other fish, the bulk of the HCO_3^- dehydrated in the branchial vasculature for CO_2 excretion may be carried in the RBC (Tufts and Boutilier, 1989; Nikinmaa and Mattsoff, 1992); Tufts *et al.* (1992) have calculated that 62% of the difference in total CO_2 content between arterial and venous blood can be accounted for by changes in the RBC total CO_2 content (versus 8% in rainbow trout). Lamprey are also unusual in actively maintaining RBC pH at a high value and over a broad range of extracellular pH values (Nikinmaa, 1986; Tufts, 1991; Tufts *et al.*, 1992). Thus, events in the plasma and RBC appear to be largely independent insofar as CO_2 excretion and pH regulation are concerned. As lamprey appear to lack gill membrane-bound CA that faces the plasma as well as extracellular CA activity (Henry *et al.*, 1993), plasma CO_2–HCO_3^-–H^+ reactions proceed at the uncatalyzed rate, and therefore slow plasma pH increases might be expected to occur in the postbranchial blood, resulting from the effect of diffusive CO_2 loss at the gills on plasma CO_2–HCO_3^-–H^+ reactions. This question undoubtedly warrants investigation, particularly in light of recent studies that suggest that CO_2 reactions in the postbranchial blood of the lamprey may be in equilibrium (Tufts *et al.*, 1996) and that plasma HCO_3^- may have access to the RBC via an anion exchanger-like protein in the RBC membrane (Kay *et al.*, 1995; Cameron *et al.*, 1996).

HYPOXIA AND HYPERCAPNIA

Exposure to hypoxia slightly increased the magnitude of both the plasma pH rise in the postbranchial blood of trout (Gilmour and Perry, 1994) and the plasma pH fall in dogfish postbranchial blood (Gilmour *et al.*, 1997) under stopflow conditions. The effect of hypoxia on the disequilibrium in rainbow trout blood may be linked to the presence of a Haldane effect in this species. Bohr protons derived from the oxygenation of hemoglobin provide a significant contribution to RBC HCO_3^- dehydration in teleosts, and hence to CO_2 excretion (Klocke, 1988; Jensen, 1991; Perry and Gilmour, 1993; Perry *et al.*, 1996). The contribution of oxylabile protons to HCO_3^- dehydration will be reduced under hypoxic conditions (Jensen, 1986; Brauner *et al.*, 1996), resulting in a relatively greater depletion of RBC protons in comparison to plasma protons than under control conditions

(Brauner and Randall, 1996). Hence, a larger H^+ imbalance across the RBC membrane will ensue, and the correction of this imbalance by the Jacobs–Stewart cycle in the postbranchial blood (see Section II,A) will lead to an increase in the magnitude of the pH rise under stopflow conditions. In elasmobranch fish, on the other hand, the Haldane effect appears to be absent (Albers and Pleschka, 1967; Pleschka *et al.*, 1970; Butler and Metcalfe, 1988; Jensen, 1989; Graham *et al.*, 1990; Wood *et al.*, 1994), but a significant alkalinization of the plasma occurred as a result of exposure to hypoxia (Gilmour *et al.*, 1997). Although RBC intracellular pH would presumably also increase during hypoxia, the effect might be expected to be more pronounced in the plasma because of its lower buffering capacity. As in rainbow trout, then, the increased magnitude of the disequilibrium in dogfish blood under hypoxic conditions (a larger fall in plasma pH) would reflect the greater H^+ imbalance across the RBC membrane in blood leaving the branchial vasculature resulting, in this case, from the relatively greater depletion of plasma protons in comparison to RBC protons (see Section II,A). The effects of hypoxia on the disequilibria in trout and dogfish blood underline the importance of RBC and plasma nonbicarbonate buffer capacities in determining the size and direction of the acid–base disequilibrium. Not only do differences between the RBC and plasma compartments in proton availability, which is a function of the non-HCO_3^- buffer capacity, and the availability of CA activity setup the conditions that result in an acid–base disequilibrium (see Sections II,A), but theoretical and experimental studies on saline-perfused mammalian lungs suggest that the nonbicarbonate buffer capacity of the perfusate has significant, nonlinear effects on the size of the postcapillary pH disequilibrium (Gray, 1971; Bidani and Heming, 1991; Heming and Bidani, 1992). This prediction has not yet been tested in fish.

In contrast to the similar effect of hypoxia on the size of the disequilibria in trout and dogfish blood, hypercapnia caused very different responses in the two species. Whereas the magnitude of the disequilibrium was unaffected by hypercapnia in dogfish (Gilmour *et al.*, 1997), the disequilibrium in hypercapnic trout was approximately double that observed under normocapnic conditions (Gilmour and Perry, 1994). This difference is particularly striking when the different levels of hypercapnia used in the two experiments are considered: a final arterial P_{CO_2} of about 9 torr was achieved in dogfish, but only 5 torr in trout. Correspondingly, the respiratory acidosis resulting from the exposure to environmental hypercapnia was more severe in dogfish (arterial pH during hypercapnia of 7.29 versus control pH of 7.80) than in trout (arterial pH during hypercapnia of 7.65 versus control pH of 7.93). The dramatic increase in the magnitude of the disequilibrium in trout blood under hypercapnic conditions may have resulted from a combination of an increased plasma H^+ ion concentration, the Haldane effect, and the mobilization of catecholamines into the circulation. Because of the respiratory acidosis, the O_2-carrying capacity of hemoglobin will be lowered via the Root effect, which may, in turn, lead to a reduction in the release of Bohr protons during oxygenation (Brauner *et al.*, 1996). The increase in plasma proton availability coupled with a relative reduction in RBC proton availability will magnify the H^+ imbalance across the RBC membrane in blood leaving the branchial vasculature, requiring a greater transfer of protons from plasma to RBC to reestablish equilibrium and thereby resulting in an enlargement of the disequilibrium. At the same time, the mobilization of catecholamines into the circulation may have provided a second factor contributing to the increase in the magnitude of the disequilibrium. Catecholamine concentrations were not measured in the study of Gilmour and Perry (1994), but other studies have demonstrated catecholamine release at similar levels of hypercapnia (Thomas *et al.*, 1994). Large, significant increases in the magnitude of the disequilibrium in post-branchial blood were measured in rainbow trout following either the administration of exogenous catecholamines or the stimulation of endogenous catecholamine release by exposure to hypoxia, presumably owing to the increased plasma proton availability stemming from catecholamine-mediated extrusion of protons from the RBC (Gilmour and Perry, 1996).

Clearly, much of the preceding discussion of the effects of varying the environmental gas tensions on the acid–base disequilibria in fish blood is speculative, and numerous additional parameters, including ventilation status and cardiac output, may contribute to the differences observed. It is evident that further work is necessary to clarify the complex array of factors that may influence the acid–base disequilibria in the postbranchial blood of fish.

III. PREBRANCHIAL DISEQUILIBRIA

The circumstances that result in the creation of an acid–base disequilibrium in blood leaving the gas exchange organ should, in theory, also lead to the generation of an acid–base disequilibrium in blood leaving any organ in which CO_2 has been added to the blood, such as muscle. That is, in the absence of CA activity available to plasma reactions, the addition of CO_2 to the blood might be expected to cause slow plasma pH decreases downstream of the site of CO_2 addition. This aspect of the disequilibrium issue has received little attention to date, perhaps at least in part owing to the technical challenge of selecting an appropriate measurement site from among the relatively extensive network of vessels leaving many organs.

Two investigations of acid–base status following gas exchange in a systemic vascular bed are reported in the literature for saline-perfused isolated organ preparations from mammals (O'Brasky et al., 1979; O'Brasky and Crandall, 1980). Extension of these results to an in vivo situation should be treated with caution, given the discrepancies between results for saline-perfused isolated lung preparations and in vivo preparations (see Section II,A). A disequilibrium in which pH decreased was found to exist in the effluent perfusate from an isolated guinea pig liver preparation (O'Brasky et al., 1979). The disequilibrium was unaffected by treatment with acetazolamide but abolished by addition of CA to the perfusate, results that are in agreement with the situation predicted to exist when plasma reactions are not catalyzed by CA. The magnitude of the disequilibrium measured in the effluent perfusate from isolated rabbit hindlimb or liver preparations, on the other hand, was increased by addition of acetazolamide to the perfusate (O'Brasky and Crandall, 1980), suggesting that plasma reactions in these preparations had access to some CA activity, albeit not enough to establish an equilibrium condition. An indicator-dilution technique was utilized to confirm the presence of extracellular CA in skeletal muscle (Zborowska-Slus et al., 1974; Effros and Weissman, 1979; Geers and Gros, 1984; Geers et al., 1985); the technique involves comparing the washout from the perfused tissue of ^{14}C-labeled HCO_3^- and/or CO_2 to that of labeled indicators for the vascular, extracellular, and water compartments, in the presence and absence of CA inhibitors. Extracellular CA activity in muscle is believed to accelerate CO_2 transport from the muscle to the blood as well as enhance CO_2 uptake by the blood (Gros and Dodgson, 1988; Henry, 1996). The precise cellular location of the extracellular CA in muscle tissue is uncertain, and additionally may be species- and organ-dependent (Gros and Dodgson, 1988). Some histochemical and biochemical evidence (rabbit: Geers et al., 1985; rat: Dermietzel et al., 1985; Waheed et al., 1992) and studies making use of the CA inhibitor prontosil conjugated to dextran molecules to vary its tissue permeability (rabbit: Geers et al., 1985) have indicated that the extracellular CA in skeletal muscle is extravascular, associated with the sarcolemma of the muscle fibers and oriented toward the interstitial space. An extravascular site for the extracellular CA in muscle would contrast with the capillary endothelial location of lung extracellular CA activity, and would certainly necessitate a rapid flux of HCO_3^- and H^+ across the capillary endothelium for its participation in plasma CO_2–HCO_3^-–H^+ reactions. Given this requirement, it is difficult to conceive of an effective role for extravascular, extracellular CA activity in the establishment of an equilibrium state in the plasma CO_2–HCO_3^-–H^+ reactions of muscle capillaries. Immunohistochemical techniques, however, have been used to demonstrate a muscle capillary endothelium membrane-bound location for CA activity (rat: Sender *et al.*, 1994), a location consistent with the moderate acceleration of CO_2–HCO_3^-–H^+ reactions detected in the effluent perfusate from isolated rabbit hindlimb or liver preparations (O'Brasky and Crandall, 1980).

Recent work by Henry *et al.* (1997b) provides evidence for the presence of a corresponding extracellular CA in rainbow trout white muscle that may facilitate both CO_2 and NH_3 transport out of the muscle. Differential centrifugation of muscle samples to isolate subcellular fractions revealed microsomal (membrane-associated) CA activity, while a reduction in the CO_2 efflux from a saline-perfused isolated tail preparation following addition of the CA inhibitor QAS to the perfusate implied an extracellular orientation for this membrane-associated CA activity (Henry *et al.*, 1997b). By analogy with the results for mammalian hindlimb preparations, then, a small disequilibrium that comprises a decrease in plasma pH and that can be eliminated by treatment with CA or increased in magnitude by treatment with CA inhibitors might be expected to exist in the venous blood of rainbow trout. This prediction was tested using a stopflow technique and extracorporeal circulation in which blood was withdrawn from a cannula implanted in the caudal vein (Perry *et al.*, 1997). The results were somewhat unexpected: a small disequilibrium was detected, but pH *increased* (by about 0.015 unit) during the stopflow period. Furthermore, a small disequilibrium (pH increase of about 0.01 unit) was also measured during similar experiments carried out on rainbow trout in which blood was withdrawn from the afferent branchial artery into the extracorporeal loop (C. J. Brauner, K. M. Gilmour, and S. F. Perry, unpublished data). Although the basis of the venous disequilibrium in rainbow trout is at present unclear, a role for slow uncatalyzed HCO_3^- dehydration may be deduced from the effect of CA treatment in eliminating the disequilibrium (Perry *et al.*, 1997). The membrane-associated extracellular CA activity of rainbow trout muscle (Henry *et al.*, 1997b) does not, however, accelerate plasma CO_2–HCO_3^-–H^+ reactions since the magnitude of the venous disequilibrium was insensitive to treatment with benzolamide (Perry *et al.*, 1997). The basis of the venous disequilibrium observed in rainbow trout is clearly an area open for further research.

IV. CONSEQUENCES OF ACID–BASE DISEQUILIBRIA

The consequences of acid–base disequilibria in fish blood can be considered in terms of their physiological significance as well as with respect to their practical implications for making physiological measurements. Although this discussion will focus on the consequences of postbranchial disequilibria due to the paucity of information on prebranchial disequilibria in fish, similar arguments can in many cases be applied to venous disequilibria. From the practical standpoint, the implications of the acid–base disequilibrium in the postbranchial blood are fairly straightforward: the acid–base status of blood withdrawn from the fish will change during and after sampling and will not reflect exactly the acid–base status of the blood leaving the gill. Given that the discrepancy between the measured pH and the pH of blood leaving the branchial vasculature will depend upon a variety of factors, including the time required for sampling and measurement, the respiratory and acid–base status of the fish at the time of sampling, recent activity of the RBC Na^+/H^+ exchanger if present, and any drug treatment the fish has received, it is imperative that the effect of the experimental conditions on the acid–base disequilibrium be taken into consideration in making comparisons in arterial pH among different experimental groups. To ensure that the measured pH is actually an equilibrium value, a delay of at least 8 min should routinely be imposed between withdrawing the blood sample and measuring pH. It is advantageous to measure the equilibrium pH because the extent to which this measurement will deviate from the pH of blood leaving the gills can be predicted or determined. A more difficult issue is whether the equilibrium pH is actually the measurement of interest to the researcher. Although our knowledge of the relationship between the equilibrium pH and the pH of blood leaving the gills is limited, even less is known of the relationship between the equilibrium pH and the pH of blood entering any particular tissue. Indeed, whether the CO_2–HCO_3^-–H^+ reactions in fish blood ever reach equilibrium *in vivo* remains a matter for speculation.

The physiological significance of the acid–base disequilibria in the postbranchial blood of trout and dogfish is also a topic of speculation, one that it seems logical to discuss within the context of the different CA distribution patterns between teleost and elasmobranch fish. Plasma CO_2–HCO_3^-–H^+ reactions in the gills of teleost fish do not have access to CA activity, and consequently a slow increase in plasma pH occurs in the postbranchial blood, downstream of the CO_2 loss at the gills. The lack of plasma-accessible branchial CA activity in teleosts has been attributed to a requirement to avoid short-circuiting the action of catecholamines on RBC pH (Randall and Perry, 1992; Lessard *et al.,* 1995). Protons removed from the RBC interior by the action of a catecholamine-activated Na^+/H^+ antiporter slowly reequilibrate across the RBC membrane via the Jacobs–Stewart cycle (reviewed by Randall and Perry, 1992; Thomas and Perry, 1992). It has been argued that in the presence of plasma-accessible CA activity, the reequilibration of protons across the RBC membrane would occur too rapidly for catecholamine-induced RBC pH regulation to be possible. Although a short-circuit effect was demonstrated following the addition of high concentrations of CA to blood samples *in vitro* (Nikinmaa *et al.*, 1990), catecholamine-mediated regulation of RBC pH has been found to occur *in vitro* and *in vivo* even in the presence of CA levels similar to those available to plasma reactions in the mammalian lung (Motais *et al.*, 1989a,b; Lessard *et al.*, 1995). Lessard *et al.* (1995) accordingly proposed that the inaccessibility of CA activity to plasma reactions in teleost fish acts to protect the RBC from a plasma acidosis, such as that elicited by burst exercise. In the absence of plasma-accessible CA activity, the rate of transfer of protons from plasma to RBC is limited by the speed of the uncatalyzed plasma HCO_3^- dehydration step in the Jacobs–Stewart cycle, which will, in turn, minimize any Root effect reductions in hemoglobin–O_2 binding. The blood of elasmobranch fish does not exhibit a Root effect (Lenfant and Johansen, 1966, reviewed by Butler and Metcalfe, 1988) and may therefore be more tolerant of RBC acidosis, explaining the presence of extracellular CA activity in this group.

The concept that the lack of plasma-accessible CA activity in the gills of teleost fish acts to protect intracellular events from extracellular acidoses was also explored by Randall and Brauner (1997), but within the context of the requirements for ionic regulation in freshwater teleosts. Randall and Brauner (1997) suggested that the distribution of CA within gill tissue reflects the need to provide the apical H^+-ATPase with protons, while ensuring that the activity of this pH-sensitive proton pump, which functions in sodium uptake, is not constrained by the plasma pH. Hence, CA activity within gill tissue is concentrated in the apical region (Rahim *et al.*, 1988), where it catalyzes the hydration of CO_2 to supply protons to the H^+-ATPase, and is unavailable to plasma reactions, preventing rapid transfer of protons from plasma to gill cell interior (Randall and Brauner, 1997).

Like the mammals, reptiles, and amphibians studied to date, elasmobranch fish appear to possess CA activity that is available to plasma reactions at the gas exchange surface. Such CA activity could, in theory, participate in CO_2 excretion by providing an extracellular pathway for rapid HCO_3^- dehydration in addition to the standard route through the RBC. However, it has been argued that the low nonbicarbonate buffer capacity of plasma relative to that of the RBC will restrict the contribution of pulmonary capillary endothelial CA activity to $<10\%$ of CO_2 excretion in mammals because HCO_3^- dehydration in the plasma will be limited by proton availability (Crandall and Bidani, 1981; Bidani and Heming, 1991; Heming and Bidani, 1992). Evidence in support of this argument was provided by experimental results from a saline-perfused isolated rat lung preparation showing that selective inhibition of only the pulmonary capillary endothelial CA activity using dextran-coupled prontosil did not cause a reduction in CO_2 excretion (Heming *et al.*, 1986). Nonetheless, it should be recognized that measurements have yet to be made in either mammals or fish of the actual contributions of the extracellular and RBC pathways to HCO_3^- dehydration *in vivo*, i.e., in a situation in which local P_{CO_2} gradients and access limitations may be important.

An alternative role that has been advanced for lung capillary endothelial CA activity is in the promotion of an equilibrium condition for CO_2–HCO_3^-–H^+ reactions in the pulmonary vasculature (Heming *et al.*, 1993, 1994). This hypothesis proposes that pulmonary capillary endothelial CA activity aids in linking ventilation to gas transfer by minimizing changes between the end-capillary CO_2 tension, which is governed by the efficiency of gas transfer, and the arterial P_{CO_2}, which is detected by chemoreceptors and used in setting ventilation levels. Henry *et al.* (1997a) noted the parallel between the patterns of CA distribution in mammals and those in dogfish and suggested that the presence of CA activity that is available to plasma reactions at the gas exchange surface in dogfish is associated with the existence of a CO_2/H^+-sensitive ventilatory drive in elasmobranchs (Heisler *et al.*, 1988; Graham *et al.*, 1990; Wood *et al.*, 1990; Perry and Gilmour, 1996). The absence of branchial membrane-bound or plasma CA activities in teleost fish would, by this theory, be related to the primarily O_2-driven ventilatory drive (e.g., Perry and Wood, 1989) in this group. One difficulty with this explanation of the physiological significance of gill membrane-bound and extracellular CA activity in dogfish is that such CA activity does not actually result in an equilibrium state for CO_2–HCO_3^-–H^+ reactions in the postbranchial blood (Wilson, 1995; Gilmour *et al.*, 1997). The disequilibria measured in dogfish under control conditions and following selective inhibition of the extracellular and gill membrane-bound CA activities were, in fact, of similar magnitude (but opposite direction) (Gilmour *et al.*, 1997). The size of the disequilibrium in dogfish postbranchial blood (about 0.03 pH unit; Gilmour *et al.*, 1997) also falls within the range of disequilibria measured for trout under control conditions (0.02 to 0.06 pH unit, Gilmour *et al.*, 1994; Gilmour and Perry, 1996). Based on these results, it is conceivable that the role of branchial CA activity available to plasma reactions is in determining the direction of the acid–base disequilibrium rather than its presence or absence. The physiological significance of the postbranchial acid–base disequilibrium itself, however, remains unclear.

ACKNOWLEDGMENTS

Thanks are extended to C. M. Wood, B. L. Tufts, R. P. Henry, C. J. Brauner and particularly S. F. Perry for their enthusiastic participation in disequilibrium projects, and to

R. P. Henry, E. K. Stabenau, C. J. Brauner and J. M. Wilson for access to unpublished work. The volume editors as well as T. A. Heming and E. K. Stabenau are gratefully acknowledged for their comments on the manuscript. Original research reported here was funded by NSERC of Canada operating and equipment grants to S. F. Perry and C. M. Wood.

REFERENCES

Albers, C., and Pleschka, K. (1967). Effect of temperature on CO_2 transport in elasmobranch blood. *Respir. Physiol.* **2,** 261–273.

Bhargava, V., Chin Lai, N., Graham, J. B., Hempleman, S. C., and Shabetai, R. (1992). Digital image analysis of shark gills: Modeling of oxygen transfer in the domain of time. *Am. J. Physiol.* **263,** R741–R746.

Bidani, A., Crandall, E. D., and Forster, R. E. (1978). Analysis of postcapillary pH changes in blood in vivo after gas exchange. *J. Appl. Physiol.* **44,** 770–781.

Bidani, A., Mathew, S. J., and Crandall, E. D. (1983). Pulmonary vascular carbonic anhydrase activity. *J. Appl. Physiol.* **55,** 75–83.

Bidani, A., and Crandall, E. D. (1978). Slow postcapillary pH changes in blood in anesthetized animals. *J. Appl. Physiol.* **45,** 674–680.

Bidani, A., and Crandall, E. D. (1982). Analysis of the effects of hematocrit on pulmonary CO_2 transfer. *J. Appl. Physiol.* **53,** 413–418.

Bidani, A., and Crandall, E. D. (1988). Velocity of CO_2 exchanges in the lungs. *Annu. Rev. Physiol.* **50,** 639–652.

Bidani, A., and Heming, T. A. (1991). Effects of perfusate buffer capacity on capillary CO_2–HCO_3^-–H^+ reactions: Theory. *J. Appl. Physiol.* **71,** 1460–1468.

Brauner, C. J., Gilmour, K. M., and Perry, S. F. (1996). Effect of haemoglobin oxygenation on Bohr proton release and CO_2 excretion in the rainbow trout. *Respir. Physiol.* **106,** 65–70.

Brauner, C. J., and Randall, D. J. (1996). The interaction between oxygen and carbon dioxide movements in fishes. *Comp. Biochem. Physiol. A* **113,** 83–90.

Butler, P. J., and Metcalfe, J. D. (1988). Cardiovascular and respiratory systems. *In* "Physiology of Elasmobranch Fishes" (T. J. Shuttleworth, ed.), pp. 1–47. Springer-Verlag, Berlin.

Cameron, B. A., Perry, S. F., Wu, C., Ko, K., and Tufts, B. L. (1996). Bicarbonate permeability and immunological evidence for an anion exchanger-like protein in the red blood cells of the sea lamprey, *Petromyzon marinus. J. Comp. Physiol. B.* **166,** 197–204.

Cameron, J. N. (1978). Chloride shift in fish blood. *J. Exp. Zool.* **206,** 289–295.

Cameron, J. N., and Polhemus, J. A. (1974). Theory of CO_2 exchange in trout gills. *J. Exp. Biol.* **60,** 183–194.

Chakrabarti, M. K., Cobbe, S. M., Loh, L., and Poole-Wilson, P. A. (1983). Measurement of pulmonary venous and arterial pH oscillations in dogs using catheter tip pH electrodes. *J. Physiol.* **336,** 61–71.

Crandall, E. D., Bidani, A., and Forster, R. E. (1977). Postcapillary changes in blood pH in vivo during carbonic anhydrase inhibition. *J. Appl. Physiol.* **43,** 582–590.

Crandall, E. D., Mathew, S. J., Fleischer, R. S., Winter, H. I., and Bidani, A. (1981). Effects of inhibition of RBC HCO_3^-/Cl^- exchange on CO_2 excretion and downstream pH disequilibrium in isolated rat lungs. *J. Clin. Invest.* **68,** 853–862.

Crandall, E. D., and Bidani, A. (1981). Effects of red blood cell HCO_3^-/Cl^- exchange kinetics on lung CO_2 transfer: Theory. *J. Appl. Physiol.* **50,** 265–271.

Crandall, E. D., and O'Brasky, J. E. (1978). Direct evidence for participation of rat lung carbonic anhydrase in CO_2 reactions. *J. Clin. Invest.* **62,** 618–622.

Dermietzel, R., Leibstein, A., Siffert, W., Zamboglou, N., and Gros, G. (1985). A fast screening method for histochemical localization of carbonic anhydrase: Application to kidney, skeletal muscle, and thrombocytes. *J. Histochem. Cytochem.* **33,** 93–98.

Dimberg, K. (1994). The carbonic anhydrase inhibitor in trout plasma: Purification and its effect on carbonic anhydrase activity and the Root effect. *Fish Physiol. Biochem.* **12,** 381–386.

Effros, R. M., Chang, R. S. Y., and Silverman, P. (1978). Acceleration of plasma bicarbonate conversion to carbon dioxide by pulmonary carbonic anhydrase. *Science.* **199,** 427–429.

Effros, R. M., and Weissman, M. L. (1979). Carbonic anhydrase activity of the cat hind leg. *J. Appl. Physiol.* **47,** 1090–1098.

Fain, W., and Rosen, S. (1973). Carbonic anhydrase activity in amphibian and reptilian lung: A histochemical and biochemical analysis. *Histochem. J.* **5,** 519–528.

Forster, R. E., and Crandall, E. D. (1975). Time course of exchanges between red cells and extracellular fluid during CO_2 uptake. *J. Appl. Physiol.* **38,** 710–718.

Geers, C., Gros, G., and Gärtner, A. (1985). Extracellular carbonic anhydrase of skeletal muscle associated with the sarcolemma. *J. Appl. Physiol.* **59,** 548–558.

Geers, C., and Gros, G. (1984). Inhibition properties and inhibition kinetics of an extracellular carbonic anhydrase in perfused skeletal muscle. *Respir. Physiol.* **56,** 269–287.

Gilmour, K. M., Randall, D. J., and Perry, S. F. (1994). Acid–base disequilibrium in the arterial blood of rainbow trout. *Respir. Physiol.* **96,** 259–272.

Gilmour, K. M., Henry, R. P., Wood, C. M., and Perry, S. F. (1997). Extracellular carbonic anhydrase and an acid–base disequilibrium in the blood of the dogfish, *Squalus acanthias.* *J. Exp. Biol.* **200,** 173–183.

Gilmour, K. M., and Perry, S. F. (1994). The effects of hypoxia, hyperoxia or hypercapnia on the acid-base disequilibrium in the arterial blood of rainbow trout. *J. Exp. Biol.* **192,** 269–284.

Gilmour, K. M., and Perry, S. F. (1996). Effects of metabolic acid–base disturbances and elevated catecholamines on the acid–base disequilibrium in the arterial blood of rainbow trout. *J. Exp. Zool.* **274,** 281–290.

Graham, M. S., Turner, J. D., and Wood, C. M. (1990). Control of ventilation in the hypercapnic skate *Raja ocellata:* I. Blood and extradural fluid. *Respir. Physiol.* **80,** 259–277.

Gray, B. A. (1971). The rate of approach to equilibrium in uncatalyzed CO_2 hydration reactions: The theoretical effect of buffering capacity. *Respir. Physiol.* **11,** 223–234.

Gros, G., and Dodgson, S. J. (1988). Velocity of CO_2 exchange in muscle and liver. *Annu. Rev. Physiol.* **50,** 669–694.

Hanson, M. A., Nye, P. C. G., and Torrance, R. W. (1981). Studies on the localization of pulmonary carbonic anhydrase in the cat. *J. Physiol.* **319,** 93–109.

Heisler, N., Toews, D. P., and Holeton, G. F. (1988). Regulation of ventilation and acid–base status in the elasmobranch *Scyliorhinus stellaris* during hyperoxia-induced hypercapnia. *Respir. Physiol.* **71,** 227–246.

Heming, T. A., Geers, C., Gros, G., Bidani, A., and Crandall, E. D. (1986). Effects of dextran-bound inhibitors on carbonic anhydrase activity in isolated rat lungs. *J. Appl. Physiol.* **61,** 1849–1856.

Heming, T. A., Vanoye, C. G., Stabenau, E. K., Roush, E. D., Fierke, C. A., and Bidani, A. (1993). Inhibitor sensitivity of pulmonary vascular carbonic anhydrase. *J. Appl. Physiol.* **75,** 1642–1649.

Heming, T. A., Stabenau, E. K., Vanoye, C. G., Moghadasi, H., and Bidani, A. (1994). Roles of intra- and extracellular carbonic anhydrase in alveolar-capillary CO_2 equilibration. *J. Appl. Physiol.* **77,** 697–705.

Heming, T. A., and Bidani, A. (1990). In situ characterization of carbonic anhydrase activity in isolated rat lungs. *J. Appl. Physiol.* **69,** 2155–2162.

Heming, T. A., and Bidani, A. (1992). Influence of proton availability on intracapillary CO_2–HCO_3^-–H^+ reactions in isolated rat lungs. *J. Appl. Physiol.* **72,** 2140–2148.

Henry, R. P. (1991). Techniques for measuring carbonic anhydrase activity *in vitro:* The electrometric delta pH and pH stat assays. *In* "The Carbonic Anhydrases: Cellular Physiology and Molecular Genetics" (S. J. Dodgson, R. E. Tashian, G. Gros, and N. D. Carter, eds.), pp. 119–126. Plenum, New York.

Henry, R. P. (1996). Multiple roles of carbonic anhydrase in cellular transport and metabolism. *Annu. Rev. Physiol.* **58,** 523–538.

Henry, R. P., Smatresk, N. J., and Cameron, J. N. (1988). The distribution of branchial carbonic anhydrase and the effects of gill and erythrocyte carbonic anhydrase inhibition in the channel catfish *Ictalurus punctatus. J. exp. Biol.* **134,** 201–218.

Henry, R. P., Tufts, B. L., and Boutilier, R. G. (1993). The distribution of carbonic anhydrase type I and II isozymes in lamprey and trout: Possible co-evolution with erythrocyte chloride/bicarbonate exchange. *J. Comp. Physiol. B.* **163,** 380–388.

Henry, R. P., Boutilier, R. G., and Tufts, B. L. (1995). Effects of carbonic anhydrase inhibition on the acid–base status in lamprey and trout. *Respir. Physiol.* **99,** 241–248.

Henry, R. P., Gilmour, K. M., Wood, C. M., and Perry, S. F. (1997a). Extracellular carbonic anhydrase activity and carbonic anhydrase inhibitors in the circulatory system of fish. *Physiol. Zool.* **70,** 650–659.

Henry, R. P., Wang, Y., and Wood, C. M. (1997b). Carbonic anhydrase facilitates CO_2 and NH_3 transport across the sarcolemma of trout white muscle. *Am. J. Physiol.* **41,** R1754–R1761.

Henry, R. P., and Kormanik, G. A. (1985). Carbonic anhydrase activity and calcium deposition during the molt cycle of the blue crab, *Callinectes sapidus. J. Crust. Biol.* **5,** 234–241.

Hill, E. P., Power, G. G., and Longo, L. D. (1973). Mathematical simulation of pulmonary O_2 and CO_2 exchange. *Am. J. Physiol.* **224,** 904–917.

Hill, E. P., Power, G. G., and Gilbert, R. D. (1977). Rate of pH changes in blood plasma in vitro and in vivo. *J. Appl. Physiol.* **42,** 928–934.

Hill, E. P. (1986). Inhibition of carbonic anhydrase by plasma of dogs and rabbits. *J. Appl. Physiol.* **60,** 191–197.

Hughes, G. M., Horimoto, M., Kikuchi, Y., and Kakiuchi, U. (1981). Blood-flow velocity in microvessels of the gill filaments of the goldfish (*Carassium auratus*). *J. Exp. Biol.* **90,** 327–331.

Jensen, F. B. (1986). Pronounced influence of Hb–O_2 saturation on red cell pH in tench blood in vivo and in vitro. *J. Exp. Zool.* **238,** 119–124.

Jensen, F. B. (1989). Hydrogen ion equilibria in fish haemoglobins. *J. Exp. Biol.* **143,** 225–234.

Jensen, F. B. (1991). Multiple strategies in oxygen and carbon dioxide transport by haemoglobin. *In* "Physiological Strategies for Gas Exchange and Metabolism" (A. J. Woakes, M. K. Grieshaber, and C. R. Bridges, eds.), pp. 55–78. Cambridge Univ. Press, Cambridge.

Jensen, F. B., and Brahm, J. (1995). Kinetics of chloride transport across fish red blood cell membranes. *J. Exp. Biol.* **198,** 2237–2244.

Kay, M. M. B., Cover, C., Schluter, S. F., Bernstein, R. M., and Marchalonis, J. J. (1995). Band 3, the anion transporter, is conserved during evolution: Implications for aging and vertebrate evolution. *Cell Mol. Biol.* **41,** 833–842.

Klocke, R. A. (1978). Catalysis of CO_2 reactions by lung carbonic anhydrase. *J. Appl. Physiol.* **44,** 882–888.

Klocke, R. A. (1988). Velocity of CO_2 exchange in blood. *Annu. Rev. Physiol.* **50,** 625–637.

Lenfant, C., and Johansen, K. (1966). Respiratory function in the elasmobranch *Squalus suckleyi* G. *Respir. Physiol.* **1,** 13–29.

Lessard, J., Val, A. L., Aota, S., and Randall, D. J. (1995). Why is there no carbonic anhydrase activity available to fish plasma? *J. Exp. Biol.* **198,** 31–38.

Motais, R., Fievet, B., and Garcia-Romeu, F. (1989a). Effect of Na$^+$/H$^+$ antiport activation on pH for erythrocytes suspended in a HCO$_3^-$-containing saline. *Studia Biophys.* **134**, 121–126.

Motais, R., Fievet, B., Garcia-Romeu, F., and Thomas, S. (1989b). Na$^+$-H$^+$ exchange and pH regulation in red blood cells: Role of uncatalyzed H$_2$CO$_3$ dehydration. *Am. J. Physiol.* **256**, C728–C735.

Nikinmaa, M. (1986). Red cell pH of lamprey (*Lampetra fluviatilis*) is actively regulated. *J. Comp. Physiol. B.* **156**, 747–750.

Nikinmaa, M., Tiihonen, K., and Paajaste, M. (1990). Adrenergic control of red cell pH in salmonid fish: Roles of the sodium/proton exchange, Jacobs–Stewart cycle and membrane potential. *J. Exp. Biol.* **154**, 257–271.

Nikinmaa, M., and Mattsoff, L. (1992). Effects of oxygen saturation on the CO$_2$ transport properties of *Lampetra* red cells. *Respir. Physiol.* **87**, 219–230.

Nikinmaa, M., and Railo, E. (1987). Anion movements across lamprey (*Lampetra fluviatilis*) red cell membrane. *Biochim. Biophys. Acta.* **899**, 134–136.

O'Brasky, J. E., Mauro, T., and Crandall, E. D. (1979). Postcapillary pH disequilibrium after gas exchange in isolated perfused liver. *J. Appl. Physiol.* **47**, 1079–1083.

O'Brasky, J. E., and Crandall, E. D. (1980). Organ and species differences in tissue vascular carbonic anhydrase activity. *J. Appl. Physiol.* **49**, 211–217.

Obaid, A. L., Critz, A. M., and Crandall, E. D. (1979). Kinetics of bicarbonate/chloride exchange in dogfish erythrocytes. *Am. J. Physiol.* **237**, R132–R138.

Parrott, A., Ponte, J., Purves, M. J., and Stephenson, T. (1979). Changes in carbon dioxide and pH in pulmonary post-capillary blood in cats. *J. Physiol.* **296**, 23P-24P.

Perry, S. F. (1986). Carbon dioxide excretion in fishes. *Can. J. Zool.* **64**, 565–572.

Perry, S. F., Wood, C. M., Walsh, P. J., and Thomas, S. (1996). Fish red blood cell carbon dioxide transport *in vitro:* A comparative study. *Comp. Biochem. Physiol. A* **113**, 121–130.

Perry, S. F., Brauner, C. J., Tufts, B. L., and Gilmour, K. M. (1997). Acid–base disequilibrium in the venous blood of rainbow trout (*Oncorhynchus mykiss*). *Exp. Biol. Online* **2**, 1.

Perry, S. F., and Gilmour, K. M. (1993). An evaluation of factors limiting carbon dioxide excretion by trout red blood cells *in vitro. J. Exp. Biol.* **180**, 39–54.

Perry, S. F., and Gilmour, K. M. (1996). Consequences of catecholamine release on ventilation and blood oxygen transport during hypoxia and hypercapnia in an elasmobranch (*Squalus acanthias*) and a teleost (*Oncorhynchus mykiss*). *J. Exp. Biol.* **199**, 2105–2118.

Perry, S. F., and Laurent, P. (1990). The role of carbonic anhydrase in carbon dioxide excretion, acid–base balance and ionic regulation in aquatic gill breathers. *In* "Animal Nutrition and Transport Processes. 2. Transport, Respiration and Excretion: Comparative and Environmental Aspects" (J.-P. Truchot, and B. Lahlou, eds.), Vol. 6, pp. 39–57. Karger, Basel.

Perry, S. F., and Wood, C. M. (1989). Control and coordination of gas transfer in fishes. *Can. J. Zool.* **67**, 2961–2970.

Pleschka, K., Albers, C., and Spaich, P. (1970). Interaction between CO$_2$ transport and O$_2$ transport in the blood of the dogfish *Scyliorhinus canicula. Respir. Physiol.* **9**, 118–125.

Ponte, J., and Purves, M. J. (1980). Changes in pH and P_{CO_2} with time in pulmonary post-capillary blood in cats. *In* "Biophysics and Physiology of Carbon Dioxide" (C. Bauer, G. Gros, and H. Bartels, eds.), pp. 315–320. Springer-Verlag, Heidelberg.

Rahim, S. M., Delaunoy, J.-P., and Laurent, P. (1988). Identification and immunocytochemical localization of two different carbonic anhydrase isoenzymes in teleostean fish erythrocytes and gill epithelia. *Histochemistry.* **89**, 451–459.

Randall, D. J. (1982). The control of respiration and circulation in fish during exercise and hypoxia. *J. Exp. Biol.* **100**, 275–288.

Randall, D. J., and Brauner, C. J. (1997). Interactions between ion and gas transfer in freshwater teleost fish. *Comp. Biochem. Physiol.* A **119**, 3–8.

Randall, D. J., and Perry, S. F. (1992). Catecholamines. *In* "The Cardiovascular System" (W. S. Hoar, D. J. Randall, and A. P. Farrell, eds.), Vol. XIIB, pp. 255–300. Academic Press, San Diego.

Rispens, P., Oeseburg, B., Zock, J. P., and Zijlstra, W. G. (1980). Intra-aortic decrease in blood plasma pH. *Pflügers Arch.* **386**, 97–99.

Roughton, F. J. W. (1935). Recent work on carbon dioxide transport by the blood. *Physiol. Rev.* **15**, 241–296.

Roush, E. D., and Fierke, C. A. (1992). Purification and characterization of a carbonic anhydrase II inhibitor from porcine plasma. *Biochemistry* **31**, 12536–12542.

Sender, S., Gros, G., Waheed, A., Hageman, G. S., and Sly, W. S. (1994). Immunohistochemical localization of carbonic anhydrase IV in capillaries of rat and human skeletal muscle. *J. Histochem. Cytochem.* **42**, 1229–1236.

Stabenau, E. K. (1994). Pulmonary CO_2 excretion and postcapillary CO_2–HCO_3^-–H^+ equilibration in the turtle. Ph.D. thesis. Univ. of Texas.

Stabenau, E. K., Bidani, A., and Heming, T. A. (1996). Physiological characterization of pulmonary carbonic anhydrase in the turtle. *Respir. Physiol.* **104**, 187–196.

Swenson, E. R. (1984). The respiratory aspects of carbonic anhydrase. *Ann. N. Y. Acad. Sci.* **429**, 547–560.

Swenson, E. R. (1990). Kinetics of oxygen and carbon dioxide exchange. *In* "Advances in Comparative and Environmental Physiology" (R. G. Boutilier, ed.), Vol. 6, pp. 163–210. Springer-Verlag, Berlin.

Swenson, E. R., Lippincott, L., and Maren, T. H. (1995). Effect of gill membrane-bound carbonic anhydrase inhibition on branchial bicarbonate excretion in the dogfish shark, *Squalus acanthias. Bull. MDI Biol. Lab.* **34**, 94–95.

Swenson, E. R., Taschner, B. C., and Maren, T. H. (1996). Effect of membrane-bound carbonic anhydrase (CA) inhibition on bicarbonate excretion in the shark, *Squalus acanthias. Bull. MDI Biol. Lab.* **35** (In press).

Swenson, E. R., and Maren, T. H. (1987). Roles of gill and red cell carbonic anhydrase in elasmobranch HCO_3^- and CO_2 excretion. *Am. J. Physiol.* **253**, R450–R458.

Takahashi, E., and Phillipson, E.A. (1991). Effect of changing venoarterial pH difference on in vivo arterial pH. *J. Appl. Physiol.* **70**, 1586–1592.

Thomas, S., Fritsche, R., and Perry, S. F. (1994). Pre- and post-branchial blood respiratory status during acute hypercapnia or hypoxia in rainbow trout, *Oncorhynchus mykiss. J. Comp. Physiol.* B **164**, 451–458.

Thomas, S., and Perry, S. F. (1992). Control and consequences of adrenergic activation of red blood cell Na^+/H^+ exchange on blood oxygen and carbon dioxide transport in fish. *J. Exp. Zool.* **263**, 160–175.

Tufts, B. L. (1991). Acid–base regulation and blood gas transport following exhaustive exercise in an agnathan, the sea lamprey *Petromyzon marinus. J. Exp. Biol.* **159**, 371–385.

Tufts, B. L., Bagatto, B., and Cameron, B. (1992). *In vivo* analysis of gas transport in arterial and venous blood of the sea lamprey *Petromyzon marinus. J. exp. Biol.* **169**, 105–119.

Tufts, B. L., Currie, S., and Kieffer, J. D. (1996). Relative effects of carbonic anhydrase infusion or inhibition on carbon dioxide transport and acid–base status in the sea lamprey *Petromyzon marinus* following exercise. *J. Exp. Biol.* **199**, 933–940.

Tufts, B. L., and Boutilier, R. G. (1989). The absence of rapid chloride/bicarbonate exchange in lamprey erythrocytes: Implications for CO_2 transport and ion distributions between plasma and erythrocytes in the blood of *Petromyzon marinus. J. Exp. Biol.* **144**, 565–576.

Tufts, B. L., and Boutilier, R. G. (1990). CO_2 transport in agnathan blood: Evidence of erythrocyte Cl^-/HCO_3^- exchange limitations. *Respir. Physiol.* **80,** 335–348.

Waheed, A., Zhu, X. L., Sly, W. S., Wetzel, P., and Gros, G. (1992). Rat skeletal muscle membrane associated carbonic anhydrase is 39-kDa, glycosylated, GPI-anchored CA IV. *Arch. Biochem. Biophys.* **294,** 550–556.

Wilson, J. M. (1995). The localization of branchial carbonic anhydrase in the shark, *Squalus acanthias.* M.Sc. thesis. Univ. of British Columbia.

Wood, C. M., Turner, J. D., Munger, R. S., and Graham, M. S. (1990). Control of ventilation in the hypercapnic skate *Raja ocellata:* II. Cerebrospinal fluid and intracellular pH in the brain and other tissues. *Respir. Physiol.* **80,** 279–298.

Wood, C. M., Perry, S. F., Walsh, P. J., and Thomas, S. (1994). HCO_3^- dehydration by the blood of an elasmobranch in the absence of a Haldane effect. *Respir. Physiol.* **98,** 319–337.

Wood, C. M., and Munger, R. S. (1994). Carbonic anhydrase injection provides evidence for the role of blood acid–base status in stimulating ventilation after exhaustive exercise in rainbow trout. *J. Exp. Biol.* **194,** 225–253.

Wood, C. M., and Perry, S. F. (1991). A new *in vitro* assay for carbon dioxide excretion by trout red blood cells: Effects of catecholamines. *J. Exp. Biol.* **157,** 349–366.

Zborowska-Slus, D. T., L'Abbate, A., and Klassen, G. A. (1974). Evidence of carbonic anhydrase activity in skeletal muscle: A role for facilitative carbon dioxide transport. *Respir. Physiol.* **21,** 341–350.

Zock, J. P., Rispens, P., and Zijlstra, W. G. (1981). Calculated changes in pH and P_{CO_2} in arterial blood plasma assuming absence of ion and water exchange between plasma and erythrocytes during their equilibration with alveolar gas. *Pflügers Arch.* **391,** 159–161.

INDEX

A

Acetazolamide, 86, 90–1, 100
Acid–base balance, 244, 266, 269–71, 302
  carbonic anhydrase, 81, 101
Acid–base disequilibria
  Cl^-/HCO_3^-, 332
  consequences, 339–42
  dogfish, 327–30
  trout, 324–7
Acid load, 169
Acidosis, 63, 168–70, 174, 204, 215
  metabolic, 67, 117, 120, 122, 131, 244–5, 327
  respiratory, 67, 167, 171, 302, 337
  respiratory and metabolic, 171
β-Adrenergic receptors, 62, 166, 307
β-Adrenergic stimulation, 21, 62–8, 147, 160, 174, 236, 240–9
Adrenergic stimulus, 49
Alanine, 60–1
Alkalosis
  metabolic, 67
  respiratory, 166, 239
β-Amino acid, 60
Amino acids
  rbc transport, 60–2
  transporters
    Na^+ dependent, 61
    Na^+ independent, 61–2
Amiloride, 91, 166
Ammonia, 87–8
  carbonic anhydrase, 89
  excretion, 87–9, 236, 241
Ammonium, 88, 90, 339

Anemia, 173, 186, 210–11, 240, 305–6
ATP, 6, 20–2, 26, 42, 65, 144–9, 167, 174, 305
  catecholamines, 147
  RBC turnover, 44
  Root effect, 117

B

Band-3, 11–2, 30–1, 57–8, 63, 154, 238, 241–2, 308–9
  carbonic anhydrase, 79
Benzolamide, 251, 329–30
Bicarbonate
  dehydration, 94, 98, 133, 237–43, 250–2, 255, 297, 303, 307, 326–7, 341
  plasma, 83, 230, 236–7, 244, 248, 259, 264, 269, 311
  RBC, 83, 153, 161, 248–9, 261, 263, 334
  Root effect, 129
Blood buffering capacity, 232–4
Bohr effect, 6, 8, 10, 15–6, 19, 21, 113–4, 116–7, 156, 159, 165, 283–6,292
  Haldane effect, 284, 299–303, 306–7, 309
    air-breathing, 309–312
    coefficient, 289–92, 295, 304
    non linear, 289
    reverse, 23, 26
Bohr groups, 8
Bohr proton , 12, 19
Boyle's law, 120

C

cAMP, 63, 166, 172
Carbamino, 230, 233, 235, 255

Carbon dioxide
  anemia, 173
  blood capacitance, 231–2, 235, 255
  branchial epithelium, 86
  carbonic anhydrase 239–41, 246, 251, 265,
    270–1
  carriage, 254–5, 261
  disequilibrium, 244, 246, 260
  dissociation curve, 255, 261
  exercise, 267–9, 271
  excretion, 11, 83–91, 236–250, 303
    agnathans, 252, 264–7, 272
    elasmobranchs, 251–2
    hagfish, 260–4, 271
    lamprey, 254–60, 267–71
    teleost, 236–250
  Haldane effect, 255–6, 262–4
  hydration, 11, 76, 86, 90–91, 97, 157
  loading, 85, 258
  production, 64, 83–91, 245
  Root effect, 120–124, 128–130, 133
  transport, 230–6, 258–60, 264
    catecholamines, 248
Carbonic anhydrase
  acid base disequilibrium, 321
    post-branchial, 322–8
    pre-branchial, 337–42
  assays
    electrometric-delta pH, 96–97
    electrometric-pH stat, 96–98
    manometric, 98–99
    spectrophotometric, 98
    stable isotope, 100
  Band-3, 79
  CO_2 excretion, 83–91, 236–8, 246
  CO_2 hydration reaction, 76–77
  distribution
    branchial epithelium, 86
    erythrocyte, 75, 79
    gill, 81–83, 86–88, 91
    mucus, 82–83, 91
    muscle, 77
    plasma, 81
  hypoxia, 304
  inhibitors, 85–87, 102, 322
    acetazolamide, 86, 90–91, 100
    other sulfonamides, 100–101
    plasma (pICA), 91–96
  isozymes, 77–83
  kinetic properties

  pH disequilibrium, 101–103
  respiratory gas exchange
  Root effect, 129, 133–134
Catecholamines, 42–43, 244, 306, 337
  β-adrenergic stimulation, 62–68
    RBC glucose uptake, 57
    monocarboxylate, 58
    rbc pHi regulation, 62, 64–66, 131–132,
      160, 174
    O_2 consumption, 65–66
  CO_2 transport, 248–51
  hypoxia, 166
  Na^+/H^+ exchanger activation, 63–4, 94,
    153, 160–1, 165, 167, 171
  Root effect, 131
Chemoreceptors, 85
Chloride
  acid base disequilibria, 332–5
  bicarbonate exchanger, 11, 61, 63, 152,
    157–158, 237, 240–3, 251–2, 297, 299,
    304–5, 307-8, 324, 329
  Root effect, 118
  shift, 12
Choroid rete mirabile
  circulation, 126–7
  O_2 supply, 128–31, 144
  Root effect, 116
α-CIM, 58
Cortisol, 148
Cytochalasin B, 55, 57

                           D

DIDS, 58–59, 158
2,3 DPG, 13, 22

                           E

Erythrocytes, *see also* Red blood cells
  cell volume regulation, 57, 60, 149–55,
    170
    RVD, 151–4
    RVI, 151–3
  swelling, 63, 151, 195, 206–7
    adrenergic stimulation, 153, 170
Erythropoiesis, 209–11
Exercise, 85, 170–1, 174, 204–5, 212–4, 236
  CO_2, 244–6, 250
  PCO_2, 246–8

G

Gills
  carbonic anhydrase, 81–2, 86–8
  CO_2 removal/O_2 uptake, 299, 307
  diffusion distance, 169–70
  diffusion limited, 240
  excretion pattern, 86–7, 244, 246, 251, 271, 340
  surface pH, 87
Glucose
  consumption, 52–3, 67
  Root effect, 128
  transport, 55–7, 59
Glut-1, 55, 57
  inhibitors
    cytochalasin B, 55, 57
    phloretin, 55, 57
Glycine, 61–62
GTP, 13, 20, 144–149, 167–168, 170, 305
  Haldane effect, 286
  Root effect, 117

H

H^+
  H^+-ATPase, 133–4
  H^+ binding site on Hb, 6
  H^+ equilibria, 8–9
  H^+ monocarboxylate carrier, 58
H_2O_2, 29
Haldane effect, 10, 159, 165, 243, 245, 255, 262–3, 268, 283, 286–8, 293–4, 297–9, 301, 304, 311, 337
  Bohr effect, 284, 287–8
    optimal coefficient, 294
  Root effect, 132
Hematocrit
  effects of sampling, 187–95
  interspecific diversity, 206–9
  intraspecific regulation, 209–15
  optimal Hct theory, 186, 215–9
  serial sampling, 205–6
  values, 188–93, 197–203, 206–7, 246
Heme group, 3, 5, 119

Hemoglobin
  autoxidation, 27–29, 163
  Bohr/Haldane effect, 291
  buffering capacity, 133–4, 157, 161, 218, 293
  cellular concentration, 149, 188–95, 197–203, 206, 209, 212–3
  functional adaptation, 20
  larval, 26
  ligand binding, 5–8
    H^+, 6, 9, 155–157
    CO, 19
    CO_2, 12, 230
    O_2, 6
    organic phosphate, 13, 145
  membrane proteins, 30–31
  multiplicity, 22–27
  nitrite induced oxidation, 29
  O_2 affinity, 20–1, 30, 143–4, 149, 155–7, 162–3, 165–7, 170, 284–5, 290
  O_2 carrying capacity, 114, 125, 132, 185
  oxidation vs. O_2 affinity, 29–30, 162–3
  Root effect, 116, 118–9, 122, 125, 129, 131–2
  structure, 2–5, 25, 118
Hepatocytes, 92, 147
Heterotropic effector, 7, 144
Hypercapnia, 167–8, 215, 306, 335
  hypoxia, 167–8, 170
Hyperoxia, 167–8, 215, 306
  in choroid rete, 128
Hyperventilatory response, 143, 165, 168, 173, 248
Hypoventilatory response, 168
Hypoxia, 63, 165–7, 171, 174, 204, 211–2, 214, 303–6, 309–10, 335–6
  catecholamines, 166
  chronic, 167

I

IPP, 22
Isoprenaline, 63
Isoproterenol, 61, 162

J

Jacobs–Stewart cycle, 63, 134, 157, 264, 329, 334–5, 341

K

Kidney
K$^+$
  K$^+$/Cl$^-$ cotransport, 31, 154–5, 161
  rbc concentrations, 65

L

Lactate, 144, 270
  inhibitors
    α-CIM, 58–59
    DIDS, 58–59
    p-chloromercuriphenylsulfonic acid, 58
    PCMBS, 58
    SITS, 58
  rbc consumption, 67
  Root effect, 118, 128
  transport, 58, 60, 64
Lactic acid, 117, 122–3, 128
L-Leucine, 61

M

MCHC (mean cell hemoglobin
  concentration), 188–95, 197–203
Metabolic acidosis, 67, 117, 120, 122, 131,
  244–5, 327
Metabolic alkalosis, 67
Methemoglobin, 4, 163–4
Monocarboxylate transport, 57–60

N

Na$^+$
  Na$^+$/H$^+$ antiporter, 63
  Na$^+$/H$^+$ exchange, 21, 31, 63–5, 91, 147,
    152–3, 158, 161, 166–8, 172, 174,
    248–9, 257–8, 262, 265, 340, 367–8
  Na$^+$/K$^+$-ATPase, 64–66
  Na$^+$-rbc concentrations, 65
NADH, 29, 163
NADPH, 48, 163–4
NH$_3$, *see* Ammonia
NH$_4$$^+$, *see* Ammonium

NTP, 13, 19, 21–2, 24, 26, 29–30, 306–8
  depletion, 66
  rbc, 44, 54, 63–6, 148, 163, 167–8, 170,
    173
  Root effect, 117
Nitrite, 174
Nucleotide phosphate/ Hb molar ratio, 148
Normocythemia, 217–9
Normoxia
  adrenergic stimulation, 65
  exercise, 172

O

O$_2$, *see* Oxygen
3-OMG, 55, 57
Organic phosphates, 24
  Hb binding, 13, 20–1, 144–5, 156, 159
  ATP, 6, 20–2, 26, 42, 44, 65, 117, 144–9,
    167, 174, 305
  GTP, 13, 20, 117, 144–149, 167–168, 170,
    286, 305
  2,3-DPG, 13, 22
  NTP, 13, 19, 21–2, 24, 26, 29–30, 44, 54,
    63–6, 148, 163, 167–8, 170, 173,
    306–8
Ouabain, 64
Oxygen
  affinity, 6, 144–5, 210
  arterial content, 159, 171–2
  binding sites on Hb, 6
  blood carrying capacity, 94
  consumption rates, 44, 64–5, 142, 186,
    217
  equilibrium curve, 6, 113, 144
    exercise, 171
    NTP, 144
    pH, 144, 155
    temperature, 145

P

P$_{50}$, 21
PCO$_2$ gradient, 84–85
PCMBS, 58
pIAC, 91–96
pH disequilibrium, 82, 86–87, 101–3
pH/HCO$_3$$^-$ diagram, 302

pHe, 67
  as ventilatory stimulus, 63
pH/PCO$_2$ disequilibrium, 85, 102
  ventilatory drive, 85–86
Plasma skimming, 195
Pollutants, 215
  O$_2$ transport, 174
Polycythemia, 186, 216–7
Potassium, *see* K$^+$
Propanolol, 166
Pseudobranch
  Root effect, 129–130
Pyruvate, 58–59

Q

QAS, 90, 101, 339

R

Red blood cells, *see also* Erythrocytes
  β-adrenergic stimulation, 62–8, 147, 160,
    174, 246
  CO$_2$ excretion, 240–9
  metabolism, 46–69
    anabolic pathways, 54
    catabolic pathways, 51–54
    fuel transport, 54–62
    methodology, 42–46
    regulation, 48–9
  O$_2$ consumption, 65–7, 119
  O$_2$ transport, 63
  pHi regulation, 21–2, 43, 62–4, 66–7,
    157–62, 234, 248, 258
    agnathans, 159–159
    elasmobranchs, 159
    teleosts, 159–162
  respiration rates, 53
  splenic release, 204–5
Respiratory acidosis, 67, 167, 171, 302, 337
Respiratory alkalosis, 166, 239
Rete mirabile, 120–1, 124
Root effect, 8, 14–8, 286, 292–3, 311, 341
  allosteric ligands, 117–8
  catecholamines, 131
  choroid rete, 116
  definition, 113–4
  distribution, 115–7

  Haldane effect, 132
  kinetics, 118–20
  measurement, 115
  rete mirabile, 120–1, 124
  swim bladder, 116, 119–25, 130

S

Salinity change, 171
L-Serine, 61–62
SITS, 242
Sodium, *see* Na$^+$
Sodium–proton exchanger, 21
Starvation, 210–1
Stress, 172, 185, 194
Swim bladder, 144, 284
  Root effect, 116, 119–25, 130

T

Temperature effect, 18–20, 145, 163, 165
  Bohr/Haldane effect, 306
  Hct, 211
Trimethyl amine oxide, 22

U

Urea, 22, 145

V

V-ATPase, 122
Ventilatory stimulus, 63, 85, 342

W

White blood cells
  metabolism, 64

X

Xenobiotics
  Hb,
  rbc, 68–69

OTHER VOLUMES IN THE
FISH PHYSIOLOGY SERIES

VOLUME 1   Excretion, Ionic Regulation, and Metabolism
            Edited by W. S. Hoar and D. J. Randall

VOLUME 2   The Endocrine System
            Edited by W. S. Hoar and D. J. Randall

VOLUME 3   Reproduction and Growth: Bioluminescence, Pigments,
            and Poisons
            Edited by W. S. Hoar and D. J. Randall

VOLUME 4   The Nervous System, Circulation, and Respiration
            Edited by W. S. Hoar and D. J. Randall

VOLUME 5   Sensory Systems and Electric Organs
            Edited by W. S. Hoar and D. J. Randall

VOLUME 6   Environmental Relations and Behavior
            Edited by W. S. Hoar and D. J. Randall

VOLUME 7   Locomotion
            Edited by W. S. Hoar and D. J. Randall

VOLUME 8   Bioenergetics and Growth
            Edited by W. S. Hoar, D. J. Randall, and J. R. Brett

VOLUME 9A   Reproduction: Endocrine Tissues and Hormones
            Edited by W. S. Hoar, D. J. Randall, and E. M. Donaldson

VOLUME 9B   Reproduction: Behavior and Fertility Control
            Edited by W. S. Hoar, D. J. Randall, and E. M. Donaldson

VOLUME 10A Gills: Anatomy, Gas Transfer, and Acid–Base Regulation
            Edited by W. S. Hoar and D. J. Randall

VOLUME 10B Gills: Ion and Water Transfer
            Edited by W. S. Hoar and D. J. Randall

VOLUME 11A The Physiology of Developing Fish: Eggs and Larvae
            Edited by W. S. Hoar and D. J. Randall

VOLUME 11B   The Physiology of Developing Fish: Viviparity and
Posthatching Juveniles
Edited by W. S. Hoar and D. J. Randall

VOLUME 12A   The Cardiovascular System
Edited by W. S. Hoar, D. J. Randall, and A. P. Farrell

VOLUME 12B   The Cardiovascular System
Edited by W. S. Hoar, D. J. Randall, and A. P. Farrell

VOLUME 13   Molecular Endocrinology of Fish
Edited by N. M. Sherwood and C. L. Hew

VOLUME 14   Cellular and Molecular Approaches to Fish
Ionic Regulation
Edited by Chris M. Wood and Trevor J. Shuttleworth

VOLUME 15   The Fish Immune System: Organism, Pathogen,
and Environment
Edited by George Iwama and Teruyuki Nakanishi

VOLUME 16   Deep Sea Fishes
Edited by D. J. Randall and A. P. Farrell